Use of
Earth Sciences Literature

Editor

D. N. Wood, B.Sc., Ph.D.
British Library, Lending Division

ARCHON BOOKS
1973

Library of Congress Cataloging in Publication Data

Wood, David Norris.
    The use of earth sciences literature.

    (Information sources for research and development)
    1. Earth sciences—Bibliography. I. Title.
Z6031.W67   550'.7   72–13674
ISBN 0–208–00669–9

Published in the United States of America
as an Archon Book
by The Shoe String Press, Inc.,
Hamden, Connecticut 06514
First published 1973
© Butterworth & Co. (Publishers) Ltd., 1973
Printed in England

Preface

As the increasing volume of literature makes it more and more difficult to discover previously published information, there has developed an urgent need for up-to-date guides to primary and secondary sources of knowledge. This book, which forms part of a series covering many subject fields, is an attempt to provide guidance to the users of earth sciences literature. It is offered in the hope that it will be of use to both librarians and scientists, irre-spective of whether they are learning their professions or have been practising them for some time. It is not intended as a comprehensive list of publications of potential interest to earth scientists. Its aim has been to present a general picture of the structure of the literature and to illustrate this with examples. Where the literature of particular branches of the subject is discussed, the cited items are those which the authors, in many cases practising scientists, have found to be the most useful.

Since individual chapters have been completed at different times over a period of about 18 months, some are inevitably more up to date than others. In all cases literature published up to and including 1970 has been covered and in some cases 1971 material has also been considered. A few 1972 items have been added at proof stage.

As with other books in the series, space considerations have led to the town of publication being omitted for all but a handful of lesser-known publishers.

Not surprisingly, the style of individual chapters varies considerably. This reflects the nature of the literature being discussed and also the ideas of the various authors on how the material should be presented.

As editor I should like to thank the various contributors, without whose co-operation the book would not have been possible. As editor and as an author of several chapters, my thanks are also due to Dr. D. J. Urquhart, Director of the National Lending Library, for permission to undertake the work and to my colleagues at the Library for helpful discussions on numerous occasions. Special thanks are due to Miss C. A. Bower for critically reading much of the manuscript and to Miss D. Howard and Mrs. M. Redfearn for secretarial assistance. The authors of Chapter 16 would like to

PREFACE

acknowledge the valuable comments and advice of their colleagues Mr. R. I. Davies, Dr. R. G. Wyn-Jones and in particular of Mr. G. V. Jacks.

Thanks are due to the following organisations who have given permission for parts of their publications to be reproduced: E. Schweizerbart'sche Verlagsbuchhandlung (*Zentralblaat für Mineralogie*); Institute for Scientific Information (*Science Citation Index*); Geological Society of America (*Bibliography and Index of Geology Exclusive of North America*); Mineralogical Society (*Mineralogical Abstracts*); Chemical Abstracts Service (*Chemical Abstracts*); Centre National de la Recherche Scientifique (*Bulletin Signalétique*). Thanks are also due to the Copeland-Chatterson Co. Ltd. for permission to reproduce one of the edge-notched cards marketed by them.

D.N.W.

Note added in proof. By the time this book is published, the National Lending Library for Science and Technology, to which reference is made several times, will have been combined with the British Museum Library, the National Reference Library of Science and Invention, the National Central Library and the British National Bibliography to form the British Library. The National Lending Library and the National Central Library will constitute the British Library, Lending Division (BLL).

D.N.W.

Contributors

D. A. BASSETT, B.SC., PH.D., F.G.S., Keeper of Geology, National Museum of Wales, Cardiff

O. R. BRADLEY, B.SC., PH.D., Assistant Keeper, National Reference Library of Science and Invention, London

J. W. CLARKE, B.A., PH.D., Geologist, US Geological Survey, Washington

K. M. CLAYTON, M.SC., PH.D., Professor of Environmental Sciences, University of East Anglia, Norwich

A. P. HARVEY, Librarian, Department of Palaeontology, British Museum (Natural History), London

C. H. JAMES, B.SC., PH.D., A.R.S.M., D.I.C., Senior Lecturer in Economic Geology, Department of Geology, University of Leicester

D. A. JENKINS, B.A., PH.D., Department of Biochemistry and Soil Science, University College of North Wales, Bangor

E. L. MARTIN, B.SC., F.R.E.S., Department of Mines, Hobart, Tasmania

A. G. MYATT, B.SC., M.I.BIOL., A.I.INF.SC., National Lending Library, Boston Spa, Wetherby

W. A. S. SARJEANT, B.SC., PH.D., F.G.S., Department of Geology, University of Nottingham (Present address, University of Saskatchewan, Saskatoon, Canada)

P. W. G. TANNER, B.SC., PH.D., D.I.C., F.G.S., Research Institute of African Geology, Department of Earth Sciences, University of Leeds

R. I. J. TULLY, F.L.A., The Library, University College of North Wales, Bangor

CONTRIBUTORS

A. M. TYLER, formerly Librarian, Geological Society, London

D. N. WOOD, B.SC., PH.D., National Lending Library, Boston Spa, Wetherby

Contents

CONTENTS

1

Introduction

D. N. Wood

It is a desirable feature of scientific life that scientists should communicate their results to one another. In the first place it is only in this way that unnecessary duplication of scientific effort can be avoided, and in the second a knowledge of what other workers are doing can be of enormous benefit in stimulating thought and provoking new ideas.

One useful and long-practised means of interchanging ideas is by private communication either orally or by letter. Indeed, one of the few acknowledged benefits of attending present-day conferences is that they give one an opportunity to meet and discuss with colleagues working in the same field. There are so many scientists in existence, however (over 90 per cent of those who have ever lived are living today), that, unless one's interest lies in a very restricted subject area, personal contact cannot be relied upon as the only source, or even the principal source, of information. It is to the published literature that one must turn.

The first scientific periodicals, *Journal des Sçavans* and the *Philosophical Transactions of the Royal Society*, were published in 1665 as an aid to the dissemination of information. It is interesting to note the first paragraph of the first issue of the *Philosophical Transactions*: 'There is nothing more necessary for promoting the improvement of philosophical matters than the communicating to such as apply their studies and endeavours that way, such things as are discovered or put in practice by others. It is therefore thought fit to employ the press as the most proper way to gratify those whose engagement in such studies and delight in the advancement of learning and profitable discoveries doth entitle them to the knowledge of what this Kingdom or other parts of the world do from time to time afford.'

Since the seventeenth century and for a variety of motives it has become traditional for the scientist to publish his thoughts and results of experiments and investigations. This tendency has been reflected in a steady increase in the number of scientific and technical periodicals. The growth of the literature has itself been the subject of many papers (e.g. Urquhart, 1964) and appears to have been exponential. The doubling time has been between 10 and 15 years and compares with 30–50 years for non-scientific activities. The result of this literature explosion, as it has been termed, is that over 26 000 scientific and technical journals are currently being produced (Barr, 1967) and an average of three new journals are published each day.

Periodical publications, of course, are not the only sources of scientific information. Other types of literature exist and, like the periodical, have increased in number at a rate which reflects the general growth of scientific activity. These other types of publication include textbooks, monographs, handbooks, encyclopaedias, government and institutional reports, conference proceedings, trade literature, standard specifications, patent specifications, theses and maps.

Altogether it has been estimated that 60 million pages of technical literature are published yearly throughout the world or, put another way, enough technical literature is produced every 24 hours around the globe to fill seven sets of the 24-volume *Encyclopaedia Britannica*.

The information contained in this body of literature may be of use to the scientist in a number of tasks: keeping up to date with current work; locating specific facts such as the optical properties of a mineral, the chemical composition of a certain rock or the manufacturer of a particular instrument; and carrying out an exhaustive search for information on a particular subject. Whatever the reason for using it, the literature is probably the most expensive tool that the scientist has at his disposal. The cost of producing, acquiring and storing published material is infinitely greater than even the most sophisticated electronic equipment known to man.

Despite its cost and potential value, few scientists are taught about the structure and use of their literature. There is a tendency in most universities to assume that a knowledge of scientific literature is gained intuitively, and most students obtain degrees and even higher degrees without ever really discovering how to use this valuable research tool effectively. On the other hand, considerable time and effort is devoted to teaching students how to use relatively inexpensive equipment such as a petrological microscope, an X-ray fluorescence spectrograph or a flame photometer.

Much has been written on the need to teach scientists about scientific literature, and there is evidence that the students themselves desire such instruction (Wood, 1968). The topic has been the subject of numerous official recommendations—The Royal Society of London, the UK Committee of Vice-Chancellors and Principals and the UK Committee on University Libraries are but three bodies who have urged that library education for scientists should be introduced into British universities. Similar sentiments have been expressed in the United States and in other countries.

As a result of these recommendations, courses on the structure of scientific and technical literature have been organised in the United Kingdom. In 1962 the first of a series of such courses, each lasting two weeks, was offered by the National Lending Library (NLL) to postgraduate research students (Wood and Barr, 1966), and since 1962 similar courses have been organised for university librarians and academic staff. Largely as a result of the NLL's courses, seminars on the use of scientific literature have been arranged in a number of British universities, and technical colleges. Similar courses have been organised in some American universities (Power, 1964) and in Europe (Bottle, 1967).

It is probably true to say that the majority of these courses have been concerned with chemical literature. Certainly chemists appear to be more aware of the information problem than are most other scientists. Earth scientists in particular have been slow to appreciate the problems posed by the literature explosion and only recently have there been moves to try and control the effects of the explosion (see, for example, Smith, Creager and Sayer, 1967). This is partially understandable in view of the fact that the growth of literature has not been quite as spectacular in the earth sciences as in other subject fields. It has been upwards nevertheless, and for this reason, and because of certain characteristics displayed by the literature, the earth scientist cannot afford to remain indifferent to the information problem. The characteristics in question include the interdisciplinary nature of much of the geological literature and the facts that it is more international than the literature of any other subject and that it remains current for a much greater period than the literature of other scientific disciplines.

This book has been prepared with a view to its being used as the basis for a course on the literature of the earth sciences but it will also be of use to students and practising geologists who wish to undertake for themselves the task of becoming familiar with the literature. Library school students should also find it of use. The book is not intended to be a comprehensive reference tool: the more important literature is of course mentioned, but for details of additional publications the reader will find himself referred to other reference works. In the interest of readability and following the precedent established by earlier books in the Information Sources for Research and Development series, certain bibliographic details (such as the place of publication and length of book) have been omitted. These are readily obtainable from standard reference works.

Chapter 2 concerns itself with libraries where earth science literature can be obtained and provides some useful background information regarding the classification systems in use in such libraries. Chapters 3–7 deal with the various types of publications which contain information of potential interest to the earth scientist, and Chapter 8 concludes the first half of the book with a section on the technique of literature searching and sections on personal record keeping and report writing. In the second half of the book the literature of each branch of the earth sciences is considered in turn and the concluding chapter presents a series of questions which can be used for practical work in the type of course mentioned above.

The need for the present book arises out of the considerable changes and growth which have taken place in the earth science literature since 1951, i.e. the date of publication of *Guide to Geologic Literature*, by R. M. Pearl (McGraw-Hill). This was the first real attempt to introduce geologists to the various forms of scientific publication. The book has the merit of being eminently readable and, although it is now somewhat out of date, it presents a vast amount of still relevant information.

Since 1951 other guides to earth sciences literature have been published but, unlike the present work and that by Pearl, they have been mainly reference works. Their function has not been to provide an introduction to the whole subject of searching the earth sciences literature but to help the practising scientist locate a book, periodical or other source of information on a particular aspect of the earth sciences. Since these guides, particularly if used in conjunction with the present work, will be of undoubted value to the reader who is intent on becoming familiar with the literature of the earth sciences, details are given below.

The Literature of Geology, by B. Mason (Edwards Bros., 1953). This consists of two parts, the first of which lists items according to the form of the publications, e.g. abstracting journals, bibliographies, dictionaries, glossaries, etc., and according to broad subjects, e.g. economic geology, geophysics, etc. In the second part entries are arranged on a regional basis.

'Sciences de la Terre', by P.-M. Guelpa, in *Les Sources du Travail Bibliographique*, Tome III, *Bibliographies Spécialisées (Sciences Exactes et Techniques)*, by L.-N. Malcles (Droz, 1958). A select list of books, bibliographies and journals arranged according to broad subject groups. This is now mainly of historical interest.

Guide to Information Sources in Mining Minerals and Geosciences, Vol. 2 of *Guide to Information Sources in Science and Technology*, edited by S. R. Kaplan (Interscience, 1965). This is a comprehensive publication produced in two parts. Part I lists, in country order, the full name, address, telephone number, details of function and organisational structure, membership and publications of over 1000 worldwide organisations. Part II deals with the literature and lists over 600 worldwide publications, including bibliographies, dictionaries, abstracting journals, primary periodicals, directories and yearbooks. Entries are arranged in broad subject groups.

'Geologic Reference Sources', by D. C. Ward (*University of Colorado Studies, Series in Earth Sciences*, No. 5, 1967). The book is worldwide in scope and lists over 1000 publications of all types together with brief explanatory notes. It is divided into three sections—general, subject and regional. A revised and enlarged edition by D. C. Ward and M. J. Wheeler was published under the same title by Scarecrow Press in 1972.

Information Sources in the Earth Sciences, by J. B. Watkins (Syracuse University Libraries, 1967). This is a select list of earth science publications held by the Syracuse Natural Sciences Library. The entries, of which there are just over 1000, are arranged by form of publication, i.e. abstracting journals are dealt with in one section, handbooks in another and so on. Within each section, titles are listed in alphabetical order and in many instances are accompanied by annotations. Primary periodicals are not included.

An Introductory Guide to Sources of Information for the Literature of Geology, by J. W. Mackay (University College London, Department of Geology, 1971). This is a duplicated pamphlet, 63 pages long, aimed at research workers. It deals briefly with reference works and in considerable detail with abstracting and indexing services.

In addition to these general guides, there are many others which have been prepared for more specialised purposes. These include such publications as *Directory of Geological Material in North America*, by J. V. Howell (American Geological Institute, 1957); *Oceanography Information Sources*, compiled by R. C. Vetter

(National Academy of Sciences, 1970); *Oceanography in Print—A Selected List of Educational Resources*, by L. Forbes (Town of Falmouth, Massachusetts, 1968); 'Oceanographic Research Literature—A Bibliography of Selected Guides and Periodicals', compiled by A. P. Harvey, in *Ocean Research Index* (Francis Hodgson, 1969); *Sources of Information on Geology and Mining in the Western States* (Stanford Research Institute, 1957); *Selected References for Earth Science Courses*, by W. H. Mathews III (Prentice-Hall, 1964); and a host of guides to the publications of particular bodies such as national and state geological surveys and professional institutes. Many of these will be mentioned in subsequent chapters.

Because of the dependence of the earth sciences on work done in other fields, the reader will find much useful information in guides to the literature of other disciplines as well as in guides to the literature of science as a whole. A list of those most likely to be of value is given in the Appendix at the end of this chapter.

REFERENCES

Bottle, R. T. (1967). *Courses in the Use of Scientific and Technical Literatific and Technical Periodicals'. *Journal of Documentation*, **23** (2), 110–116

Barr, K. P. (1967). 'Estimates of the Number of Currently Available Scienture in Central and Eastern Europe*. University of Bradford

Power, E. (1964). 'Instruction in the Use of Books and Libraries: A Preliminary Report to the International Association of Technical University Libraries', *Libri*, **14** (1), 253–263

Smith, F. D., Creager, W. A. and Sayer, J. S. (1967). *Developing a Coordinated Information Program for Geological Scientists in the United States*. American Geological Institute

Urquhart, D. J. (1964). 'Rising Tide of Paper'. *The Advancement of Science*, **21** (91), 279–285

Wood, D. N. (1968). 'Library Education for Scientists and Engineers'. *Bulletin of Mechanical Engineering Education*, **8** (1), 1–8

Wood, D. N. and Barr, K. P. (1966). 'Courses on the Structure and Use of Scientific Literature'. *Journal of Documentation*, **22** (1), 22–32

APPENDIX: SOME 'GUIDES TO THE LITERATURE' OF OTHER SUBJECT FIELDS

Anthony, L. J. *Sources of Information on Atomic Energy*. Pergamon (1966)

Bottle, R. T. (ed.). *The Use of Chemical Literature*, 2nd edn. Butterworths (1969, revised 1971)

Bottle, R. T. and Wyatt, H. V. (eds.). *The Use of Biological Literature*, 2nd edn. Butterworths (1971)

Blanchard, J. R. and Ostvold, H. *Literature of Agricultural Research*. University of California Press (1958)

Burkett, J. (ed.). 'Concise Guide to the Literature of Geography'. *Ealing Technical College, School of Librarianship, Occasonal Paper*, No. 1 (1967)

Carey, R. P. J. *Finding and Using Technical Information*. Edward Arnold (1966)

Goldman, S. *Guide to the Literature of Engineering, Mathematics and the Physical Sciences*, 2nd edn. Johns Hopkins University (1964)

Gould, R. F. (ed.). 'Searching the Chemical Literature'. *Advances in Chemistry Series*, No. 30. American Chemical Society (1961)

Grogan, D. *Science and Technology—An Introduction to the Literature*. Clive Bingley (1970)

Houghton, B. *Mechanical Engineering—The Sources of Information*. Clive Bingley (1970)

Jenkins, F. B. *Science Reference Sources*, 5th edn. M.I.T. Press (1969)

Malinowsky, H. R. *Science and Engineering Reference Sources*. Libraries Unlimited (1967)

Minto, C. S. *How to Find Out in Geography—A Guide to Current Books in English*. Pergamon (1966)

Parke, N. G. *Guide to the Literature of Mathematics and Physics including Related Works on Engineering*, 2nd edn. Dover Publications (1958)

Roberts, E. G. *The Literature of Science and Engineering*, 2nd edn. School of Information Science, Georgia Institute of Technology (1969)

Smith, R. C. and Painter, R. H. *Guide to the Literature of the Zoological Sciences*, 7th edn. Burgess Publishing Company (1966)

Swift, L. H. *Botanical Bibliographies: A Guide to Bibliographic Materials Applicable to Botany*. Burgess Publishing Co. (1970)

Whitford, R. S. *Physics Literature: A Reference Manual*. Scarecrow Press (1968)

Yates, B. *How to Find out About Physics*. Pergamon (1965)

2

Libraries and their use

Alice M. Tyler

Interest in the earth sciences has increased rapidly in the last 25 years, with a resulting increase in the information available. A particular piece of information may be stored in the mind of one man, included in a paper or book, or recorded on microfilm or magnetic tape. The methods used differ but their function is the same—to hold the piece of information in readiness for the next user. At present most scientific information is still stored in the form of the written word, both published and unpublished. Libraries have developed as storehouses for these written records and as a means of access to the information contained in them.

'A place for everything and everything in its place' is an excellent motto for any library. In pursuit of this ideal, various schemes have been devised to classify knowledge so that information, and the literature containing it, can be arranged systematically. The three most widely used schemes, in the English-speaking countries, are the Dewey Decimal Classification, the Universal Decimal Classification and the Library of Congress Classification. The Dewey Decimal Classification scheme (DDC) was developed in the USA and has, naturally, a somewhat North American bias. It was for this reason that, about 70 years ago, two Belgians, Paul Otlet and Henry la Fontaine, obtained Melville Dewey's permission to adapt the 5th edition of his scheme to make it more applicable to the European scene and, more importantly, to expand its coverage of science and technology to include all the new developments. The adaptation became the Universal Decimal Classification (UDC), which is kept up to date through constant revision by subject experts. Development and further improvement takes place under the auspices of the International Federation for Documentation (FID) and the most up to date section on the earth sciences has been published under the title *Special-Subject Tables for Geology, Surveying, Cartography and Related Branches of Science and Engineering* (FID No 431, Israel Program for Scientific Translations, 1970). The Library of Congress Classification (LC) was, as its name implies, developed for the Library of Congress in Washington. Since it was developed for an existing library and not based on an abstract division of knowledge, it has certain advantages, for the librarian, over the other schemes. The different histories of the three schemes may account for many public libraries using DDC, for some scientific and technical libraries using UDC and many academic libraries using LC.

The three schemes divide knowledge in the following ways. DDC and UDC use Arabic numerals, while LC uses the Roman alphabet, to symbolise the major divisions. Thus:

Decimal		Library of Congress	
000	General works	A	General works
100	Philosophy	B	Philosophy–Religion
200	Religion	C	History—Auxiliary sciences
300	Social sciences		(D–P, see Library of Congress
400	Language		schedules)
500	Pure science	Q	Science
600	Technology	R	Medicine
700	The Arts	S	Agriculture
800	Literature	T	Technology
900	History	U	Military science
		V	Naval science
		Z	Bibliography

UDC retains the same major divisions as DDC but drops the three-digit minimum, e.g. in DDC Earth Sciences is 550, in UDC 55; similarly Palaeontology is symbolised 560 and 56. UDC uses letters and arbitrary symbols in addition to the Arabic numerals, so that the scheme can be both more expansive and more flexible, without the necessity for very long decimal numbers, such as are found in DDC. Thus it would be unwise to assume that a thorough knowledge of DDC would enable one to use a library classified by UDC, or vice versa, without help from the staff.

Within the DDC Classes 550 and 560 the Earth Sciences and Palaeontology are broken down into ten major divisions, as follows:

550 Earth sciences
551 Physical and dynamic geology
552 Petrology
553 Economic geology

554 to 559	Regional geology	554	Europe
		555	Asia
		556	Africa
		557	North America
		558	South America
		559	Other parts of the world

| 560 | Palaeontology |
| 561 | Palaeobotany |

562 to 569	Taxonomic palaeontology	562	Invertebrate
		563	Protozoa, Parazoa, Metazoa
		564	Mollusca and Molluscoidea
		565	Other invertebrates
		566	Vertebrate palaeozoology
		567	Anamnia
		568	Saropsida
		569	Mammalia

Each of the above divisions is further divided: e.g. Petrology (552) is divided into Igneous rocks (general) (552.1), Volcanic rocks (552.2), Plutonic rocks (552.3), etc.

The Library of Congress scheme divides the Earth Sciences into six major divisions, as follows:

QE		Geology
QE	1–350	General
QE	351–399	Mineralogy
QE	420–499	Petrology
QE	500–625	Dynamic and structural geology
QE	651–700	Stratigraphic geology
QE	701–996	Palaeontology. Palaeozoology. Palaeobotany

Variations on all three themes, both official and unofficial, are to be found in use. There are other recognised schemes, though none is as widely used as the three described. In addition, some home-made schemes are still used; and some libraries have no scheme—the books are simply shelved in the order in which they are added to the library.

In an ideal library the reader would find everything the library possesses relevant to his enquiry shelved at the same place. Reference to an index to the classification used and to a plan of the library would tell him at which point on the shelves to look. In life this ideal is not possible. A book (or journal article or pamphlet, etc.) may deal with more than one subject, or several aspects of a single subject, but it can only occupy one point in space. The obvious solution of providing duplicate copies to be placed at every relevant point in the scheme is ruinously wasteful of both money and space. Thus the librarian must decide the most appropriate place for each publication. The decision will be influenced by the type of library as well as by the subject of the book. For example, a general library would place a book on the Geology of Scotland in its geology section at 544.1 whereas a library specialising in geology would place it in the section on Scotland at 914.1. This is why the same book may appear in several different places in different libraries, even when they use the same classification scheme. It is a good idea, therefore, when visiting a library for the first time, to ask the librarian to explain the classification scheme and to show one round.

The contents of a library and their location on the shelves are given in the catalogue, which can be thought of as a directory giving the bibliographic details (full name) and the classification (zip or postal code) of each item in the library's stock. Additional information—for example, that the book is too large to be placed at its expected point on the shelves and is therefore placed at a special location—is also given. There may also be various other symbols which are of a purely administrative nature, such as price, date of accession or bookseller, which can be ignored by the reader. But if in doubt as to the importance of a symbol in the catalogue, ask the librarian.

The catalogue may appear in various forms, the most usual being a card index on 5×3 in cards (125×75 mm) (see *Figure 2.1*). However, sheaf, guard-book and 'visible index' catalogues may also be encountered.

```
WOOD, Alan (ed.)
    The Pre-Cambrian and Lower Palaeozoic
    Rocks of Wales. Cardiff, University
    of Wales Press. 1969.

    x + 461 pp. figs. tbls. sk. maps.
    Report of a symposium. . .Jan., 1967.

                              551. 714 29
```

Figure 2.1

Within the catalogue the entries are usually arranged in two main sequences, by author and by subject.

The 'author' can be a person, or an institution responsible for the publication of a document. There are various rules and regulations governing the way an author catalogue is compiled. The two in most common use are the Anglo-American Cataloguing Code compiled by the American Library Association in conjunction with the Library Association, London, and the British Museum Cataloguing Rules. The codes differ in the methods used to deal with main entries for authors with double-barrelled names or pseudonyms or where the author has changed his name, etc. In these cases, however, there should be a reference from the rejected form to the chosen main entry form. There may also be problems with alternative forms of spelling, e.g. Frič or Fritch, but, again, cross-references will refer the reader to the main entry. The Main Entry is the full description of an item in the library. The fullness of the description varies; it may simply identify a publication or, in the case of valuable books, it may be so detailed that each copy can be distinguished from every other copy of the same work. However, since books and papers are not written to conform with the rules of librarianship, frequent adaptation and bypassing of these rules may be found in the catalogue of a large library.

The subject catalogue can be in dictionary form or in classified order. Under each heading are listed the books, etc., dealing with that particular subject. A classified catalogue has an index to refer the reader to the appropriate section of the scheme. Since there can be an excellent section or chapter in a book that deals mostly with another subject, there may well be cross-references, as, for example, the instance when some very useful information on iron was found in a book on non-ferrous metals. This detailed subject searching and indexing, however, is only possible in the high-powered research libraries, where the ratio of librarians (or information officers) to research workers may be as high as 1:20 or even higher.

In the UK the usual form of the catalogue is an alphabetical author catalogue with a separate classified subject catalogue. In the USA the subject catalogue is usually in dictionary form and is interfiled with the author catalogue in one alphabetical sequence. There are two major methods for filing the entries in an alphabetical catalogue: 'word-by-word' and 'letter-by-letter'. These are illustrated below.

Word-by-word	*Letter-by-letter*
New Amsterdam	New Amsterdam
New York	Newark
New Zealand	Newcastle
Newark	Newnham
Newcastle	New York
Newnham	New Zealand

Periodicals may be listed by title in the main sequence or formally catalogued under the societies or institutions publishing them. However, because of the increasing number of commercially published journals the rules tend to break down. It is becoming more common for the periodical holdings of a library to be shown in a separate list, which may or may not have the same physical form as the main catalogue. The periodicals themselves are often shelved in a particular section of the library. Their arrangement may be by subject, by country of publication, by title, or purely arbitrary. There is often a special display of the latest numbers, and very early issues may be stored at some distance from the current volumes. This may involve the reader in ordering a particular part a day or two before it is required, if the back numbers are in a warehouse on the other side of the town. All the information required to locate a particular part is usually given either on the catalogue card or in a special periodicals list.

Pamphlets or offprints from periodicals may be catalogued and classified or simply filed in boxes in author order. There is no standard practice for dealing with such material, so the advice of the staff should be sought. This also applies to theses, manuscripts and other material which will not fit neatly into the categories of book or periodical.

Of particular interest to geologists are the maps published by the various surveys, and other bodies, throughout the world. There is, again, no standard practice for the cataloguing and classification of maps but interested bodies on both sides of the Atlantic are discussing the problems involved. It is thus necessary to ask the staff for advice on the maps held by a particular library and their availability. In the majority of cases maps are strictly for reference only.

Some libraries hold material in micro form. This, again, will be out of the main sequence, since it requires special storage and equipment for its use. Photographs, illustrations and lantern slides may also be available. There may even be specimens: after all, a museum is simply a 'library' of objects rather than words. For the geologist, particularly the palaeontologist, reference to the type specimens can be as important as reference to the standard works.

Obviously, no one library can hold all the material its readers may ever require. To overcome this difficulty, co-operative schemes have grown up between libraries in related fields. Some of this co-operation is purely informal on a mutual benefit basis, but there are, in many areas, regional interlending schemes. In Britain these are often based on the local Public Library headquarters with the National Central Library (NCL) acting as a national co-ordinator for inter-regional borrowing. The NCL can also arrange to borrow material from abroad. Another national library is the National Lending Library (NLL) at Boston Spa, Yorkshire. The NLL collects and makes available for loan the scientific and technical literature of the whole world. It currently subscribes to over 36 000 periodicals and has a comprehensive collection of recent English language and Russian books, semi-published report literature, conference proceedings and translations. Its by-return-of-post loans service is restricted to the UK and can be utilised through the library of one's own organisation. The NLL also offers a rapid photocopy and microfilm service to organisations and individuals throughout the world. A public reading room is available at Boston Spa and any scientist or librarian is welcome to visit the library and make use of its stock.

The ideal situation, in which the reader would find everything relevant to his enquiry 'shelved at the same place', may be achieved with the establishment of data banks based on computer storage and retrieval. Ultimately, perhaps, the data bank would contain data from all the work done by every earth scientist, whatever his field of activity. Anyone requiring to know what was known on any topic would simply 'set the dial' and the relevant facts would be produced, like sausages out of a machine. Before the computer can be built, however, the terminological, legal and political problems involved must be solved. Computers are currently being used to trace the literature in which the data have been reported, in, for example, the MEDLARS scheme (for medical literature), and the new GEO-REF system operated by the Geological Society of America and the American Geological Institute. However, finding that the information exists is only the beginning; it is sometimes very difficult to place it in the reader's hands. Perhaps it should be made obligatory that a bibliographic service guarantees to supply on demand every article cited, to any reader, either in his own language or in one of those recommended by UNESCO.

In the meantime the reader can study the literature in the libraries accessible to him. Most countries have established geological surveys and their libraries are often open to the public. Even if they are not 'public' libraries, no reader need hesitate to ask if it might be possible for him to make use of their facilities, if they are essential to his research and are unobtainable elsewhere. If it is not possible to arrange a personal visit, some form of copying service may be available. The making of copies of material required for research is now universal, but copyright law and customs differ widely. One may be required to fill in four forms in triplicate, or none at all. Many countries are reviewing their copyright law, in the light of the tremendous increase in copying with the development of modern techniques. There may well be major changes in the future.

In the UK one of the main geological libraries is that of the Institute of Geological Sciences (IGS) in South Kensington, London. It serves the Institute as a whole and the Geological Survey of England and Wales in particular. The Institute's offices in Edinburgh (19 Grange Terrace) house the Geological Survey of Scotland and its library, and another library is maintained at the Survey's offices in Leeds (Ring Road, Halton). The Institute's libraries are public reference libraries requiring no ticket or prior appointment. Northern Ireland has its own survey, based in Belfast. The United States Geological Survey is based in Washington, with branch offices throughout the country. Again, its libraries are available for public reference. There are also the state geological surveys, whose libraries specialise in literature relating to the particular state. These state surveys are often based on the state universities. The situation in France is similar; a university is responsible for the geological survey work in its area, with national responsibility held by the Bureau de Recherches Géologiques et Minières.

University libraries generally, whether or not they include that of a geological survey, are sources of earth sciences literature. Departments often specialise in a particular branch of the science— for example, sedimentology at Reading, England. The ease of access, and formalities to be completed, before a non-member of the university may use a departmental library are up to the individual department, so it is wise to introduce oneself by letter or by telephone. The same procedure is advised when planning a visit to other 'private' libraries such as those run by government research bodies or those owned by learned societies. Apart from the formalities, one might travel many miles only to discover the place closed for six months for redecoration!

The national libraries—for example, the British Museum Library or the Library of Congress—will be listed in guide-books, with their times of opening (and closures) and the procedure for using them. Other libraries, of every description, are included in special directories (see Bibliography) which are available for consultation in most university or local public library reference sections.

It is worth mentioning here, to illustrate the range of sources available to an earth scientist, the major libraries to be found in and around London. The British Museum (Natural History), in South Kensington, holds the national collections of minerals and fossils and its library is therefore of major importance to any earth scientist. Also in South Kensington is the Science Museum, whose library, in addition to being one of the major scientific libraries in the world, acts as a 'back-up' for the NLL, supplying material published prior to the latter's existence. The library belonging to the Geological Society of London (Burlington House, Piccadilly) is the oldest geological library in the world. Specialist aspects of the earth sciences are covered by material in the libraries of the National Institute of Oceanography (Godalming), the Institution of Mining and Metallurgy (Portland Place) and the Institute of Petroleum (New Cavendish Street). The stock of the Chemical Society's library (Burlington House) and that of the Institution of Civil Engineers (Great George Street) are of great importance to the mineralogist and engineering geologist, respectively. The Royal Astronomical Society has a geophysics section, and the extraterrestrial geologist will find its library useful. Details of other libraries in the London area can be found in a duplicated guide entitled *Libraries for the Geologist in and around London*, compiled by J. E. Hardy, the library information officer at Imperial College.

London is not unique; the libraries of New York and Washington, Moscow and Leningrad, Paris and the other major cities of the world have comparable collections.

SELECT BIBLIOGRAPHY

Anderson, B. L. (comp.). 'Special Libraries in Canada—A Directory'. *Research and Special Libraries Section. Canadian Library Association. Occasional Paper*, 73 (1968)

Dewey Decimal Classification and Relative Index, 17th edn., 2 vols. Forest Press (1965)

Directory of Geoscience Libraries in the United States and Canada. Geoscience Information Society (1966)

Directory of Information Resources in the United States. National Referral Center for Science and Technology (1967)

Lewanski, R. C. (comp.). *European Library Directory. A Geographical and Bibliographical Guide.* Leo S. Olschki (1968)

United States Library of Congress. *Classification. Class Q. Science*, 5th edn. USGPO (1950); reprinted 1963

Universal Decimal Classification. Abridged English Edition, 3rd edn. British Standards Institution (1961)

Wilson, B. J. (ed.). *Aslib Directory, Vol. 1. Information Sources in Science, Technology and Commerce*, 3rd edn. Aslib (1968)

World of Learning: Directory of the World's Universities, Colleges, Learned Societies, Libraries, Museums, Art Galleries and Research Institutes. Europa Publications, annual editions

3

Primary literature

D. N. Wood

Earth sciences literature, in common with the literature of other scientific disciplines, can conveniently be divided into primary and secondary. The primary literature contains the first reports of laboratory studies and field investigations, details of new hypotheses, descriptions of new equipment and so on. In this category are periodicals, theses, research reports, conference proceedings, government and international publications, patents and standards. The secondary literature is compiled from these primary sources and its object is to distil the information and present it in a convenient form. Various ways of doing this have been invented and as a result numerous types of secondary literature exist. Among these are included one-off bibliographies, abstracting and indexing periodicals, review journals, monographs, textbooks and numerous reference publications such as dictionaries, handbooks, encyclopaedias and data compilations. A third class of literature (tertiary) is sometimes recognised. This includes such things as location lists of periodicals, lists of books, lists of indexing and abstracting publications and guides to the literature such as those listed in the Appendix to Chapter 1.

In reality these categories are rather arbitrary and the dividing lines between them are far from distinct. Many periodicals, for instance, contain a mixture of primary and secondary information. Both theses and reports may contain substantial reviews of previous work in the field, while the original work they contain may be written up subsequently for publication in a periodical. Similarly textbooks sometimes contain facts, or more usually hypotheses, which have never been reported or advanced before.

Notwithstanding the rather indefinite boundaries, however, the divisions do exist and this chapter will concern itself with the primary literature—periodicals, government publications, theses, reports and conference proceedings. Patents and standards are not dealt with, since few earth scientists will have cause to use them and in any case they are covered adequately in other guides to the literature, e.g. *Technical Information Sources*, by B. Houghton (Clive Bingley, 1967).

PERIODICALS

The term 'periodical' is used here as a blanket term to include serials, journals, memoirs, bulletins, transactions, newsletters, proceedings, circulars or indeed anything which is issued regularly or irregularly as part of a numbered series.

The first scientific periodicals, *Philosophical Transactions of the Royal Society* and *Journal des Sçavans*, were published in 1665 and developed out of personal correspondence and meetings—the two main means used by scientists of the day to communicate with one another. Many of the early journals were short-lived but by 1800 over 100 titles were currently being produced. By 1900 this figure had increased to 5000 and at the present time it is over 30 000. The value of the periodical as a means of disseminating and acquiring information is evidenced in a recent description of it as the 'most successful and ubiquitous carrier of scientific information in the entire history of science'.

The over-all growth in the number of periodicals has been reflected in the earth sciences field and it has been estimated that information of interest to geoscientists is to be found scattered through about 4000 current titles. These generate about 50 000 individual papers per annum. In common with the periodical literature of other fields, earth sciences journals have become progressively more specialised. In the early years periodicals were general in their coverage and contained geological and mineralogical material alongside that of other disciplines. The first issue of the *Philosophical Transactions of the Royal Society*, for instance, contained a paper on 'A peculiar lead-ore of Germany . . .' alongside 'an account of a very odd monstrous calf' and 'an account of the improvement of optick glasses'. Although some general journals, such as *Nature*, continue to exist, the majority are now devoted to narrower branches of science. This tendency to specialise developed towards the end of the eighteenth century and during the early part of the nineteenth century. The first journal to be produced which was primarily of interest to earth scientists was *Journal des Mines* (1795–1815) and the first purely geological journal appears to have been the *Transactions of the Geological Society of Cornwall* (1818–32). During the nineteenth century such titles as *Mineralogical Magazine* (1876–   ) appeared; the current trend towards even further specialisation is typified by the title *Clays and Clay Minerals* (1968–   ).

Although the early periodicals were issued almost exclusively by learned societies, an examination of current periodicals shows that a wide range of organisations is involved in their production. Learned societies still figure prominently and many of what can be described as the core journals are published by them. The characteristic feature of their publications is the strict refereeing system which pertains. All submitted papers are sent to an independent referee for comment and criticism. Although this leads to delays in publication, it helps to maintain the quality of the journal and thus the reputation of the society concerned. Typical of this category of periodicals are *Bulletin of the Geological Society of America*, *Bulletin of the American Association of Petroleum Geologists*, *Proceedings of the Geologists Association* and *Palaeontology* (the journal of the Palaeontological Association).

Few, if any, disciplines depend more than the earth sciences upon the work that has been undertaken and published by governments and this fact is reflected in the number of periodicals produced under their auspices. In most cases these are products of the various national geological surveys and carry titles such as *Bulletin*, *Memoir* and *Record* of the appropriate organisation.

During the last 20 years commercial publishers have developed a keen interest in the production of learned journals. Elsevier and Pergamon are two companies that have entered this field and among the titles produced are *Tectonophysics*, *Marine Geology* and *Sedimentology* (Elsevier), and *Journal of Terramechanics*, *International Journal of Rock Mechanics and Mining Science* and *Deep Sea Research and Oceanographic Abstracts* (Pergamon). The quality of such publications is maintained by having an editor (in most cases an academic scientist) usually supported by a distinguished editorial board. In addition, many commercial publishers produce periodicals on behalf of learned societies. Such is the case with the *Scottish Journal of Geology*, published by Oliver and Boyd on behalf of the Geological Societies of Glasgow and Edinburgh.

Commercial organisations are not, of course, solely concerned with learned and research journals; indeed they are more commonly associated with the applied and trade end of the literature range. The mining industry in particular is well served by such publications as *Mining Annual Review*, *Mining Journal* and

Mining Magazine. These types of journals rely a great deal on advertising revenue to keep the price relatively low and the circulation figures correspondingly high. Compared with learned periodicals they are characterised not only by a proliferation of advertisements but by many more news items and photographs. Correspondence columns, editorials and shorter articles are other distinguishing features.

Commercial firms are also responsible for most of the 'popular' general science journals, including such titles as *Scientific American*, *Science*, *New Scientist* and *Nature*. All these contain mainly news or reviews and are therefore more secondary than primary. *Nature*, however, does contain papers reporting original work and the speed with which it publishes make it a useful vehicle for establishing 'priority' in a field.

House journals are another type of periodical. Produced by industrial firms for both prestige and advertising reasons, they contain not only news about the companies concerned but frequently articles of a more general nature. A typical example is *Optima*, the house publication of the Anglo-American Corporation of South Africa. This regularly contains articles of geological interest.

Universities and independent research institutes also frequently produce periodicals. These are often semi-administrative documents reporting on activities carried out during a particular period, or they may be truly research journals. Examples include the *Annual Report of the Research Institute of African Geology, University of Leeds* and *Polar Record*, published by the Scott Polar Research Institute at Cambridge, England.

Most periodicals are provided with annual indexes to help the user locate particular items of information. In some cases even longer indexes have been produced. Such cumulations are available for, e.g., Vols. 1–50, 51–90 and 91–118 of the *Quarterly Journal of the Geological Society of London* and for Vols. 9–100 of the *Geological Magazine*. Although now considerably out of date, a still useful guide to such cumulative indexes is D. C. Haskell's *A Check List of Cumulative Indexes to Individual Periodicals in the New York Public Library*, published in 1942.

Lists of periodicals

The identification and location of relevant periodicals is a high-priority and difficult job for any scientist. For the earth scientist, whose information is liable to come from any corner of the world and who relies to such a large extent on information from other disciplines, the task is daunting. To help, a large number of lists is available, many of which are detailed in two bibliographies, *Guides to Scientific Periodicals*, by M. J. Fowler (Library Association, 1966), and *Union Lists of Serials—A Bibliography*, by R. S. Freitag (Library of Congress, 1964). One of the most comprehensive and widely available lists is the *World List of Scientific Periodicals* (4th edn., 3 vols., Butterworths, 1963–65). This is an alphabetical listing of scientific and technical periodicals currently published after 1900. For each title the recommended abbreviation and the holdings of certain British libraries are indicated. Also published by Butterworths is the *British Union-Catalogue of Periodicals* (BUCOP), a list of all periodicals held by major British libraries. The work, published 1955–58, is in four volumes and a *Supplement* covering the period up to 1960 was issued in 1962. Since then supplements (*New Periodical Titles*) with cumulations (e.g. 1960–68) have appeared at regular intervals and keep the work up to date. During the 1960s it was decided to combine the *World List* and BUCOP and the latter now carries the sub-title 'Incorporating World List of Scientific Periodicals'.

The North American equivalent to BUCOP is the *Union List of Serials in Libraries of the United States and Canada* (2nd edn., 4 vols., H. W. Wilson, 1943). Supplements covering 1941–43 and 1944–49 were published in 1945 and 1953, respectively. The list records the holdings of approximately 500 libraries and is updated by the Library of Congress publication *New Serial Titles*. This appears eight times a year with quarterly and annual cumulations. Larger cumulations cover 1950–60, 1961–65 and 1966–68. A *Subject Index to New Serial Titles 1950–65* was published by Pierian Press in 1968.

Union lists are available for many other countries. Two published relatively recently in English-speaking countries are the *Union List of Scientific Serials in Canadian Libraries* (2nd edn., National Science Library–National Research Council of Canada, 1967) and *Scientific Serials in Australian Libraries* (Commonwealth Scientific and Industrial Research Organisation, Australia, 1967).

Union lists record the holdings of several libraries. In addition, many individual libraries publish lists of their own periodical holdings and in the case of the larger ones these can be of great value in helping one determine the available material, correct titles and where to consult them. Among such lists can be cited (a) *Periodical Publications in the National Reference Library of Science and Invention. Part 2, List of Slavonic and East European Titles in the Bayswater Division.* Part 1, *List of Non-Slavonic Titles in the*

Bayswater Division. Part 3, *List of Current Titles in the Holborn Division* (British Museum, 1969, 1970, 1971)—most of the earth sciences material is in Part 1; (b) *Current Serials Available in the University Library and in Other Libraries Connected with the University, 1970* (Cambridge University Library, 1971)—this contains 27 000 titles; (c) *List of Serial Publications in the British Museum (Natural History) Library* (British Museum (Natural History), 1968)—a list of 12 500 titles held by the various departments, including mineralogy and palaeontology, of the museum; (d) *Current Serials Received by the NLL, March 1971* (HMSO, 1971)—this contains nearly 36 000 entries.

Some lists such as *Ulrich's International Periodicals Directory* (14th edn., 2 vols., Bowker, 1971) and its sister publication *Irregular Serials and Annuals* (Bowker, 1972) provide more information than just the title. In these two cases entries are arranged under subject headings and contain details of starting date, frequency, price and publisher's address. More or less similar information is given in hundreds of nationally based lists (see M. J. Fowler, above). Typical of these are the *Guide to Current British Periodicals*, edited by M. Toase (Library Association, 1962), which contains 3800 titles; and *Directory of Canadian Scientific and Technical Periodicals: A Guide to Currently Published Titles* (4th edn., National Science Library, Ottawa, 1969), which lists 890 titles.

For details of old titles reference should be made to 'Catalogue of Scientific and Technical Periodicals 1665–1895', by H. C. Bolton, which was published as *Smithsonian Miscellaneous Collections*, **40**, 1898; and to *Catalogue of Scientific Serials . . . ,* by S. H. Scudder (Harvard University Library, 1879; reprinted by Kraus, 1965). It is important to note that Bolton's work does not include the transactions of learned societies. Of more specific interest to the earth scientist is the list of periodicals which appears in Volume 1 of *Bibliographia Zoologiae et Geologiae. A Catalogue of all Books, Tracts and Memoirs on Zoology and Geology,* by L. Agassiz, corrected, enlarged and edited by H. E. Strickland (4 vols., Ray Society, 1848). This was reprinted by Johnson as *The Sources of Science*, No. 20, 1968.

Probably the most useful list of periodicals for earth scientists is 'Geoserials 1969', a world list of over 1500 current geoscience serial publications plus publishers' addresses, which was issued as *Geoscience Documentation*, **1** (1) (1969), and which is updated regularly in the same journal. Also of value are the various lists put out by abstracting and indexing services to indicate coverage. One of the best of these is *Serial Publications Commonly Listed in Technical Bibliographies of the United States Geological Survey* (US

Geological Survey, 1967), which lists over 1800 periodicals covered by *Geophysical Abstracts, Abstracts of North American Geology, Bibliography of North American Geology* and *Bibliography of Hydrology of the United States.*

The Geological Society of London has issued *List of Periodicals Taken in the Library*, by A. M. Paddick (1962). Of interest to geographers on the one hand and geochemists, petrologists and mineralogists on the other are 'Annotated World List of Selected Current Geographical Serials in English, French and German: including serials in other languages with supplementary use of English or other International languages', by C. D. Harris (3rd edn., *University of Chicago, Dept. of Geography, Research Paper*, 137, 1971), 'International List of Geographical Serials', by C. D. Harris and J. D. Fellman (2nd edn., *University of Chicago, Dept. of Geography, Research Paper*, 138, 1971) and *Access* (Chemical Abstracts Service, 1969). The last is a list of 21 000 periodicals, monographs and conference proceedings, including virtually all publications abstracted by *Chemical Abstracts* since 1907 and *Chemisches Zentralblatt* since 1830 and those cited in *Beilstein's Handbuch der Organischen Chemie* before 1907.

Finally it is worth noting that the American Geological Institute has commissioned H. R. Malinowsky to undertake an 'Inventory Analysis of the Geoscience Serials of the World'. This will include map series and will contain all material current in 1960 or later.

Guides to government publications

As has been pointed out earlier, earth scientists rely heavily on work published under government auspices. The majority of this work is issued in periodical form as numbered circulars, papers, bulletins, memoirs, reports, etc., of national or state geological surveys. Details of many of these publications are given in Chapter 9 and in the periodical lists mentioned above. The individual issues in the various series are usually concerned with the geology, mineral resources, hydrology, etc., of a particular area or with a particular group of fossils. In this respect, since they are dealing with a single subject, they are more like books than the typical periodical issue.

In addition to using purely earth science publications, readers may occasionally have recourse to other types of government publications—agricultural, statistical, legislative and so on.

Accounts of government publishing activities, particularly in the UK and USA, will be found in any large library. A useful summary appears on pp. 66–83 of *The Use of Biological Literature* (2nd edn., Butterworths, 1971) and, although it is now somewhat out of date, considerable space is devoted to official publications in *Guide to Geological Literature*, by R. M. Pearl. For those who wish to delve even deeper into this complicated field a list of useful books is given in the Appendix to this chapter.

\This section will deal only with the major indexes to government publications.

As far as British material is concerned, a *Daily List of Government Publications* appears on Mondays to Fridays inclusive and gives details of both Parliamentary and non-Parliamentary publications which are available through HMSO. The *Daily List* cumulates monthly and annually under the title *Government Publications*. Both cumulations have an alphabetical and Standard Book Number index. *Government Publications* includes details not only of the publications of British Government Departments but also those of the Research Councils (including the Natural Environment Research Council, of which the Institute of Geological Sciences is a part), and international organisations such as the United Nations, for which HMSO is the British agent. In the case of the annual cumulations, non-British material is excluded but is listed in a supplement entitled *International Organisations and Overseas Agencies Publications*. Since 1936 quinquennial indexes to *Government Publications* have been produced. The latest covers the period 1961–65.

Besides these general lists, HMSO issues *Sectional Lists*. These are produced at irregular intervals and record the publications of a particular Department, in a particular series or on a particular subject. Successive editions of a particular 'List' always carry the same number. Those most likely to be of interest to earth scientists are Nos. 3 (Department of Trade and Industry), 37 (Meteorological Office), 45 (Institute of Geological Sciences), 63 (Atomic Energy) and 67 (Overseas Affairs). Many, but by no means all, British government publications are also listed in the *British National Bibliography* (see p. 66).

The principal guide to United States government publications is the *Monthly Catalog of United States Government Publications*, which started life as *Catalogue of the United States Public Documents* in 1895. The December issue each year contains a cumulated index and a *Decennial Cumulative Index* is available for 1941–50 (USGPO, 1953) and 1951–60 (USGPO, 1967).

The coverage of the *Monthly Catalog* is not entirely complete as a record of US government documents, since not all such publications are produced by the US Government Printing Office and many consequently go unlisted. In an effort to overcome this problem, the Library of Congress has made a special effort to bring this mainly ephemeral material under bibliographical control by publishing *Non-GPO Imprints Received in the Library of Congress*. The first issue (1970), which covers July 1967–December 1969, includes a number of US Geological Survey leaflets, open-file reports and circulars.

In addition to the *Monthly Catalog*, the USGPO issues *Price Lists*. These give details of available government publications in different subject fields and are updated from time to time. Like the HMSO *Sectional Lists*, they are numbered and those primarily of interest to earth scientists are Nos. 15 (Geology), 46 (Soils and Fertilizers), 48 (Weather, Astronomy, Meteorology) and 58 (Mines).

Also important on the American scene are the individual state publications. These are particularly valuable to geologists, and a useful guide to the various series is *An Index to State Geological Survey Publications Issued in Series*, by J. B. Corbin (The Scarecrow Press, 1965). Also helpful in this area is the *Monthly Checklist of State Publications* produced by the Library of Congress. This lists monographs (including annuals and monographs in series) by state and issuing agency. Periodicals are listed semi-annually in June and December.

Space considerations preclude the possibility of discussing all the guides to government publications issued in other countries. A few of the more important titles, however, are listed below.

Australia:	*Australian Government Publications* (1952–)
	Australian National Bibliography (1961–)
Canada:	*Daily Checklist of Government Publications* (1952–)
	Canadian Government Publications, Monthly Catalogue (1953–)
	Canadian Government Publications, Annual Catalogue (1953–)
	Canadiana (1950–)
France:	*Bibliographie de la France* (1811–). F, 'Publications Officielles'
German Democratic Republic:	*Deutsche Nationalbibliographie* (1931–)
German Federal Republic:	*Deutsche Bibliographie* (1947–)
USSR:	*Knizhnaya Letopis* (1907–)

For a more detailed treatment and listing, readers should consult *The Use of Biological Literature*; 'Government Publications', by J. B. Childs, *Library Trends*, **15**, 378–397 (1967); and *Bibliographical Services Throughout the World 1960–64*, by P. Avicenne

(UNESCO, 1969). The last reviews the major bibliographical tools in each country and includes details of national bibliographies and any special guides to government publications. A volume covering 1965–69 by the same author was published in 1972.

Besides the official national lists of government publications, many government and state agencies produce their own lists of publications. The United States Geological Survey, for instance, has produced *Publications of the Geological Survey 1879–1961* (USGPO, 1964), which is updated by the monthly and annual *New Publications of the Geological Survey*. Similarly, the Bureau of Mines has prepared an annotated *List of Publications Issued by the Bureau of Mines from July 1st 1910 to January 1st 1960*. This is updated by a *List of Bureau of Mines Publications and Articles, January 1st 1960 to December 31st 1964*, which is itself supplemented by annual volumes issued under the same title. Canadian Geological Survey publications are listed in the *Index of Publications of the Geological Survey of Canada 1845–1958* and *1959–1969*, by A. G. Johnston (1961, 1970). It is intended that annual supplements will be issued. Examples of state lists include those of the Quebec Department of Natural Resources, *Catalogue of Publications since 1883* (1968), and Missouri Geological Survey and Water Resources, *List of Publications* (1970).

Publications from outside North America include the *Catalogue des Publications* of the Bureau de Recherches Géologiques et Minières (published annually) and the *Publications of the Bureau of Mineral Resources, Geology and Geophysics* (Dept. of National Development, Australia, 1967).

Publications of interest to earth scientists and particularly to those concerned with natural resources are frequently issued by the United Nations. As mentioned earlier, details of some UN publications appear in the HMSO lists, but a more complete record is to be found in the *United Nations Documents Index*, issued monthly since 1950.

Geographic origin of periodicals

Reflecting the universal interest in, and economic importance of, geology and related subjects, earth sciences periodicals are produced in practically every country of the world. This fact can be confirmed by scanning any of the lists mentioned earlier in this chapter. Two such lists—*Serial Publications Commonly Cited in Technical Bibliographies of the United States Geological Survey*, and the 'List of Serials' in *Bibliography and Index of Geology*—contain between them 2594 different titles from 80 countries. Not surprisingly, a detailed analysis of the lists reveals that a few countries are responsible for a large number of titles. The United States alone publishes over 1000 periodicals and with the USSR, United Kingdom, Germany, Canada, France and Japan is responsible for 70 per cent of the titles. A list of those countries producing 10 or more periodicals is given below.

United States	1002	Switzerland	26
USSR	199	Norway	21
United Kingdom	141	Spain	21
Germany	140	Czechoslovakia	20
Canada	131	Denmark	20
France	106	Mexico	20
Japan	101	Finland	18
Italy	67	South Africa	18
Netherlands	49	Argentina	17
India	46	Austria	15
Australia	43	Portugal	15
Belgium	32	Hungary	14
Brazil	32	New Zealand	14
Romania	30	Bulgaria	13
Sweden	29	Chile	11
Poland	27	Colombia	10

These figures, however, are only indicative of the volume and geographic distribution of earth sciences periodicals. The American bibliographic services are by no means comprehensive and there are many journals, particularly in peripheral subject fields, that occasionally carry relevant material but are not covered. Furthermore, a comparison of the figures with those published by Hawkes (see Chapter 6) suggests that the coverage of Russian journals in particular is less than satisfactory.

Use of periodicals

There are several characteristics of use of periodicals which are worth considering: (1) the relative use of periodicals and other forms of literature, (2) the use made of material in different languages, (3) the use made of periodicals of different ages, (4) the range of titles used and (5) the use made of periodicals from other disciplines. Information regarding these characteristics can be obtained by analysing the citations in earth sciences articles. The results of citation studies in the earth sciences field have been published by Gross and Woodford (1931), Brown (1956), Woodford (1969) and Craig (1969). An additional study was undertaken by the author in 1966; and since the results of this have never previously been published, they will be used as the basis of this section. Comparisons with other results will be made where appropriate.

The citation analysis was carried out on the 1963 issues of nine British geological journals. The form (i.e. periodical, book, etc.), title and date of publications cited in articles by British-based authors were recorded. The results indicate in a broad way the range of literature used by British geologists. The journals used for the analysis were *Quarterly Journal of the Geological Society of London, Journal of Petrology, Palaeontology, Proceedings of the Geologists' Association, Geological Magazine, Mineralogical Magazine, Proceedings of the Yorkshire Geological Society, Liverpool and Manchester Geological Journal* and *Bulletin of the Geological Survey of Great Britain.*

In all, 3504 citations were examined. Of these, 82 per cent were to periodicals, a figure virtually identical with Woodford's (82.5 per cent) and very similar to Craig's (79.2 per cent). Books accounted for 12.5 per cent and the remainder were to theses, maps and reports.

A total of 536 periodicals were referred to, of which 287 were cited once only. Ninety-two titles were responsible for roughly 75 per cent of the citations, 20 for 50 per cent and 3 for 25 per cent. In order of frequency of citation, the top 20 periodicals were as listed in *Table 3.1.*

Alongside each title in *Table 3.1* is the number of citations it received (the figure in parentheses is the number of self-citations) and the ranked position in similar lists published by Craig (C) and Woodford (W).

A comparison of the three lists suggests that both British and American geologists are somewhat parochial in their interests and consequent reading habits. Only four periodicals, all American, appear in the top 10 titles of each survey—*Journal of Geology, American Mineralogist, Bulletin of the Geological Society of America* and *American Journal of Science.*

An interesting feature of the study is the number of titles other than earth sciences ones which are used by geologists. Of the 536 titles referred to, 88 (16.5 per cent) were from fields other than the earth sciences and ranged from astronomy, acoustics and agriculture to odontology, pharmacy and urology. This dependence on the literature of other subject fields, however, although considerable, does not appear to be as great as suggested by Earle and Vickery (1969), who found that only 52 per cent of the citations in geological periodicals and books were to literature that could

Table 3.1

	C	W
1. *Quarterly Journal of the Geological Society of London*	241(26)	23	19
2. *Geological Magazine*	211(56)		22
3. *Proceedings of the Geologists' Association*	156(106)		
4. *Memoirs of the Geological Survey of the United Kingdom*	135		
5. *Mineralogical Magazine*	73(34)	10	18
6. *Journal of Geology*	68	6	6
7. *American Mineralogist*	66	10	3
8. *Bulletin of the Geological Society of America*	64	1	1
9. *American Journal of Science*	58	6	4
10. *Transactions of the Royal Society of Edinburgh*	45		
11. *Monographs. Palaeontographical Society*	43		83
12. *Philosophical Transactions of the Royal Society. B.*	43	23	37
13. *Journal of Paleontology*	36		7
14. *Palaeontology*	33		50
15. *Bulletin of the American Association of Petroleum Geologists*	31		2
16. *Nature*	30	10	17
17. *Summary of Progress of the Geological Survey of Great Britain*	28		
18. *Journal of Sedimentary Petrology*	25		13
19. *Memoirs of the Geological Society of America*	22	23	15
20. *Proceedings of the Yorkshire Geological Society*	21(10)		

be classed as geology. This figure compares with 86 per cent in botany, 79 per cent in mathematics, 75 per cent in physics and 73 per cent in chemistry.

The language distribution is as might be expected and largely confirms the results of a previous survey of British geologists (see Chapter 6). Over-all, English is the dominant language with 78.5 per cent of the citations, while German, French and Russian account for 8.0 per cent, 5.0 per cent and 2.0 per cent, respectively. If the citations from individual journals are considered separately, some slightly different patterns emerge. In *Palaeontology* only 68 per cent of the citations are to English-language material and 15 per cent are to German literature, and in *Mineralogical Magazine* 70 per cent of the references are English. If the citations from these two journals are omitted from the over-all figures, the proportion of English references rises to 83.5 per cent.

The over-all date distribution of the journal citations reveals a 'half-life' (the time during which one-half of all the currently active literature was published) of 11.5 years. Allowing for delays in publication, the real 'half-life' is probably nearer 10 years. This is almost identical with the figure obtained by Brown and is much greater than the equivalent figures (also obtained by Brown) for physics (4.6), chemical engineering (4.8), metallurgy (3.9), mechanical engineering (3.9) and chemistry (8.1). If the citations in *Palaeontology* are considered separately, they indicate that palaeontologists rely even more than other geologists on the older literature. Only 41.5 per cent of the *Palaeontology* citations were published from 1950 onwards, compared with 57.5 per cent for the rest of the sample.

The study reported here did not consider the use of oceanographic, hydrological, meteorological and geomorphological literature. Information regarding the temporal distribution of currently active literature in these fields is given on p. 361.

Some problems with periodicals

Although the periodical is probably the best means available for disseminating and acquiring information, this form of publication is not without its problems. To a large extent periodicals have become the victim of their own success. There are so many current journals and their rate of increase is so great that libraries are finding difficulty in providing space for them even if they are able to afford the high prices charged for many titles. Abstracting and indexing services have problems in handling the output and individual scientists find themselves unable to locate information effectively. Allied to these problems is the fact that the periodical is a fairly slow means of transmitting information. To a large extent this is due to the refereeing system used to control quality but it is also a reflection of the increasingly large number of scientists who feel obliged to publish in order to demonstrate their ability. As a result, many of the more important journals have long lists of papers waiting to be published.

Various ways of overcoming these problems have been devised. Because of the delays in publication, the news-carrying function of many journals has been taken over by special publications such as 'newsletters' and 'letters' journals. These contain pure news items such as dates of conferences, details of new books, etc., and unrefereed preliminary communications on scientific research. Many of the newsletters are sponsored by societies, although some are issued by research organisations and at least one by a geological survey. Typical of this type of publication are *Geotimes* (American Geological Institute), monthly; *Gondwana Newsletter* (Centro de Investigação de Gondwana, Universidade Federal do Rio Grande do Sol, Brazil), irregular; *Newsletter of the Cave Research Group of*

Great Britain, quarterly; and *Newsletter of the Arctic Institute of North America*, approximately every 2 months. Representative of the 'letters' type of journal is *Earth and Planetary Science Letters*, published monthly by North-Holland. To facilitate rapid publication, contributions are typically short, the use of photographs and complex maps is restricted and proof correction is undertaken by editorial staff rather than authors.

In an effort to reduce the amount of material published and so keep costs down, some publishers are exploring the possibility of asking authors to omit from their papers data which the average reader will find unnecessary and indigestible. The NLL operates a scheme whereby such data can be deposited at the Library and lent to whomever wishes to make use of it. The availability of the data is indicated in the journal article with which it is associated.

The increasing problem of locating relevant articles in the thousands of journals produced might ultimately lead to the death of many periodicals, particularly those carrying purely research material. It is conceivable that all papers will in future be stored in a few central depositories which would abstract, classify and make available for loan the manuscripts concerned. Subscribers could either borrow these on a one-off basis or submit a subject interest profile which could be matched mechanically against titles and keywords. Relevant manuscripts would be sent to subscribers on a regular basis. A further development along these lines might see the complete text of the manuscripts stored in a computer, with users having access to the files through remote consoles.

The twin problems of storage and cost of periodicals are also being tackled by making material available in microform (microfilm or microfiche). As yet there are only a few journals which are available *only* in microform but an increasing number of titles are optionally available in this format. The American Geophysical Union, for instance, offers microform versions of *Journal of Geophysical Research*, *Water Resources Research* and *Reviews of Geophysics and Space Physics*. *Chemical Abstracts* is also available on microfilm. Many of the back issues of journals can be obtained in microform and a leading producer in this field is NCR Microcard Editions, who produce an annual *Guide to Microforms in Print*. Recent advances in micro-reading and printing equipment have made this type of publication more acceptable to scientists and it is likely that this transitional step from print-based to computer-based systems is likely to be taken by more and more publishers.

REPORT LITERATURE

Another form of primary literature, and one which is often over-
looked as a source of information, is the report. A report can be
defined as a document which states the results of, or the progress
made with, a research or development investigation. Typically it
is submitted direct to the person for whom the work was done and
so the contents escape any formal refereeing. The majority of
reports are subsequently made available to the general public,
although some may remain classified.

Characteristic features of reports are that they are mainly issued
in semi-published (duplicated) form or in microform and usually
carry an identifying alphanumeric code. The quality of report
literature varies a great deal. Some reports are progress reports and
as such are produced more for administrative than for scientific
reasons. On the other hand, many contain a great deal of valuable
scientific and technical information and are far more detailed than
anything which might subsequently be published on the same topic
in the periodical literature. They also have the advantage that they
may precede formal publication by anything up to several years.

The annual output of unclassified reports is estimated at about
150 000. These contain information on a wide range of subjects
from social science to agriculture and space technology. A rough
estimate puts the number in the earth sciences field at around 4000
per year. Reports emanate from many different types of organisation—government departments, universities, industrial firms, independent or sponsored research laboratories and international
organisations. Although most countries in the world produce some
reports, the largest source is the USA, where over three-quarters of
the material is generated. The next most important sources are
Canada, France, West Germany and the UK.

Since by far the greater proportion of research in the USA is
carried out under government contract, it is natural that the principal outlets of report literature should be government agencies.
Among the most important of these are the US Atomic Energy
Commission (USAEC), the National Aeronautics and Space
Administration (NASA) and the Department of Defense. All three
organisations sponsor a great deal of work in the earth sciences
field.

Locating details of report literature requires access to special
bibliographies, since the majority of subject-oriented guides to the
literature (see Chapter 5) do not cover this type of material in anything like a comprehensive fashion. Among these special biblio-

graphies there are three (all American) which are of particular importance. These are *Scientific and Technical Aerospace Reports (STAR), Nuclear Science Abstracts (NSA)* and *Government Reports Announcements (GRA)*.

Scientific and Technical Aerospace Reports is issued semi-monthly by NASA and includes details not only of American reports in the space and aeronautics field but also of reports from other countries. Citations and abstracts are arranged in broad subject groups and six indexes are included in each issue: subject, personal author, corporate source, contract number, report number and accession number. Cumulative index volumes are published semi-annually and annually. At least three of the 34 subject groups are of potential interest to earth scientists: 'Geophysics', which includes oceanography, cartography and geodesy; 'Meteorology', which includes climatology; and 'Space sciences', which includes lunar exploration.

Nuclear Science Abstracts is a semi-monthly guide to the world's nuclear science literature. It covers scientific and technical reports of the USAEC, other US Government agencies, other governments, international organisations, universities and industrial organisations. Non-report literature is also covered. Abstracts are arranged in 11 major categories, one of which is 'Earth sciences'. This in turn is subdivided into 'Geology and hydrology', 'Meteorology', 'Mineralogy and exploration', 'Oceanography' and 'Seismology'. Each issue of *NSA* contains a subject, personal author, corporate source and report number index. Annual and quinquennial indexes are also available. During the last few years there have been moves to reduce the international coverage of *NSA* and to replace it by an International Nuclear Information System (INIS) under the auspices of the International Atomic Energy Agency (IAEA). The latter already produces *INIS Atomindex*, a bi-monthly list of atomic energy reports of all nations.

The third guide, *Government Reports Announcements* (formerly *US Government Research and Development Reports*), is also issued semi-monthly. It is published by the National Technical Information Service (NTIS) and lists about 40 000 reports per year made available to the public through the Service. The majority of these reports are issued in two series: *AD* (material released by the Defense Documentation Center) and *PB* (reports from the Department of Commerce). Between them these two series account for approximately 23 000 reports per year. In addition, some NASA and USAEC reports and translations sponsored by the US Government (*JPRS* and *TT* series) are listed. As with the other two publications, entries are arranged in broad subject fields. Field 8—

'Earth sciences and oceanography'—is further subdivided into 13 groups, and contains about 3000 reports per year. Entries contain bibliographic details, subject descriptors and an abstract (see *Figure 3.1*). The indexes—corporate author, subject, personal author, contract number and report number—are published separately under the title *Government Reports Index*. Besides producing *Government Reports Announcements*, NTIS offers several subject-oriented current awareness services. For a fee, subscribers can receive information on reports in a particular field. This service is provided through frequently issued 'flash sheets' entitled *Fast Announcements*. An SDM (Selective Dissemination of Micro-

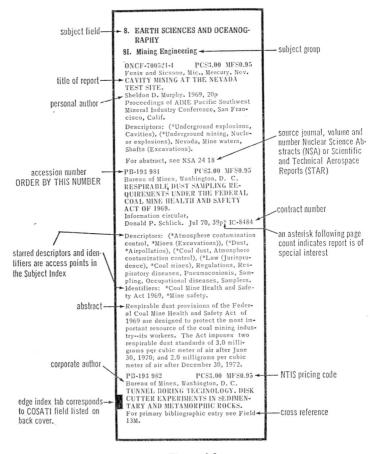

Figure 3.1

fiche) service is available for those who wish to receive the actual reports in a particular field on a regular basis.

Another group of American reports is worthy of brief mention. These are the *Rand Reports*, produced by the Rand Corporation, Santa Monica, California. The Corporation is an independent non-profit organisation engaged in research supported by the US Government, state governments, foundations and others. From time to time work is carried out in most scientific disciplines, including the earth sciences. About 500 reports per year are produced, details of which appear in the quarterly *Selected Rand Abstracts* (1963–). Material produced before 1963 is listed in the *Index to Selected Publications of the Rand Corporation 1946–1962* (Rand Corporation, 1962).

Outside the US the production of reports is comparatively small and bibliographical control is correspondingly poor. Atomic energy establishments are the principal producers and many, such as the UKAEA and Euratom, produce guides to their own publications. It is probably easier, however, to make use of *NSA* to identify these.

The most comprehensive guide to British reports is the National Lending Library's publication *NLL Announcement Bulletin* (a continuation of *British Research and Development Reports*). Issued monthly, this lists under broad subject headings all the British reports, theses and translations which have been received by the Library. In 1971 approximately 7000 items were listed, of which 2500 were reports. Another British guide to reports is the Department of Trade and Industry's *R and D Abstracts*. This not only includes information on British reports but also covers a selection of US and European material.

Although most reports can be obtained as full-size paper copy, the more usual format, particularly for American reports, is microfiche. A microfiche is a sheet of film, which can be of any size but which is commonly 105 mm × 148 mm, on which up to 98 pages of a document can be arranged in seven rows. Usually the top row is devoted to an eye-readable positive or negative title. A considerable number of older reports are available on 35 mm microfilm. Microforms require special reading equipment but most universities and many of the larger firms now have microfilm and microfiche readers available.

Availability of reports

In the UK the principal collection of reports is at the NLL, where an attempt has been made to acquire all material issued in the major American government report series as well as reports produced under other government auspices and by universities, research organisations and industrial firms in Britain and abroad. This material is available on loan (UK only) or as photocopies. The National Reference Library of Science and Invention and the Technical Reports Centre of the Department of Trade and Industry also have good collections of reports, and more specialised collections are available in many public, university and special libraries.

In the USA the main source of supply is the National Technical Information Service, Springfield, Virginia. In addition, numerous university and public libraries carry depository sets of particular reports. For example, 65 libraries (see inside front cover of *NSA*) hold sets of USAEC reports.

Because of the problems of bibliographic control, it is essential when applying to any source for a report to quote all available information. This should include alphanumeric code, authors, title and issuing body.

THESES

A further source of original information is the thesis literature. Theses are usually produced in partial fulfilment of the conditions for a higher degree and submitted to a university or other award-giving body. Since acceptance of a thesis usually requires the work to be 'worthy of publication', much of the information subsequently appears in the periodical literature. The process of formal publication, however, may take years and in fact a great deal of valuable information, particularly reviews of previous work (with which most theses begin) and negative results, never gets published elsewhere. A search of the thesis literature can therefore often be a profitable exercise when a search for information is undertaken. Unfortunately, university regulations relating to theses vary from country to country and even universities within the same country differ greatly in the number and form of copies required.

In the UK it is usual for a student to provide two, three or four typewritten copies of his thesis, one of which is deposited in the university library and another kept within his department. Every university has its own special conditions for making theses available to others. In most cases they can be borrowed or copied, although authors and heads of departments usually have the right to restrict lending or copying for a period which varies from 2 to

10 years depending on the university. There have been recent moves to standardise the procedure for making theses available and the UK Standing Conference on National and University Libraries has asked Vice-Chancellors to consider the following conditions:

1. That at least one copy of every thesis accepted for higher degrees should be deposited in the university library.
2. That, subject to the author's consent, all theses should be available for inter-library loan.
3. That, subject to the author's consent, all theses should be available for photocopying.
4. That authors of theses should be asked at the time of deposit to give their consent for (2) and (3) in writing and that this consent should be inserted in the deposit copy of the thesis.

In other countries the regulations are different. Many European universities, for instance, require authors to have their theses printed. Since copies are plentiful, they are usually available in departmental, university and national libraries and are frequently used as exchange items for material from other libraries. Authors themselves usually have copies to give away as 'reprints'.

In the USA, although theses are no longer generally published, the problem of availability has been overcome to a large extent by University Microfilms. This commercial organisation has been given permission by most universities to microfilm doctoral dissertations. Authors provide typescript copies and pay a small fee to the company for the privilege of having their work included in the programme.

Although a few subject-oriented abstracting and indexing publications are attempting to cover university theses, it is still necessary to make use of a variety of specialist guides to locate details of many theses.

In the UK the most complete guide is the Aslib publication *Index to Theses Accepted for Higher Degrees in the Universities of Great Britain and Ireland and the Council for National Academic Awards* (1953–). The latest volume, published in 1971, covers 1968–69 and lists over 6000 thesis titles under broad subject headings. Theses from 38 universities as well as 64 colleges of London University and the CNAA are included. An editorial note gives details of availability.

A second British publication is more up to date but at the moment less comprehensive than the Aslib guide. This is the *NLL Announcement Bulletin*. Among other material this monthly list includes details of university theses which have been sent to the Library for filming and then returned to the universities concerned.

Copies of the theses can be obtained from the NLL, provided that a signed copyright form accompanies the request. At the time of writing seven universities are co-operating with the NLL in its attempt to promote the use of theses. It seems likely that more will follow suit.

In addition to these general lists, many individual universities produce lists of their own. These may be included in the university *Calendar*, as in the case of Dublin University; or may be published separately, e.g. *List of Theses Accepted for Higher Degrees, University of Edinburgh* (annual) and *Index of Theses Accepted for Higher Degrees 1955–1970* (University of Exeter, 1971). Birmingham University produces an annual list entitled *Research and Publications* which includes details not only of theses but also of publications of members of staff.

Occasionally special subject-oriented guides to theses are published. These may appear as periodical articles, e.g. 'Theses on Scottish Geology', by W. D. I. Rolfe (*Scottish Journal of Geology*, **6**, 401–407, 1970), or as separate publications, e.g. 'Bibliography of Unpublished Theses Relating to the Geology of Ireland' issued as *Geological Survey of Ireland Information Circular*, No. 1 (1969).

The most useful guide to American theses is *Dissertation Abstracts International* (1969–   ), published monthly by University Microfilms and containing abstracts of approximately 25 000 dissertations per year. It started in 1938 as *Microfilm Abstracts* and then changed its title to *Dissertation Abstracts*. It is issued in two parts: *A. Humanities and Social Sciences* and *B. Sciences and Engineering*. Both parts are provided with a keyword subject index. In 1971 Part *B* contained abstracts of nearly 500 earth sciences theses. All theses abstracted can be bought either as hard copy or microfilm from University Microfilms, and in the UK copies of material announced since 1970 can be borrowed from the NLL. A further service offered by University Microfilms is DATRIX (**D**irect **A**ccess **T**o **R**eference **I**nformation: a **X**erox service). This is a mechanised search service based on files containing details of almost 200 000 theses submitted to American and Canadian universities since 1938. Computer searches can be made for theses described by particular combinations of keywords. For details of American theses prior to 1938 one can consult the annual list produced by the Library of Congress—*List of American Doctoral Dissertations Printed in . . .* (USGPO, 1912–38). Overlapping with this and the University Microfilms publications is another guide, *Doctoral Dissertations Accepted by American Universities*, produced annually from 1934 to 1956 for the Association of Research Libraries by the H. W. Wilson Co.

Just as in Britain, individual American universities publish lists of theses submitted for higher degrees. A useful, although somewhat out of date, guide to these is *Guide to Bibliographies of Theses—United States and Canada*, compiled by T. R. Palfrey and H. E. Coleman (American Library Association, 1940; reprinted by University Microfilms, 1963). A similar publication covering master's theses is *Guide to Lists of Master's Theses*, by D. M. Black (American Library Association, 1966). This contains details of special subject lists and lists issued by institutions. Many of the latter include details of doctoral as well as master's theses.

A series of bibliographies covers American and Canadian geological theses. The first, containing details of 11 000 dissertations, was compiled by B. J. and H. Chronic under the title *Bibliography of Theses Written for Advanced Degrees in Geology and Related Sciences at Universities and Colleges in the United States and Canada through 1957* (Pruett Press, 1958). This was supplemented by the same authors in *Bibliography of Theses in Geology (1958–63)* (American Geological Institute, 1965). Further supplements are 'Bibliography of Theses in Geology (1965–1966)', by D. C. Ward (*Geoscience Abstracts*, **7** (12), Part 1, 1965), and *Bibliography of Theses in Geology (1965–1966)*, by D. C. Ward and T. C. O'Callaghan (American Geological Institute, 1969). The AGI intend to publish similar lists on an annual basis with quinquennial cumulations. From time to time more specialised lists appear in the periodical literature, e.g. 'Index to Graduate Theses on Californian Geology to December 31, 1961', by C. W. Jennings and R. G. Strand (*Special Report, Division of Mines, California*, No. 74, 1963), and 'Bibliography of Graduate Theses on Nevada Geology', by R. V. Wilson (*Nevada Bureau of Mines Report*, No. 8, 1965). Lists such as these can be located through such guides as *Bibliography and Index of Geology*.

Various guides exist to theses of other countries. Where theses are published, details can often be found in the appropriate national bibliography. *Bibliographie de la France*, for instance, periodically contains a thesis supplement. French theses are also listed in *Catalogue des Thèses et Ecrits Académiques* (1884–). Material is listed by subject and institution, and annual and quinquennial indexes are available. German theses are recorded in *Jahresverzeichnis der deutschen Hochschulschriften*, formerly *Jahresverzeichnis der an deutschen Universitaten erschienen Schriften* (1885–1934).

Since 1945 all Russian theses have been deposited at the State V.I. Lenin Library, Moscow. Following initial haphazard attempts to list them, a quarterly publication was started in 1957. This is *Katalog Kandidatskikh i Doktorskikh Dissertatsii Postupivshikh v Biblioteku im V.I. Lenina i Gos. Tsentral'nuyu Nauchnuyu Meditsinskiyu Biblioteku* (Catalogue of Masters and Doctoral Dissertations received by the State V.I. Lenin Library and the State Medical Library).

Other lists include *Canadian Theses* (National Library of Canada, 1962–), annual; *Union List of Higher Degree Theses in Australian University Libraries* (University of Tasmania Library, 1967), covering up to 1965; and *Union List of Theses of the Universities of New Zealand 1910–1954*, by D. L. Jenkins (New Zealand Library Association, 1956), supplements to which cover 1955–62 (1963) and 1963–67 (1969).

CONFERENCES AND THEIR PROCEEDINGS

The conference, understood here to be synonymous with meeting, colloquium, symposium, congress and seminar, has a long history as one of the principal means whereby scientists exchange information and ideas. Conferences are usually organised by branches or the main secretariats of learned or professional societies. They range in size from small specialist meetings (sometimes referred to as workshops) to such massive international gatherings as the International Geological Congresses. In frequency and regularity, too, they vary; the most common are annual but many of the large international meetings are less frequent and some are irregular.

Conferences perform at least two important functions. First, they provide a forum where papers can be read and discussed, and, secondly, they permit informal contact between people working in the same and related fields. Regarding the former function, it has been argued that, despite the obvious advantages of having interaction between an author and an audience, the formal presentation of papers is an overrated conference activity. To begin with, the amount of useful formal discussion which usually takes place after a presentation is minimal. Furthermore, much of the information which is being reported is not new or will be reported elsewhere in the future. There is evidence to show that large numbers of authors have already made some report of the main content of their presentation prior to the conference. In the case of the 1967 annual meeting of the American Geophysical Union, for instance, 69 per cent of the authors had done so, 7 per cent having published the results in a journal article (Johns Hopkins University, 1967). Those who did not publish before, usually intended to do so afterwards and in the AGU study it was revealed that 83 per cent of the authors planned future journal articles.

Most people would agree that the main benefits to be derived from a conference are the informal contacts and exchanges of ideas and information which can be made. In many cases not only is there an exchange at the conference itself but, as a result of contacts made, interaction is often continued into the future.

Conference publications

Publications of various types are associated with conferences. To begin with, these may be preprints consisting of the full papers, shortened versions of the papers or merely abstracts. These are intended to give participants a better idea of the presentations which they will find most useful. Where the full text is provided, the presentations themselves are usually shortened to allow more time for discussion. Preprints are frequently only available to participants and are often issued as separates and in duplicated format. Unfortunately, they are frequently cited by participants and others, and the result is a number of requests reaching libraries for material described in various ways, which the libraries have in any event been unable to obtain.

It is usual for the proceedings of most conferences together with associated field activities to be published in some form or other, although delays in publication of several years are not unknown. Commonly the proceedings are issued in book format published in one or more volumes by either the sponsors of the meeting or a commercial organisation. Sometimes the title page makes it quite clear that the contents are conference proceedings but in other cases this fact is subordinated to a more eye-catching title. Other information, such as the date of the conference, the place and the sponsoring organisation, may or may not appear in a prominent place or may be absent altogether.

An alternative form of publication is as articles in periodicals. The whole issue of a periodical is sometimes devoted to reporting the proceedings of a conference. On other occasions a special supplement may be issued. As with books, the fact that the articles are papers presented at a conference may or may not be prominently displayed.

References to conference proceedings appear frequently in the literature, where they may cite a combination of the official name of the meeting, the editor of the proceedings, the sponsoring body, the conference topic, the location of the meeting and the title of the book. More often than not, at least half of the information is likely to be missing from the citation and this, coupled with the difficulty in recognising conference proceedings as such, gives rise to severe bibliographic problems.

To help the librarian and scientist track down conferences and their associated proceedings, a number of published guides have made their appearance over the last ten years. For a long time the best guide to forthcoming conferences was the Library of Congress bi-monthly publication *World List of Future International Meetings*, but this ceased publication in September 1969. Three other American guides are still produced. These are the CCM twin publications *World Meetings—USA and Canada* and *World Meetings —Outside USA and Canada* and the US Special Libraries Association's *Scientific Meetings*. *World Meetings*, which is issued quarterly, includes information on name of meeting, location, date, sponsor, content, attendance, deadline for papers, proceedings to be published and exhibition arrangements. Each issue also contains indexes to date of meeting, keyword, deadline date, location and sponsor. The information regarding future availability of proceedings is particularly useful, since it allows librarians to place an early order and this obviates the problems caused by organisers underestimating the demand for the conference publications. *Scientific Meetings* is also issued quarterly. It is international in scope and very good in its coverage of American society meetings. Similar in scope but with a bias towards British meetings is *Forthcoming Scientific and Technical Meetings*, produced quarterly up to 1970 by the UK Department of Education and Science and now published by Aslib.

With regard to the published proceedings, the most comprehensive guide is the NLL's *Index to Conference Proceedings Received by the NLL*, issued monthly. This gives details of title, location and date of all types of conference proceedings received by the Library, including those which appear in periodicals. A subject keyword index is provided in each issue. In 1971 over 6000 conference publications were covered. Better indexed, but covering only about 2000 proceedings a year each, are two American publications, *Directory of Published Proceedings* and *Proceedings in Print*. The first is issued by InterDok in 10 parts per year, with annual and quinquennial cumulations. It has world-wide coverage and is particularly good in its treatment of American society meetings. *Proceedings in Print* (1964–) is published six times per year and an annual combined subject–sponsor–editor index is provided. Another US publication concerned with conference literature is *Current Index to Conference Papers*, published by CCM. As its name indicates, this lists individual papers presented at conferences. It is issued monthly, with annual cumulations, and lists approximately 100 000 items per year. Author addresses and future publication plans, where known, are also given.

Since many of the above lists have only been in existence for a short time, they are of limited value for retrospective searching. In this connection there are two general bibliographies which cover older material: *International Congresses and Conferences 1840–1937*, edited by W. Gregory (H. W. Wilson, 1938; reprinted by Kraus 1967), and *International Congresses 1681 to 1899* (Union of International Associations, 1960). A supplement to the latter covers 1900–1919 and more recent material is covered in *Bibliography of Proceedings of International Meetings 1957–59* (1964) and by the *Yearbook of International Congress Proceedings* (1st edn., 1960–67; 2nd edn., 1962–69), also published by the Union of International Associations.

All these lists are multidisciplinary and contain details of most earth sciences meetings. In addition, many subject-oriented abstracting and indexing publications in the earth sciences cover conference proceedings. For instance, the 'Special Publications' section of *Bibliography and Index of Geology* lists some conference literature and *Geoscience Documentation* has a section devoted to the announcement and reporting of conferences and field trips and their published proceedings. In neither case, however, is the information as comprehensive as in the specialist guides to conference literature mentioned earlier.

REFERENCES

Brown, C. H. (1956). *Scientific Serials: Characteristics and Lists of Most Cited Publications in Mathematics, Physics, Chemistry, Geology, Physiology, Botany, Zoology and Entomology.* Association of College and Research Libraries, ACRL Monograph No. 16

Craig, J. E. G. Jr. (1969). 'Characteristics of Use of Geology Literature'. *College and Research Libraries*, **30** (3), 230–236

Gross, P. L. K. and Woodford, A. O. (1931). 'Serial Literature used by American Geologists'. *Science*, **73**, 660–664

Johns Hopkins University. Center for Research in Scientific Communication (1967). *The Dissemination of Scientific Information, Informal Interaction, and the Impact of Information Associated with the 48th Annual Meeting of the American Geophysical Union.* PB176 491, CFSTI

Woodford, A. O. (1969). 'Serial Literature Used by American Geologists, 1967'. *Journal of Geological Education*, **17** (3), 87–90

APPENDIX: SOME GUIDES TO GOVERNMENT PUBLICATIONS

Directory of British Government Publications. Birmingham University Library (1966)

Guide to Official Publications. HMSO (1960)

Ford, P. and Ford, G. *A Guide to Parliamentary Papers*, 2nd edn. Blackwell (1956)

Jackson, E. *Subject Guide to Major US Government Publications.* American Library Association (1968)

Leidy, W. P. *A Popular Guide to Government Publications*, 3rd edn. Columbia University Press (1968)

Olle, J. G. *Introduction to British Government Publications.* American Library Association (1965)

Pemberton, J. E. *British Official Publications.* Pergamon (1971)

Schmeckebier, L. F. and Eastin, R. B. *Government Publications and their Use*, 2nd edn. Brookings Institution (1969)

4

Secondary literature — reference and review publications

D. N. Wood

Many problems encountered by the scientist, the answers to which can be found in the literature, do not necessarily require access to the most up to date information reported in periodicals, etc. Indeed, the primary literature, in which techniques, experiments, results and hypotheses are discussed for the first time, is often difficult and tedious to use for tracking down specific facts and for obtaining an over-all picture of a particular field. To help in the location of such information, there has grown up a body of secondary literature. Reference publications such as encyclopaedias, dictionaries, handbooks, directories and yearbooks are the most obvious types of secondary literature, but it should be remembered that the majority of textbooks, many conference proceedings and some journals perform a secondary function by bringing together and commenting on previously published information.

This chapter will briefly introduce some types of secondary publications, cite examples and indicate sources where further information can be found. The subject chapters which follow later in the book will mention those secondary sources deemed to be of especial importance in the various branches of the earth sciences. Readers are also referred to two basic sources of information on secondary publications. The first is Constance M. Winchell's *Guide to Reference Books* (8th edn., American Library Association, 1967), which deals with literature up to 1965 and covers 7500 titles in all subject fields. Annual supplements by E. P. Sheehy keep the work up to date. The second is A. J. Walford's *Guide to Reference Material* (2nd edn., 2 vols., Library Association, 1966). Volume 1 covers science and technology and contains sections on geology and palaeontology. Although the publication is international in scope, it is principally concerned with British literature.

ENCYCLOPAEDIAS

Encyclopaedias attempt to summarise existing knowledge and by their arrangement to make it as easy as possible to locate information on particular subjects or to track down specific facts. The subject matter contained is sometimes in the hands of an editor or, more often, an editorial board consisting of experts drawn from the fields covered by the encyclopaedia. The amount of information to be found in any one encyclopaedia will obviously depend on the size and subject coverage, and encyclopaedias range from multi-volume works covering all knowledge to those published as single volumes and limited to quite specialised branches of a major scientific discipline.

The sort of information contained in encyclopaedias makes them useful for a number of different purposes. They can be used as dictionaries, sources of statistics, biographical works, history texts, atlases and, because most entries contain references to other sources of information, guides to the literature.

The majority of encyclopaedias employ an alphabetical arrangement of subject headings supplemented in the larger works by a detailed index intended to guide readers to information which would otherwise be lost in the text. An alternative method, used mainly in specialised subject encyclopaedias, is to present the information in a series of major subject sections.

Most encyclopaedias are updated from time to time and new editions appear at regular or irregular intervals. However, in view of the time required to revise completely works of this magnitude, even new editions are considerably out of date at the time of publication. As a result, the publishers of some of the largest encyclopaedias have adopted a policy of continuous revision. With such a system new and improved developments can be incorporated relatively quickly and the information is made available by reprinting the whole work over one or two years. Other measures taken by publishers to keep their works up to date are the publication of irregular supplements or yearbooks. The aim of the latter is to report significant events and developments which have occurred in the year in question. Some measure of how up to date any particular encyclopaedia entry is can be gauged by scrutinising the dates of accompanying references.

Probably the most comprehensive and widely available English-language encyclopaedia is *Encyclopaedia Britannica*, published by Encyclopaedia Britannica Inc. in co-operation with the University of Chicago. Originally published in three volumes in Edinburgh in 1768, *Britannica* has had a chequered history and was acquired by American interests in 1901. Despite this it still retains an essentially British character, with British spellings and many UK-oriented entries. Following the publication of the 14th edition in 1929, the encyclopaedia has been subjected to continuous revision and has been reprinted every one or two years. The work consists of 23 volumes of alphabetised entries plus an index volume and atlas, and it is liberally illustrated with sketch maps, diagrams and figures. The fact that *Britannica* is a general encyclopaedia should not deter scientists from using it, as many of the entries are considerably more detailed than those to be found in specialised technical encyclopaedias. In this connection, it is interesting to note, for example, that in the 1969 reprint two columns are devoted to the pyroxenes, seven columns and two pages of illustrations to the Brachiopoda and four columns and two pages of illustrations to glaciers. With the proviso that much of the information is somewhat out of date, *Britannica* can be extremely useful for obtaining background information on almost any topic and for locating statistical data, physical constants and references to the more important books on a subject. The problem of reporting current information is partially overcome by the *Britannica Book of the Year*, an annual publication which started in 1938 and reports major events of the previous year, presents obituaries and biographies of personalities who died or dominated the events of the year and gives details of developments which have recently taken place in many walks of life.

When searching for information in encyclopaedias one should bear in mind that their content is largely determined by editorial policy and by the opinions and interests of individual contributors. On any particular enquiry, therefore, it is often worth-while consulting more than one encyclopaedia. There are many general encyclopaedias available besides *Britannica*, but probably the most up to date is the 4th edition of *Chambers's Encyclopaedia* (International Learning Systems, 1966). This 15-volume work has been written by over 3000 contributors and has an index with approximately 225 000 entries.

In addition to the general encyclopaedias, there are a number of more specialised works which cover science and technology as a whole, and some which are devoted to particular scientific disciplines. All can be useful in dealing with the shorter type of enquiry or for getting subject orientation prior to carrying out a lengthy literature search. Generally speaking, the more specialised the encyclopaedia the more detailed the information reported.

Of the general science encyclopaedias, the most useful English-language work is probably the *McGraw-Hill Encyclopaedia of Science and Technology* (3rd edn., McGraw-Hill, 1971). The work is published in 15 volumes, the last being a subject index to the other volumes. Each year since 1962 the publishers have also issued the *McGraw-Hill Yearbook of Science and Technology*. This has the twofold aim of providing a comprehensive record of progress and of updating the basic material in the encyclopaedia. Like the encyclopaedia, it employs an alphabetical arrangement of entries, and, in addition, each volume contains a number of review articles on topics of current general interest. In the 1967 volume, for instance, there appears an article on 'Antiquity and Evaluation of Pre-Cambrian Life' and in the 1968 volume one entitled 'World Weather Watch'.

An alternative source of general scientific information is *Van Nostrand's Scientific Encyclopaedia* (4th edn., Van Nostrand, 1968). This single-volume work of over 2000 pages contains 16 500 separate articles ranging in length from a few lines to several pages. There is a good deal of cross-referencing but no bibliography and no separate index.

More detailed background information to a problem is usually to be found in specialist encyclopaedias. Most branches of science and technology are served by one or more of these, and the earth sciences are no exception.

A great deal of basic information is to be found in the *Larousse Encyclopaedia of the Earth*, by L. Bertin (Prometheus Press, 1961). Unlike most encyclopaedias, this work does not employ an alphabetised arrangement but is more like a textbook in that information is presented under major subject headings. There are 26 such headings ranging from 'Atmosphere and Weather' through 'Oceans and Lakes' and 'Mineral Fuels' to 'The Palaeozoic Era' and 'Life in the Cenozoic'. A fairly detailed subject index aids in the location of specific information and the book is superbly illustrated.

Covering the earth sciences in more detail is a new series of encyclopaedias, known as the *Encyclopedia of Earth Sciences Series*, published by Reinhold and edited by R. W. Fairbridge. Vol. I, *The Encyclopedia of Oceanography*, was published in 1966; Vol. II, *The Encyclopedia of Atmospheric Sciences and Astrogeology*, in 1967; Vol. III, *The Encyclopedia of Geomorphology*, in 1969; and Vol. IVA, *Encyclopedia of Geochemistry and Environmental Sciences*, in 1972. Entries are arranged alphabetically, bibliographies are appended and each volume has a separate index.

Of a somewhat more specialised nature are three illustrated encyclopaedias published by Weidenfeld and Nicolson. One is *Standard Encyclopedia of the World's Rivers and Lakes*, edited by R. K. Gresswell and A. J. Huxley (1965); the second, *Standard Encyclopedia of the World's Oceans and Islands*, edited by A. J. Huxley (1962); and the third, *Standard Encyclopedia of the World's Mountains*, edited by A. J. Huxley (1962). In each case only a selection of rivers, mountains, etc., is described at any length, the remainder being dealt with briefly in gazetteers at the end of the books.

No scientific discipline is neatly circumscribed and in those areas where one subject impinges on another the practitioner in one field can find useful information in the literature of the other. There are many chemical, biological and physical encyclopaedias, for instance, that could profitably be used by earth scientists. An example is *The Encyclopedia of Physics (Handbuch der Physik)*, edited by S. Flugge (2nd edn., Springer, 1955–   ). This 54-volume work consists of review articles covering the whole field of physics. Geophysics is the subject of Vols. 47 (1956); 48 (1957); 49, Part 1 (1966); 49, Part 2 (1967); and 49, Part 3 (1971).

For details of encyclopaedias in other subject fields readers are referred to the literature guides listed in the Appendix to Chapter 1.

DICTIONARIES

Monolingual dictionaries

Typically, a monolingual dictionary is an alphabetised listing of words giving definitions and in many cases the origins of the words. Like encyclopaedias, monolingual dictionaries range from the general to the very specialised. With regard to the latter type, it is worth noting that the term 'dictionary' has in many instances been wrongly applied to a publication and that several works, though carrying the word 'dictionary' in their titles, are, in fact, more in the nature of encyclopaedias. A typical example in the earth sciences field is the *International Dictionary of Geophysics*, edited by S. K. Runcorn *et al.* (Pergamon, 1967). This two-volume work consists of alphabetised entries up to 10 pages in length and in some instances accompanying bibliographies contain over 40 references.

Dictionaries proper can be useful starting points in searching the literature of an unfamiliar subject, for in such cases sorting out the terminology should be the number one priority.

Large general works such as the multi-volume *Oxford English Dictionary* (Clarendon Press, 1933), *Webster's New International Dictionary of the English Language* (2nd edn., Merriam, 1961), and *Funk and Wagnalls New Standard Dictionary of the English Language* (Funk and Wagnalls, 1964) all contain definitions of many scientific terms, but are usually to be found only in large reference collections and may therefore not be readily at hand. More likely to be available, and a work which is extremely good in its coverage of scientific and technical terms in general is *Chambers's Technical Dictionary*, edited by C. F. Tweney and L. E. C. Hughes (3rd edn revised, W. and R. Chambers, 1958). A new edition of this popular dictionary was published in late 1971 under the title *Chambers Dictionary of Science and Technology*. Also of value is the ever-popular paperback *Dictionary of Science* by E. B. Uvarov, D. R. Chapman and A. Isaacs (3rd edn., Penguin Books, 1964).

Of more specific interest to earth scientists are the subject-oriented dictionaries concerned with geological and related nomenclature. The most comprehensive of these is the *Glossary of Geology and Related Sciences with Supplement* (2nd edn., American Geological Institute, 1960). The publication gives brief definitions of 18 000 terms and includes a useful list of additional dictionaries and glossaries. More limited in its coverage but more up to date and generally giving larger entries is *A Dictionary of Geology*, edited by J. Challinor (3rd edn., University of Wales Press, 1967). Besides definitions, this cites quotations in which the terms have been used and most entries contain references to significant books and papers on the subject. Also of limited coverage, but having the advantage of being cheap and therefore within the budget range of most individuals, are the *Dictionary of Geology*, by D. G. A. Whitten and J. R. V. Brooks (Penguin Books, 1972) and *A Dictionary of Geography*, by W. G. Moore (3rd edn., Penguin Books, 1963). Together these provide definitions of most commonly encountered terms in the earth sciences.

There are a number of earth sciences dictionaries prepared for specialists in one or other branches of the science. In this category are publications such as *Glossary of Geographical Terms*, edited by L. D. Stamp (Wiley, 1961); *International Tectonic Dictionary*, by J. G. Dennis (American Association of Petroleum Geologists, 1967); *Chambers's Mineralogical Dictionary* (Chemical Publishing Company, 1960); 'Glossary of Oceanographic Terms', by B. B. Baker *et al.* (2nd edn., *US Naval Oceanographic Office, Special Publication* No. 35, 1966); *Glossary of Meteorology*, edited by R. E.

Huschke (American Meteorological Society, 1959); and *Dictionary of Gems and Gemology*, by R. M. Shipley (Gemological Institute of America, 1946). The field best served by specialist subject dictionaries is applied geology and there are many glossaries devoted to mining, etc. The best and most up to date of these is *A Dictionary of Mining, Mineral and Related Terms*, compiled by Paul W. Thrush and published in 1968 by the Bureau of Mines of the United States Department of the Interior. Including terminology from the entire English-speaking world, 150 000 definitions appear under about 55 000 individual entry terms. Each definition is accompanied by its source. The work represents a revision of 'A Glossary of the Mining and Mineral Industry', by Albert H. Fay (*Bulletin of the US Bureau of Mines*, No. 95, 1918). Fay's work, reprinted in 1947, contains 18 000 terms with 27 000 definitions.

In so far as they list and describe stratigraphic formations and deposits, the various geological lexicons can be considered as subject dictionaries. Since the beginning of this century the US Geological Survey has paid a good deal of attention to North American stratigraphic nomenclature and has issued several *Bulletins* devoted to the subject. The first appeared in 1902 under the title 'North American Geologic Formation Names', by F. B. Weeks (*US Geological Survey Bulletin*, No. 191). In 1938 this was superseded by M. G. Wilmarth's 'Lexicon of Geologic Names of the United States—including Alaska' (*US Geological Survey Bulletin*, No. 896). This two-volume publication became known as the Wilmarth Lexicon and contained 13 090 names. It was updated by *Bulletin* No. 1056-A, 'Geologic Names of North America Introduced in 1936–1955', by D. Wilson *et al.* (1957), which contains 5 000 new names, and by *Bulletin* No. 1200, 'Lexicon of the Geologic Names of the United States for 1936–60', by G. C. Keroher *et al.* (3 vols., 1966). The latter has also been published as Vol. VII of the *Lexique Stratigraphique International*. This international lexicon was the brain child of the Stratigraphic Commission of the International Geological Congress and has been published in numerous parts by the Centre National de la Recherche Scientifique. Started in 1956, the work is still in progress and up to the end of 1969 105 parts had appeared in eight volumes (see Chapter 9). In the main, each part of the first seven volumes covers the stratigraphic nomenclature of one or more countries, but there are exceptions. For instance, England, Scotland and Wales are dealt with in 13 parts, each of which is concerned with a separate geological period. For each stratigraphic formation the lexicon gives a brief description, the type locality, age, and references to the literature wherein it is described in detail. Volume VIII consists of lengthy review articles on major stratigraphic units. Three parts have so far been published: *Infracambrian*, by R. Furon; *Aptien*, by J. Sornay; and *Pennsylvanian*, by H. R. Wanless.

Interlingual dictionaries

In addition to the monolingual dictionaries, there are other types to which attention will occasionally have to be turned. With foreign language literature, for instance, the use of bilingual or multilingual dictionaries will often be necessary. There are many such dictionaries and it would be pointless to mention all, since other published lists are available and in most cases use will be dictated by local availability. A select list of foreign language–English dictionaries is, however, given in Appendix 1 of this chapter. The list is confined to dictionaries covering those languages most likely to be encountered by earth scientists.

Details of other technical dictionaries can be located in the following guides:

> *Bibliography of Monolingual Scientific and Technical Glossaries*, by E. Wuster (2 vols., UNESCO, 1955, 1959)
> *Bibliography of Interlingual Scientific and Technical Dictionaries* (4th edn., UNESCO, 1961) and *Supplement* (1965)
> *Dictionaries of Foreign Languages; A Bibliographical Guide to the General and Technical Dictionaries of the Chief Foreign Languages*, by R. L. Collison (Hafner, 1955)
> *Foreign Language–English Dictionaries* (2nd edn., 2 vols., Library of Congress, 1955)
> 'Foreign Language and English Dictionaries in the Physical Sciences and Engineering—A Selected Bibliography, 1952–1963', by T. W. Marton (*National Bureau of Standards, Miscellaneous Publications*, No. 258, 1964)

Dictionaries of abbreviations

Apart from monolingual and interlingual dictionaries there are other types which may be useful on special occasions. For example, there is an increasing tendency in all branches of science and technology to use abbreviations and acronyms. Many of these become well known but there are many more which do not, and foreign abbreviations, in particular, can sometimes be difficult to expand. The following have been found by the author to be the most useful dictionaries of abbreviations:

Dictionary of Abbreviations in English, German, Dutch and Scandinavian Languages, by W. Bluwstein (Soviet Encyclopaedia Publishing House, 1964)

Glossary of Russian Abbreviations and Acronyms (Library of Congress, 1967)

HANDBOOKS

One of the most useful secondary sources of information is the handbook. Originally intended to be relatively small compact works containing what the author considered to be the basic principles of a subject, essential miscellaneous data, and tables, handbooks have to a certain extent lost their identity. Works carrying the title 'handbook' now range from the small pocketbook through extensive data compilations to multivolume treatises, and the only thing they really have in common is the secondary nature of the information they contain. Many of them are so detailed as to require an intimate knowledge of the subject before they can be used effectively.

Many handbooks, particularly those containing mathematical tables and physical constants, are of value to scientists in general. A classic example of such a work is the *Handbook of Chemistry and Physics*, edited by R. C. Weast and published by the Chemical Rubber Company. The 53rd edition of the 'Rubber Bible', as it is commonly known, was published in 1972, and contains a wealth of physical data and numerous mathematical tables. Another well-known work of general interest and value is Kaye and Laby's *Tables of Physical and Chemical Constants and some Mathematical Functions*, edited by N. Feather *et al.* (13th edn., Wiley, 1966). Both these works contain mathematical tables and formulae, but a more detailed treatment can be found in the *Handbook of Tables for Mathematics*, edited by S. M. Selby (4th edn., Chemical Rubber Company, 1970).

A major attempt to assemble critical data in a single place was made by the International Research Council and the National Academy of Sciences in the 1920s. On their behalf the US National Research Council compiled *International Critical Tables of Numerical Data, Physics, Chemistry and Technology* under the general editorship of E. W. Washburn (7 vols., plus index, McGraw-Hill, 1926–33). This work, which is divided into 300 sections, is an extremely valuable guide to data which appear in the primary literature up to 1924. For more up to date information it is necessary to use *Tables Annuelles de Constantes et Données Numérique de Chimie de Biologie et de Technologie*, edited by C. Marie and published by the International Union of Pure and Applied Chemistry in 12 volumes. It covers the years 1910 to 1936, and subject and formula indexes are available. References are given to the primary literature from which the data were extracted, and the mineralogist, in particular, will find much valuable information within its pages.

If the reader has a working knowledge of German, then the various editions of the Landolt-Börnstein compilations will be found useful as a source of data. The last complete edition of *Landolt-Börnstein Physikalisch-Chemische Tabellen* was the 5th, published by Springer between 1923 and 1936. It consists of eight volumes and three supplements, and a full subject index appears in the last supplement. In 1950 the 6th edition started to appear under the title *Zahlenwerte und Funktionen aus Physik, Chemie, Astronomie, Geophysik und Technik*. Published by Springer in four volumes, 24 out of 28 parts have so far been published. Again, all but one or two parts are entirely in German. A useful guide to its contents is to be found on pp. 115–118 of *The Use of Chemical Literature*, edited by R. T. Bottle (2nd edn., revised, Butterworths, 1971). Those parts likely to be of most use to earth scientists are Vol. 1, Part 4 (1955), which deals with crystallographic data; Vol. 2, Part 3 (1956), which deals with melting point equilibria; and Vol. 3 (1952), which is concerned with astronomy and geophysics and includes oceanographic, hydrographic and meteorological data.

Since 1961 a 'new series' of Landolt-Börnstein has been appearing under the title *Zahlenwerte und Functionen aus Naturwissenschaften und Technik—Neue Serie*, edited by K. H. Hellwege (Springer, 1961–). The work is being published in numerous parts as and when the appropriate data are accumulated. Each part will be issued in one of the following six basic groups:

1. *Nuclear Physics and Technology*
2. *Atomic and Molecular Physics*
3. *Crystal and Solid State Physics*
4. *Macroscopic and Technical Properties of Matter*
5. *Geophysics and Space Research*
6. *Astronomy, Astrophysics and Space Research*

The 'new series' is easier to use than the 5th and 6th editions, since the preface, introductions, contents lists and section headings are in English as well as German.

Another compilation of general physical data is the 9th edition of 'Smithsonian Physical Tables', compiled by W. E. Forsythe (*Smithsonian Miscellaneous Collections*, **120**, 1954). Reprinted for the third time in 1964, this is a much more selective and therefore smaller collection of data than those referred to above.

In addition to the general compilations already mentioned, there are many which have been prepared with the needs of particular specialists in mind. An example from the field of the earth sciences is 'Handbook of Physical Constants', edited by S. P. Clark (revised edn., *Memoir of the Geological Society of America*, No. 97, 1966). This represents a revision of the 1942 edition published by the Society as *Special Paper* No. 36, and records a 'variety of physical constants needed for geological and geophysical calculations'. The information is presented in 27 subject sections, each of which is accompanied by a lengthy list of references.

Other subject-oriented data compilations include:

'Chemical Analyses of Igneous Rocks', compiled by H. S. Washington (*US Geological Survey, Professional Papers*, No. 99, 1917)

'Handbook of Oceanographic Tables', compiled by E. L. Bialek (*US Naval Oceanographic Office, Special Publication*, No. 68, 1966)

International Tables for X-ray Crystallography (3 vols., Kynoch Press, 1952–62)

'Smithsonian Meteorological Tables' (6th edn., *Smithsonian Miscellaneous Collections*, **114**, 1951)

'The Data of Geochemistry', compiled by F. W. Clarke (*Bulletin of the US Geological Survey*, No. 770, 1924)

The mineralogical field is particularly well served by works which contain data. Most of these have been prepared as an aid to the identification of minerals. They include the following:

An Index of Mineral Species and Varieties Arranged Chemically, by M. H. Hey (2nd edn., British Museum (Natural History), 1962). An *Appendix* was published in 1963

Dana's System of Mineralogy, rewritten by C. Palache (7th edn., 3 vols., Wiley, 1944–62)

Handbuch der Mineralogie, by C. Hintze (6 vols., Walter de Gruyter, 1898–1939). *Supplements* recording details of new minerals were published in 1939, 1960 and 1968

'Microscopic Determination of the Ore Minerals', by M. N. Short (*Bulletin of the US Geological Survey*, No. 914, 1940)

Mineralogische Tabellen', by H. Strunz (2nd edn., Akademische Verlagsgesellschaft, 1949)

'The Microscopic Determination of the Non-Opaque Minerals', by E. S. Larsen and H. Berman (*Bulletin of the US Geological Survey*, No. 848, 1934)

Tables for Microscopic Identification of Ore Minerals, by W. Uytenbogaardt and E. A. J. Burke (2nd edn., Elsevier, 1970)

Many handbooks combine the functions of textbook and data compilation and in such cases tabular and graphical material and critical data are to be found interspersed with explanatory text and references to the literature. Examples include *Handbook of Applied Hydrology*, edited by V. T. Chow (McGraw-Hill, 1964); *Handbook of Subsurface Geology*, by C. A. Moore (Harper, 1963); *Handbook of Geophysics and Space Environments*, edited by S. L. Valley (McGraw-Hill, 1965); and *Handbook of Geochemistry*, edited by K. H. Wedepohl (2 vols., Springer, 1968).

Because they attempt to cover a broad subject field in great detail, many handbooks have developed into multivolume treatises. In such cases individual volumes resemble textbooks more closely than they do the traditional handbook. A good example of such a work is *Nouveau Traité de Chimie Minérale*, edited by P. Pascal (Masson, 1964). Published in 20 volumes (30 parts), each dealing with one or more elements, the treatise contains many references to the primary literature.

The Germans, in particular, have specialised in the publication of multivolume handbooks and most major disciplines are covered by at least one 'handbuch'. They are noted for their excellent bibliographies and are represented in the earth sciences by such titles as *Handbuch der Geophysik* (1942–) and *Handbuch der Paläozoologie* (1938–), both of which are being published by Gebruder Borntraeger. Another characteristic of the 'handbuch' series is that the different volumes are usually published over a long time period and very often out of turn.

This section would not be complete without some mention of *Gmelins Handbuch der Anorganischen Chemie* (8th edn., Gmelin Institute, 1924–). An indispensable reference work for generations of chemists, it contains much useful information for geochemists, mineralogists and crystallographers. The current open-ended edition covers the entire field of inorganic chemistry from the middle of the eighteenth century and is completely independent of the earlier editions. The material, selected and evaluated from the international literature (to which references are given), is arranged in terms of the chemical elements and their compounds. Each element has been assigned to one of 71 numbered systems in such a way that the anion-forming elements have smaller numbers than do the cations. The description of a particular compound can be located in the volume covering the highest-numbered constituent element. Thus the barium silicates will be found in the volume on barium, since the system number of barium is 30 and the system number of silicon 15. A list of the system numbers is to be found inside most volumes. The subject matter within each volume is presented under a series of subheadings, and in the more recent issues these and the contents lists appear in English as well as German. Each individual volume covers the literature to within

6 months of its publication and the work is kept up to date by the continuing publication of supplements for those elements last treated a long time ago.

YEARBOOKS

Typically a 'yearbook' is a serial publication appearing once a year and concerned with reporting progress and presenting statistical information for a particular subject field. In so far as their emphasis is on commercial information and production figures, the majority of earth sciences yearbooks are concerned with the mining and petroleum industries.

The following is a select list of the more important titles.

Annual Statistical Report—American Iron and Steel Institute. A compendium of statistical information relating to the production, shipment, financial, employment, raw materials usage and other aspects of the iron and steel industry in the USA and Canada. World production figures are also given.

Annual Statistical Review—Department of Statistics, American Petroleum Institute. Formerly the *Annual Statistical Bulletin*, the publication presents statistical information on the US petroleum industry. The 1971 volume covers the period 1956–70.

Annual Statistics for the United Kingdom—Iron and Steel. Published by the British Steel Corporation.

The Australian Mineral Industry. Published annually since 1948 by the Australian Bureau of Mineral Resources.

Canadian Minerals Yearbook. Produced by the Canadian Department of Mines and Technical Surveys, the publication contains annual reviews on metals, minerals and fuels.

Iron Ore. Published by the American Iron Ore Association, this is a statistical report on the iron ore industry of the world with emphasis on that of the USA and Canada. Many of the statistical tables cover a 10-year period. A directory of US and Canadian companies is included.

Metal Bulletin Handbook. Formerly *Quin's Metal Handbook* and published annually by the Metal Information Bureau since 1914, it presents statistics relating to both the ferrous and non-ferrous metal industries. Import and export data, prices and consumption are all considered.

Metal Statistics. Published annually since 1908 by the American Metal Market.

Mineral Facts and Problems. Produced every 5 years by the US

Bureau of Mines, the second revision was issued as *Bulletin No. 630* in 1965. Not strictly a 'yearbook', it performs a similar function. Reviews of all the principal minerals and mineral commodities are given together with statistical information. Details of the sources of the statistics are included.

Minerals Yearbook. Published by the US Bureau of Mines, reviews and statistical data are presented in four volumes. Volume 1 covers metals and minerals, Vol. 2 deals with fuels, and Vols. 3 and 4 are concerned with American and international area reports, respectively.

Statistical Summary of the Mineral Industry. Produced by the Mineral Resources Division of the UK Institute of Geological Sciences and published by HMSO, each annual publication covers the previous 5-year period. The 1971 volume deals with over 60 minerals, giving complete country-by-country coverage of production, exports and imports. Based on official statistical publications, a useful list of the sources used is given on pp. 404–412 of the volume covering 1957–62.

Yearbook—American Bureau of Metal Statistics. Each year since 1941 this publication has presented a complete statistical picture with respect to the economics of the non-ferrous metals on a world-wide basis.

Yearbook—International Oil and Gas Development. Published annually since 1931 by the International Oil Scouts Association, the emphasis is on US and Canadian data. Part 1 of the two-volume work covers exploration and Part 2 production.

In addition to the above, there are a number of general yearbooks which present up to date information about a particular country or group of countries. These, too, are very often sources of statistical data, although the amount of detail they contain is not always as great as that to be found in the specialised yearbooks. *The Statesman's Yearbook*, published by Macmillan, is a good example of a general yearbook. Among a mass of other data it contains a limited amount of information on the mining activities of each country, but its main use in this respect is as a guide to official statistical publications in which further information can be located. Examples of national yearbooks are the *Official Year Book of the Commonwealth of Australia* (Commonwealth Bureau of Census and Statistics, 1908–) and *Canada Year Book* (Dominion Bureau of Statistics, 1905–). Both include weather statistics and data relating to mining activities. Along with general yearbooks one can include official statistical journals such as the UK *Annual Abstract of Statistics, Statistical Abstract of the United*

States, and *United Nations Statistical Yearbook*. The first of these, prepared by the Central Statistical Office and published by HMSO, contains data on many aspects of the nation's life, including the mineral industry and the climate. In each volume statistics for the previous 10 years are presented.

Other types of publication which sometimes carry the word 'yearbook' in their titles include directories, e.g. *Mining Yearbook* (Walter R. Skinner, 1887–); annual administrative reports, e.g. *Year Book of the Royal Society of London*; proceedings of meetings, e.g. *Year Book of the American Iron and Steel Institute*; journals containing primary information, e.g. *Yearbook of the Association of Pacific Coast Geographers*; and state-of-the-art annual reviews.

DIRECTORIES

It may sometimes be necessary to locate such information as the address, first names, date of birth, publications, qualifications, or research interests of an individual scientist; or the address, telephone number, size, organisational structure, and activities, etc., of a commercial firm, university, government department or other organisation. This type of information is to be found in directories. Directories appear in many guises. Some of the more important and better-known ones are published annually; others appear irregularly; some are one-off compilations; and yet others appear as contributions in periodicals or books. Sometimes the word 'directory' will appear in the title of the publication; in other cases the work may be described as a 'handbook', 'register', 'yearbook', 'index', 'list' or 'calendar'.

There are literally thousands of directories published each year and little point would be served by attempting to list them all here. Only a selection of the more important works will be mentioned, and the reader who wishes to locate additional titles is referred to the following guides to directories:

Current British Directories, by G. P. Henderson and I. G. Anderson (6th edn., CBD Research, 1970–71)

Current European Directories, compiled and edited by G. P. Henderson (CBD Research, 1969)

Directories in Science and Technology (Library of Congress, 1963)

Directory of Scientific Directories, compiled by A. P. Harvey (2nd edn., Francis Hodgson, 1972)

Guide to American Directories, edited by B. Klein (7th edn., B. Klein, 1969)

Broadly speaking, directories can be classified into three groups: trade, personal and institutional.

Trade directories can be used either to locate information about a particular firm or to discover the manufacturer of a certain type of product. Given a particular problem, the directory to use will, in the majority of cases, be determined by the choice of directories in one's local library. These will probably range from nationally oriented general trade directories such as *U.K. Kompass* (9th edn., 2 vols., Kompass Publishers, 1971) and the *Thomas Register of American Manufacturers* (57th edn., 6 vols., Thomas Publishing Co., 1966) to such specialist subject-oriented publications as *European Metals Directory* (Quin Press, 1964–), a guide to European metals suppliers; *Geophysical Directory* (1946–), a comprehensive listing of companies and individuals connected with geophysical exploration for petroleum; and *Mining Yearbook* (Walter R. Skinner, 1887–), a directory of companies associated with the mining industry.

Details of other commercial directories can be obtained from *Trade Directories of the World* (*1964–65*), compiled and edited by U. H. E. Croner (Croner Publications, 1965), a loose-leaf guide updated through an amendment service.

Personal directories vary a great deal in their scope and in the amount of information given under each entry. The simplest type is the straightforward list of members (usually with addresses) of a particular society or association. These are published by the organisations concerned and are issued either as separate documents, e.g. *Directory of the Association of American Geographers, Geological Society of London—List, Geologists' Association—Rules and List of Members* and *Glaciological Society—List of Members and Subscribers*, or as an integral part of one of the society's regular publications. For example, a list of members appears in the October 1969 issue of the *Bulletin of the American Association of Petroleum Geologists* and from time to time in the *Bulletin of the Geological Society of America*.

In the same way large organisations such as universities and government departments often produce publications in which can be found lists of their staff members. The *Calendars* of British universities are good examples.

Some personal directories bring together the names of people with common interests. Thus we have *World Directory of Mineralogists*, by Font-Altaba (International Mineralogical Association, 1962); 'List of Ostracod Workers in the USSR' (*Ostracodologist*, **14**, 1–12, 1969); *Marine Science in the United Kingdom—A Directory of Scientists, Establishments and Facilities* (Royal Society, 1967);

Orbis Geographicus—A World Directory of Geography, edited by E. Meynen (2 vols., Franz Steiner, 1964–66); *World Directory of Crystallographers and other Scientists Employing Crystallographic Methods*, by D. W. Smith (3rd edn., International Union of Crystallography, 1965); *An International Directory of Oceanographers*, by R. C. Vetter (National Academy of Sciences—National Research Council, 1964); 'Directory of Ostracode Workers', by C. D. Wise and E. Gerry (*Micropalaeontology*, **13**, 381–384, 1967); and *Directory of Palaeontologists of the World*, by G. E. G. Westerman (McMaster University Bookstore, 1968).

A useful and up to date list of scientists and technologists has been produced annually since 1967 by the Institute of Scientific Information, the publishers of *Current Contents* (see p. 97). The list, entitled *Who is Publishing in Science*, is a cumulation of the author address directory in the weekly issues of *Current Contents*. The 1971 directory contains 236 429 names and addresses.

More detailed information on individual scientists can be obtained from various biographical directories. Generally speaking, the larger the entries the more selective the coverage. For British scientists and engineers fairly comprehensive coverage is to be found in *Directory of British Scientists* (3rd edn., 2 vols., Ernest Benn, 1966). This contains details of 54 000 scientists arranged alphabetically and has a subject index in Vol. 2. Also included are a useful list of British scientific societies, a list of British scientific journals and a list of research establishments.

Most countries are served by one of the *Who's Who* type of publication. For Britain there is *Who's Who* (Adam and Charles Black, annually) and for the United States *Who's Who in America* (Marquis–Who's Who). In the case of the former, biographies are removed on death and subsequently included in *Who Was Who*. Five volumes have so far been published, covering, respectively, 1897–1915, 1916–28, 1929–40, 1941–50, 1951–60.

Other biographical directories worthy of note include:

American Men of Science. A Biographical Directory, edited by the Jaques Cattell Press (11th edn., 6 vols., plus supplements, Bowker, 1965–69)

Chambers's Biographical Dictionary, edited by J. O. Thorne (new edn., W. and R. Chambers, 1961)

McGraw-Hill Modern Men of Science (2 vols., McGraw-Hill, 1966, 1968)

The International Who's Who (35th edn., Europa Publications, 1971)

World Who's Who in Science, edited by A. G. Debus (Marquis–Who's Who, 1968)

The primary objective of some directories is to provide a register of research projects but, since all research work is carried out by individual scientists, such directories also serve as personal directories. The UK Department of Education and Science annually produces *Scientific Research in British Universities and Colleges*, published by HMSO since 1952. The 1970–71 edition consists of three volumes: I *Physical Sciences*, II *Biological Sciences* and III *Social Sciences*. Each volume provides brief notes on research in progress and is divided into broad subject fields within which the material is arranged by institution. The *Physical Sciences* volume includes sections on crystallography; geodesy and geophysics; physical geography; geology and mineralogy; meteorology; mining; and oceanography. Each volume contains separate subject and name indexes. A similar publication, *Register of Current Scientific Research at South African Universities*, is produced annually in South Africa by the Council for Scientific and Industrial Research. Research directories of special interest to the earth sciences community include *Current Research in Geomorphology*, compiled by P. Beaumont (Geo Abstracts, 1967), and *Current Research in the Geological Sciences in Canada*, by J. F. Henderson. The latter is an annual publication sponsored by the National Advisory Committee on Research in the Geological Sciences and published in the *Paper* series of the Geological Survey of Canada. The work lists research projects, including thesis work, being carried out in Canadian institutions but excludes the work of mining and oil companies.

Although not a directory in the true sense of the word, the activities of the *Science Information Exchange* (1730 M Street, N. W., Washington, D.C. 20036) can usefully be considered at this juncture. Started in 1949 as a service for the life sciences, *SIE* is a computer-based national register of research in progress. It is administered by the Smithsonian Institution and since 1962 has also covered details of work in the earth sciences. Its object is to collect data regarding ongoing research in US federal agencies, foundations, universities, local government and industry. Over 100 000 projects are recorded annually and information can be supplied on who supports the work, who performs it, where, when and how much, etc. A nominal service fee is charged.

Institutional directories are primarily concerned with listing and describing such organisations as universities, societies, associations, museums, libraries and research centres. Very often, however, staff details accompany the entries and in such cases the publications also act as personal directories. A good example of this is the *Commonwealth Universities Yearbook*, edited by J. F. Foster and T. Craig (48th edn., Association of Commonwealth Universities,

1971), which includes the names and qualifications of all the academic staff of every university in the British Commonwealth. Performing a somewhat similar function but covering the whole world is *The World of Learning* (22nd edn., Europa Publications, 1972). In addition to giving details of the senior academic staff of the world's universities and colleges, this publication covers learned societies, research institutes and libraries. In the case of the more important societies, e.g. the Royal Society, lists of members or fellows are given. Entries are arranged by country but there is a consolidated alphabetical index of institutions at the end.

Another institutional directory of world-wide scope is the multi-volume series edited by R. J. Fifield and entitled *Guide to World Science* (20 vols., Francis Hodgson, 1968). Each volume covers a geographic region and is divided into two parts—an introductory text giving a general account of the organisation of science in the country or countries concerned and a directory of the most important scientific establishments in the region. Details are given of the membership, structure, organisation, aims and activities of each establishment. Volume 1 covers the UK and Vol. 16 the USA. The same publishers provide coverage of European organisations in *European Research Index*, edited by C. H. Williams (2nd edn., 2 vols., Francis Hodgson, 1969), a country-by-country list of research establishments and the research facilities of industrial firms.

There are a number of directories which concern themselves with particular countries. Examples from the UK include *Directory of British Associations* (3rd edn., CBD Research, 1971–72) and *Industrial Research in Britain* (6th edn., Harrap, 1968). The latter is an extremely useful guide to the research activities and senior personnel of government and government-supported organisations; industrial firms; independent research laboratories; trade and development associations; universities; technical colleges; professional and learned societies; and libraries. It also contains lists of British periodicals and abstracting journals covering industrial research. Similar publications dealing with organisations in the USA include *Encyclopedia of Associations* (6th edn., 3 vols., Gale, 1970) and *Research Centers Directory*, by A. M. Palmer (3rd edn., Gale, 1968). The latter is updated quarterly by *New Research Centers* and is arranged in subject sections. Section 12 covers the physical and earth sciences and contains 518 entries, each of which gives the name, address, telephone number, publications, research interests and organisational details of the centre concerned. A subject and institutional index is provided. For US societies there is *Scientific and Technical Societies of the U.S.* (National Academy of Sciences,

1968), which lists address, officers, history, aims and functions of 800 societies. A similar publication dealing with Canadian institutions is *Scientific and Technical Societies of Canada* (National Science Library—National Research Council of Canada, 1969).

Of a somewhat more specialised nature are those directories concerned with particular types of institutions, such as libraries and museums. These are often subject-classified and can be used by the earth scientist to locate appropriate collections of literature or exhibits. In this category is the *Directory of Special Libraries and Information Centers*, edited by A. T. Kruzas (2nd edn., Gale, 1968), an annotated list of 13 000 US organisations, of which 650 are recorded as centres of geological information. Another guide to US libraries is the *American Library Directory*, compiled by E. F. Steiner-Prag (27th edn., Bowker, 1970), which lists nearly 25 000 libraries. For British libraries the best guide is the *Aslib Directory*, Vol. 1, *Information Sources in Science, Technology and Commerce*, edited by B. J. Wilson (3rd edn., Aslib, 1968). Entries are in alphabetical order of the towns in which the organisations are situated and a subject index is provided.

Museums are obvious sources of information and can be located in such publications as *Museums Directory of the US and Canada* (American Association of Museums, 1961); *Museums Calendar* (Museums Association, 1969), which lists British and European museums; and, also covering the UK, *The Libraries, Museums and Art Galleries Year Book*, edited by E. V. Corbett (Clarke, 1968).

The institutional directories so far mentioned are general in their subject coverage. A number of guides exist, however, which are concerned exclusively with earth sciences organisations. One of the most useful is *Guide to Information Sources in Mining, Minerals and Geosciences*, edited by S. R. Kaplan (Interscience, 1965), which gives details of over 1000 government agencies, scientific organisations, institutes and associations.

For information on US earth sciences organisations, the *Directory of Geological Material in North America*, by J. V. Howell and A. I. Levorsen (2nd edn., American Geological Institute, 1957), is still worth consulting. The publication also covers institutions in Canada, Central America, West Indies and Mexico.

Earth sciences directories of a more specialised nature include those dealing with particular branches of the field, e.g. *World Wide Directory of Mineral Industries, Education and Research*, edited by H. Wohlbier *et al.* (Gulf, 1968), *Arctic Research in Western Europe—A Directory of Institutions*, by D. H. Wood (Arctic Institute of North America, 1967), and *A Directory of Meteorite*

Collections and Meteorite Research (UNESCO, 1968); and those dealing with particular types of organisations, e.g. *Directory of Geoscience Departments in the Colleges and Universities of the United States and Canada* (American Geological Institute, 1968) and *Directory of Geoscience Libraries in the United States and Canada* (Geoscience Information Society, 1966).

TEXTBOOKS AND MONOGRAPHS

Although it is convenient to consider them under a separate heading, there is no clear-cut division between textbooks and many of the reference works referred to in the previous sections of this chapter. Similarly, textbooks grade imperceptibly into monographs.

Typically, a textbook contains more detailed information and is more specific in subject coverage than a reference work. As the subject becomes more specific, the treatment more exhaustive and the observations, comments and conclusions better documented, the work becomes a monograph. Neither textbooks nor monographs usually contain original information but instead are written around information previously published in the primary literature. Both textbooks and monographs, particularly the latter, are useful for keeping in touch with progress in those branches of science outside one's own special field. They are also of value at the start of a comprehensive literature survey when one is searching for background information or for a bibliography.

There are thousands of books produced throughout the world each year. In the UK alone during the last 6 years an average of 150 books per year have appeared in the earth sciences field.

The time-honoured method used by most scientists when looking for books is either to scan the shelves or, more rarely, to search the subject catalogue in their local libraries. Since the majority of such libraries are only able to buy a proportion of the total output of books, and since at any one time much of the stock may be off the shelves, these methods of identifying potentially useful books are somewhat inefficient. If, therefore, one wishes to locate the most appropriate title from all the available books, a more reliable search procedure is to scan the various published guides. Such guides usually include details not only of textbooks and monographs but also of the whole range of reference material so far discussed in this chapter.

Some libraries do have very comprehensive collections and in many cases the catalogues of these libraries have been published. These catalogues, particularly those of the world's national libraries, are invaluable guides to books, for, apart from collecting material on a world-wide basis, many national libraries are entitled by law to receive one copy of every book copyrighted in their respective countries. One of the major published library catalogues in the English-speaking world is the British Museum's *General Catalogue of Printed Books* (Trustees of the British Museum, 1931–), which covers the literature from the fifteenth century to 1955 and is updated by the *Ten Year Supplement 1956–1965* and the annual *Additions*. In the USA the *Library of Congress Catalog* fulfils a similar function. Quinquennial cumulations are available for 1950–54 (20 vols., Edwards, 1955) and 1955–59 (22 vols., Pageant, 1960) and these are updated by quarterly and annual supplements.

Not only national but also special libraries publish their catalogues. Of particular use to earth scientists is the *Catalog of the US Geological Survey Library* (25 vols., G. K. Hall, 1965), an author, title and subject catalogue of over 500 000 publications.

Because of initial cataloguing and subsequent composition and printing delays, most published catalogues are considerably out of date when they appear. Furthermore many of them are organised in such a way (e.g. entries arranged by author) as to make subject searching very difficult. When trying to locate books in a particular field and when wanting information on current titles, therefore, one should also consult the various weekly, monthly or annual bibliographies. Typical of such guides is the *British National Bibliography (BNB)* (1950–), a weekly list of books received by the Agent for the British copyright libraries. Entries are arranged according to the Dewey Decimal Classification and each issue is accompanied by an author/title index and the final issue each month contains a cumulated index of authors, titles and subjects. Quarterly and annual cumulations are also produced and retrospective searching is simplified by the availability of quinquennial indexes for 1950–54, 1955–59 and 1960–64, and a subject catalogue for 1951–54 (2 vols.), 1955–59 (3 vols.), and 1960–64 (3 vols.). The *BNB* includes many government publications and new periodicals as well as non-British books where the publishers have British distribution centres.

The nearest American equivalent to the *BNB* is the *American Book Publishing Record (ABPR)* (1960–), a monthly list, with annual cumulations, of books as catalogued by the Library of Congress. Entries are arranged by the Dewey Decimal Classification. Specifically excluded are government publications, dissertations, new printings, periodicals and pamphlets. Non-US books are included where the listed source is the sole US distribution point. Entries for scientific books are selectively extracted and made the subject of a separate annual publication, *American Scientific Books*, by the R. R. Bowker Company.

Most countries produce similar national bibliographies based on material deposited in their national libraries. Apart from those published in English-speaking countries, e.g. the *Australian National Bibliography* (1961–   ), those most likely to be useful are the weekly *Bibliographie de la France* (cumulated annually under the title *Les Livres de l'Année*), the West German *Deutsche Bibliographie* (published in three parts, covering, respectively, books on sale, books not on sale, and maps and charts), and the East German *Deutsche Nationalbibliographie*.

In addition to official national bibliographies, numerous publications are available to help potential readers track down books on specific subjects and to locate bibliographical information. An example is *British Books in Print*, published annually by Whitaker and Sons and containing details of all British books in print and on sale at the time of publication. Paperbacks are included and are also dealt with in a sister publication *Paperbacks in Print*, the 1970 edition of which lists 37 000 paperbacks on sale in Britain. Another example is *English Catalogue of Books*, produced annually since 1837 by The Publishers Circular and containing details of all books published in a particular year in Great Britain. Performing the same function but issued quarterly and covering periods up to 5 years is *Whitaker's Cumulative Book List*. An equivalent publication in the USA is *Books in Print*, produced annually by R. R. Bowker and updated by the bi-monthly *Forthcoming Books*. The 1971 edition of *Books in Print* contains 333 000 titles available from American publishers. Entries are arranged in author and title order. Subject searching is catered for by its sister publication, *Subject Guide to Books in Print*, in which entries appear under subject headings derived from Library of Congress data. This is a particularly useful publication for people who are unfamiliar with library classification systems, since the subject headings appear in alphabetical order and there are abundant cross-references.

Probably the most comprehensive list of English-language books is H. W. Wilson's *Cumulative Book Index*, published monthly (except in August) with semi-annual, bi-annual and occasionally 4-yearly cumulations since 1898. Bibliographic details appear under author, title and subject in a single alphabetical arrangement and the work includes English-language books published anywhere in the world. Performing a similar function for French-language books is *Biblio*, an annual publication produced by Service Bibliographique de la Libraire Hachette. Also covering French-language books but restricted to those still in print is *Le Catalogue de l'Edition Française. 1970* (4 vols., Paris Publications, 1971). An equivalent publication dealing with German books is *Verzeichnis lieferbarer Bücher* (2 vols., Buchandler-Vereinigung, 1971).

A number of guides to books are concerned solely with scientific and technical literature. *Technical and Scientific Books in Print*, published annually by Whitaker, lists 34 000 books in the 1972 edition. All are in print and on sale in the UK. Another guide is the *Aslib Book List* published quarterly from 1937 and monthly since 1948. It is a select list of recommended scientific and technical books grouped by main subjects. For retrospective searching another Aslib publication *British Scientific and Technical Books* is probably of more use. Two volumes have been published; the first covers 1935–52 and the second 1953–57. Details of older American scientific books can be found in *Scientific, Medical and Technical Books Published in the United States of America: A Selected List of Titles in Print*, edited by R. R. Hawkins (2nd edn., National Research Council, 1958). It covers literature published up to 1956.

There are a number of useful book lists concerned solely with earth sciences literature. The Book Development Council, for instance, has started issuing inter-publisher lists in different subject fields including one on *Soil and Earth Sciences*. Full bibliographic details and extracts of reviews are given. The first list appeared in the spring of 1970.

A fairly general list in the earth sciences field is the British Library Association's publication *Reader's Guide to Books in Geology* (2nd edn., 1967). More specialised lists are to be found from time to time in the periodical literature. 'Paperback Books for Earth Science Teachers', by C. V. Proctor (*Journal of Geological Education,* **15**, 29–55, 1967) and 'A Reading Course in Marine Geology' (*Journal of Geological Education,* **14**, 63–65, 1966) are cases in point. An even more specialised and comprehensive list was published by the International Union of Crystallography, Commission on Crystallographic Teaching in 1965. Entitled *Crystallographic Book List*, it was edited by H. D. Megaw and a *Supplement* was issued in 1966.

Before reading, and certainly before buying, a book it can help to read a review. Most scientific and technical journals produce reviews of books in their field, but finding a review to a specific work can be somewhat time-consuming. Help in locating such reviews is provided by the *Technical Book Review Index* published monthly (except July and August) by the American Special Library Association. It consists of review extracts located in scientific, technical and trade journals. Arrangement is by author and an annual cumulative author index is available.

Although not primarily produced as such, *Science Citation Index* (see Chapter 5) is a good guide to reviews. Book reviews are a form of citation and such references are denoted in the *Citation Index* by the letter 'B'.

A serially produced index to book reviews started publication in Germany in 1971, entitled *Internationale Bibliographie der Rezensionen Wissenschaftlicher Literatur (IBR)*. The layout is identical with that of its sister publication *IBZ* (see Chapter 5).

A useful guide to reviews of earth sciences books is the section 'Review Locator' in the bi-monthly journal *Geoscience Documentation*.

REVIEW ARTICLES

As the primary literature of science continues to grow exponentially, the problem of keeping up to date with progress in one's specialised field, and of trying to keep abreast of developments in related fields, has become almost impossible. As an aid to solving this problem the review article has become increasingly important. The object of review articles is to assess critically the primary literature which has been published in a particular field during a given period. In so doing, these articles perform not only an updating function but also enable someone who is just starting to work in a field to become acquainted with the truly useful data and concepts in that field. Furthermore, in so far as most reviews are comprehensive, critical and well-documented, they constitute invaluable guides to the majority of worth-while relevant literature that has preceded them. Thus the discovery of a good review can save much time and effort when one conducts a literature survey.

The value and importance of review articles is emphasised in a paper by Craig (1969). He reports the results of a small survey in which 30 geologists were asked to list the 10 most important articles in their subject and the three most important papers in geology published since 1950. About a third of the latter were major reviews. Craig remarks that 'this relatively high proportion of review papers among the most important geological papers provides a marked contrast to the known number of review papers in the total number of geological papers produced annually. It is estimated by the American Geological Institute (Smith *et al.*, 1967) that some 1400 review articles are to be found among the 35 000 to

100 000 geological papers produced each year. In other words, review articles are held to be more important (33 per cent) than their abundance (2–5 per cent) would seem to indicate.'

The usefulness of reviews seems to have been recognised since the publication of the first scientific journals. The earliest issues of the *Philosophical Transactions of the Royal Society*, for instance, frequently contained such reviews and the first journal devoted exclusively to reviews, *Berlinischer Jahrbuch für die Pharmacie*, started publication as long ago as 1795. During the nineteenth century a number of review journals (all German) appeared which covered geological and related subjects. Among these were *Geographisches Jahrbuch* (1866–1942), which included geophysics, geomorphology and regional geology; *Jahresbericht uber die Fortschritte der Chemie und Mineralogie* (1822–51); and *Jahresbericht über die Fortschritte der Reinen, Pharmaceutischen und Technischen Chemie, Physik, Mineralogie und Geologie* (1847–1910). An extensive key to the contents of *Geographisches Jahrbuch* is given on pp. 53–57 of *Aids to Geographical Research*, by J. K. Wright and E. A. Platt (2nd edn., American Geographical Society, 1947).

Developments in certain fields during the first half of the century are reviewed in the *Annual Reports on the Progress of Chemistry*, published by the Chemical Society. Crystallography is covered in 1908, 1917, annually 1919–29, 1931, 1933, and annually 1935–42; geochemistry in 1930 and 1932; and mineralogical chemistry annually 1904–07, 1909, 1915, 1917, 1921, 1923, 1925, 1927 and 1929.

Currently, review articles are published from time to time in most earth sciences periodicals and the proceedings of conferences (see Chapter 3) often include a number of papers which summarise developments in particular fields. References to reviews can be located in abstracting and indexing journals or in special lists such as that produced by H. E. Hawkes (1967). Mineralogists, petrologists and crystallographers will find *Chemical Abstracts* a useful guide to reviews since the subject index especially indicates such articles. Between 1958 and 1963 Chemical Abstracts Service published a separate annual index to reviews entitled *Bibliography of Reviews in Chemistry*. There is a possibility that this journal may be resurrected.

There are a number of journals which are devoted entirely to the publication of reviews. These journals, which have titles such as *Advances in . . .* , *Progress in . . .* , *Annual Reviews of . . .* , and *Developments in . . .* , have proliferated over the last 10 years. They are usually annuals, although some appear more and some less frequently. Many contain cumulative indexes. Such serials are a valuable source of 'state-of-the-art' information and the majority of articles they contain are accompanied by comprehensive bibliographies.

A useful guide to review serials is the UNESCO publication *List of Annual Reviews of Progress in Science and Technology* (2nd edn., 1969). The UK National Lending Library also issues a frequently updated list entitled *KWIC Index to Some of the Review Serials in the English Language Held at the NLL*.

A list of review serials in the earth sciences is to be found in Appendix 2 of this chapter.

REFERENCES

Craig, G. Y. (1969). 'Communication in Geology'. *Scottish Journal of Geology*, **5** (4), 305–321

Hawkes, H. E. (1967). 'Recent Review Articles in Geology'. *Earth Science Reviews*, **3** (3), 135–155

Smith, F. D., Creager, W. A. and Sayer, J. S. (1967). *Developing a Coordinated Information Program for Geological Scientists in the United States.* American Geological Institute

APPENDIX 1 : A SELECT LIST OF FOREIGN-LANGUAGE/ENGLISH DICTIONARIES

GENERAL TECHNICAL DICTIONARIES

Modern Chinese–English Technical and General Dictionary, 3 vols. McGraw-Hill (1963)

Alford, M. H. T. and Alford, V. L. *Russian–English Scientific and Technical Dictionary*, 2 vols. Pergamon (1970)

Blum, A. *Concise Russian–English Scientific Dictionary for Students and Research Workers*. Pergamon (1965)

De Vries, L. and Herrmann, T. M. *English–German, German–English, Technical and Engineering Dictionary*, 2 vols., 2nd–3rd edn. McGraw-Hill (1967), (1970)

Dorian, A. F. *Dictionary of Science and Technology, English–German, German–English*. Elsevier (1967), (1970)

Kettridge, J. O. *French–English, English–French Dictionary of Technical Terms and Phrases*, 2 vols. Routledge and Kegan Paul (1955), (1956)

Walther, R. *Polytechnical Dictionary, English–German, German–English*. Pergamon (1967)

EARTH SCIENCES DICTIONARIES

Bargilliot, A. *Vocabulaire Pratique Anglais–Français et Français–Anglais des Térmes Techniques Concernant la Cartographie*. Institut Géographique National (1944)

Baulig, H. *Vocabulaire Franco–Anglo–Allemand de Géomorphologie*. Strasbourg University (1956)

Bradley, J. E. S. and Barnes, A. C. *Chinese–English Glossary of Mineral Names*. Consultants Bureau (1963)

Brazol, D. *Dictionary of Meteorological and Related Terms* (*English–Spanish–English*). Libreria Hachette (1955)

Burgunker, M. E. *Russian–English Dictionary of Earth Sciences*. Telberg Book Corp. (1961)

Cooper, S. A. *Concise International Dictionary of Mechanics and Geology* (*English, French, German, Spanish*). Philosophical Library (1958)

Davies, G. M. *A French–English Vocabulary in Geology and Physical Geography*. John Mann (1932)

Mitchell, G. S. *Russian–English Dictionary—Glossary in Geomorphology and Related Sciences*. Ohio State University (1957)

Rutten, L. *Geologische Nomenclator* (*Dutch, German, English, French*). Geologisch-Mijnbouwkundig Genootschap voor Nederland en Klonien (1929)

Sarna, A. and Telberg, V. G. *Russian–English Dictionary of Geographical Terms*, 2 vols. Telberg Book Corp. (1962), (1965)

Sofiano, T. A., Lebedev, A. P. and Khain, V. E. *Russian–English Geological Dictionary*. Fizmargiz (Moscow) (1960)

Telberg, V. G. *Russian–English Dictionary of Paleontological Terms*. Telberg Book Corp. (1966)

Telberg, V. G. *Russian–English Dictionary of Geological Map Terms*. Telberg Book Corp. (1964)

Telberg, V. G. and Deruguine, T. *Basic Russian–English Geological Dictionary*. Telberg Book Corp. (1960)

APPENDIX 2 : SOME REVIEW PUBLICATIONS IN THE EARTH SCIENCES

Advances in Geology (Academic Press, in preparation)

Advances in Geophysics (Academic Press, 1952–); annual

Advances in Hydroscience (Academic Press, 1964–); annual

Developments in Sedimentology (Elsevier, 1964–); irregular; each volume deals with a separate subject

Developments in Solid Earth Geophysics (Elsevier, 1964–); irregular; each volume deals with a separate subject

Earth-Science Reviews (Elsevier, 1966–); bi-monthly

Oceanography and Marine Biology (George Allen and Unwin, 1963–); annual

Physics and Chemistry of the Earth (Pergamon, 1956–); approximately biennial

Progress in Geography (Edward Arnold, 1969–); annual

Progress in Oceanography (Pergamon, 1963–); annual

Progress in the Science and Technology of the Rare Earths (Pergamon, 1964–); approximately biennial

Reviews in Engineering Geology (Geological Society of America, 1962–); irregular

Reviews of Geophysics and Space Physics (American Geophysical Union, 1963–); quarterly

5

Secondary literature — bibliographies, abstracts and indexes

D. N. Wood

Bibliographies, abstracts and indexes are lists of published materials which are designed to lead the potential reader to those items of literature which might be of use to him in his search for information. They have arisen out of a need to control the vast amount of journal literature, books, dissertations, conference proceedings, etc., referred to in other chapters of this book. For the purposes of this chapter, bibliographies on the one hand and abstracting and indexing publications on the other are distinguished on the grounds that the former are mainly retrospective guides to the literature, while the latter, at the time of publication, are intended to be guides to the current literature. It should be noted, however, that some abstracts and indexes carry the title 'bibliography' and vice versa.

BIBLIOGRAPHIES

As its name indicates, a bibliography is, strictly speaking, a list of books, but in fact it may be a list containing details of almost any type of publication—journal articles, reports, theses, patent specifications, government publications and so on. Bibliographies usually concern themselves with any literature on a particular subject published during a specified period. Some, however, deal only with material published in a particular country or language while others concentrate on listing only certain types of literature.

The arrangement of entries within a bibliography varies a great deal. They may be listed in chronological order, alphabetically by author or subject, alphabetically by author within subject or chronologically within author. To facilitate their use, most major bibliographies are provided with an author and/or subject index.

In view of the volume of literature being published, all modern scientific bibliographies are necessarily restricted in one way or another. In the eighteenth and nineteenth centuries, however, it was still possible to contemplate the production of a fairly comprehensive bibliography covering the whole field of science. Two such bibliographies were produced. The first, compiled by D. J. Reuss, was published by Dietrich in Göttingen between 1801 and 1821 and is entitled *Repertorium Commentationum a Societatibus Litteraris Editarum*. It is a 16-volume subject index to articles in the publications of scientific societies from 1665 to 1800. Its task was taken over by the Royal Society of London in their *Catalogue of Scientific Papers*, a 19-volume work covering the scientific periodical literature of the nineteenth century. It is an author index of articles in over 1500 journals and was published in four series between 1866 and 1925. Series 1 covers the period 1800–63 in Vols. 1–6; Series 2 covers 1864–73 in Vols. 7–8; Series 3 covers 1874–83 in Vols. 9–11; Series 3 (supplementary volume) covers 1800–83 in Vol. 12; and Series 4 covers 1884–1900 in Vols. 13–19. A *Subject Index* to the *Catalogue* was planned but only the mathematics, mechanics and physics sections were ever completed.

The *Catalogue* was superseded by the *International Catalogue of Scientific Literature*, published by the Royal Society on behalf of the International Council of Scientific Workers. Both books and articles were listed annually for the years 1901 (published in 1902)–1914 (completed in 1921). Material is arranged in 17 subject sections, and annual author and subject indexes are available. The following are the sections most likely to be of interest to earth scientists: F—'Meteorology', G—'Mineralogy', H—'Geology', J—'Geography', and K—'Palaeontology'.

Another historically important bibliography of particular interest to geologists is *Bibliographia Zoologiae et Geologiae: A General Catalogue of all Books, Tracts and Memoirs on Zoology and Geology*, compiled by L. Agassiz and corrected, enlarged and edited by H. E. Strickland (4 vols., Ray Society, 1848). This work was reprinted by Johnson in their *Sources of Science* series in 1968. The bibliographic entries are arranged in author order and Vol. 1 contains a list of periodicals arranged according to their place of publication.

Modern bibliographies can be found in various forms, the largest usually being published as books. In this category are such broadly based lists as the *McGraw-Hill Basic Bibliography of Science and Technology* (McGraw-Hill, 1966) and *A Selective Bibliography in Science and Engineering*, compiled by the North Eastern University Library Staff (G. K. Hall, 1964). Designed as a supplement to the *McGraw-Hill Encyclopedia of Science and Technology*, the former contains annotated references to recent books and monographs. Entries appear under 7000 subject headings. The North Eastern University bibliography is a reproduction of selected catalogue cards arranged in Dewey Decimal Classification order. Over 15 000 books and monographs are listed, of which about 700 are concerned with earth sciences.

Most bibliographies published as books are more restricted in subject coverage. Examples are *Bibliography and Index of Geology and Allied Sciences for Wales and the Welsh Borders 1536– 1896*, by D. A. Bassett (National Museum of Wales, 1963); *A Bibliography of British Geomorphology*, edited by K. M. Clayton (George Philip, 1964); *Bibliography of the Ionosphere*, by L. A. Manning (Stanford University Press, 1962); and *KWIC Index to Rock Mechanics Literature 1870–1969*, compiled by the Rock Mechanics Section, Imperial College, London (2 vols., American Institute of Mining, Metallurgical and Petroleum Engineers, 1969).

Some bibliographies although complete in themselves are published as part of a series. Such series may consist entirely of bibliographies, e.g. *International Atomic Agency Bibliographic Series* and the *Science Museum Library Bibliographic Series*; or may be report, bulletin or monograph series, only some issues of which are bibliographies. Thus 'Groundwater in Permafrost Regions—An Annotated Bibliography', by J. R. Williams, was published as *US Geological Survey Water Supply Paper*, No. 1792 (1965), and the two-volume work 'Bibliography of Fossil Vertebrates Exclusive of North America 1509–1927', by A. S. Romer *et al.*, was issued as *Memoir of the Geological Society of America*, No. 87 (1961).

Sometimes bibliographies appear as articles in periodicals. For instance, 'Selected Bibliography of South American Geology', by H. R. Cramer, was published in *Tulsa Geological Society Digest*, **31**, 213–239 (1963) and 'Bibliography and Index of Conodonts 1949–1958' appeared in *Micropaleontology*, **7**, 213–244 (1961).

Most scientific and technical books contain bibliographies or at least lists of items referred to by the author. For example, *Vertebrate Paleontology*, by A. S. Romer (3rd edn., University of Chicago Press, 1966), contains a bibliography with 675 entries; *Sedimentary Rocks*, by F. J. Pettijohn (2nd edn., Harper and Row, 1957), has a 700-item bibliography; and *Deserts of the World: An Appraisal of Research into their Physical and Biological Environ-*

ment, edited by W. G. McGinnies *et al.* (University of Arizona Press, 1969), contains a bibliography with 5000 references. *Handbook of Paleontological Techniques*, edited by B. Kummel and D. M. Raup, goes a stage further and besides having a 1000-item bibliography on palaeontological techniques contains a very useful list of other palaeontological and stratigraphic bibliographies.

In addition to the above, most periodical articles contain lists of references, and even entries in encyclopaedias, handbooks and dictionaries frequently give details of books and articles useful for further reading.

In view of the time they take to prepare and to avoid duplication of effort, the tracking down of existing bibliographies should be one of the first priorities when one conducts a literature search. Those bibliographies published as books can usually be located fairly easily in the guides mentioned in Chapter 4, while those issued as articles can be traced via abstracting and indexing journals, where they can usually be located in the subject indexes under such headings as 'bibliography' or 'indexes'.

As additional aids to the location of bibliographies, a number of bibliographies of bibliographies have been published. These range in coverage from the very general to the relatively specialised. The best example of the former is *A World Bibliography of Bibliographies*, by Theodore Besterman (4th edn., 5 vols., Societas Bibliographica, 1965–66). This work lists 117 187 bibliographies under 15 829 subject headings. Under the heading 'geology' there appears a 21-page list of bibliographies arranged by country.

No comprehensive up to date list of geological bibliographies exists, although one is planned by the Geoscience Information Society. This will update two works which cover the older earth sciences literature: *Catalogue des Bibliographies Géologiques*, by E. de Margerie (Gauthier-Villars, 1896), which contains details of 4000 bibliographies published between 1726 and 1895 and is especially useful for biographical material; and 'Catalogue of Published Bibliographies in Geology 1896–1920', by E. B. Mathews, in *Bulletin of the National Research Council*, **6** (5), No. 36 (1923), which contains 3699 titles. Published about the same time as the latter was the much smaller 'List of Manuscript Bibliographies in Geology and Geography', by H. P. Little, *National Research Council, Reprint and Circular Series*, No. 27 (1922). There have, of course, been more recent lists of bibliographies in the earth sciences but these have covered relatively narrow fields; for example, 'Bibliography of Bibliographies on the Geology of the States of the United States', by H. K. Long, which appeared in *Geoscience Abstracts*, **7** (7) (1965). Geological bibliographies can also be found in such works as *A Bibliography of Latin American Bibliographies*, by A. E. Gropp (Scarecrow Press, 1968); *A Bibliography of New Zealand Bibliographies* (New Zealand Library Association, 1967); and *Bibliography of Canadian Bibliographies*, by R. Tanghe (University of Toronto Press, 1960). Covering a fairly limited period but containing details of many recently published earth sciences bibliographies is 'A Checklist of Natural History Bibliographies and Bibliographical Scholarship, 1966–1970', by G. Bridson and A. P. Harvey, in *Journal of the Society for the Bibliography of Natural History*, **5** (6), 428–467 (1971).

One of the best guides to the most recent bibliographies is the serial publication *Bibliographic Index*. Produced twice yearly (with an annual cumulation) since 1937 by the H. W. Wilson Company, this lists, under alphabetically arranged subject headings, books and articles which are either bibliographies or which contain bibliographies. Also of value as a guide to current material is *Geoscience Documentation*. Published every two months since 1969, this periodical contains four sections, one of which, 'Geodoc Index', includes details of new bibliographies in the earth sciences field.

It is inevitable that by the time most bibliographies are published they are out of date. To keep abreast of current research work, attention must be turned to the abstracting and indexing publications.

ABSTRACTING AND INDEXING PUBLICATIONS

Typically, an abstracting or indexing publication is a journal produced at regular intervals and designed to guide the reader to items of interest appearing in the primary literature. An indexing journal consists of a list giving enough bibliographic information about each item for it to be traced. The list may be arranged by subject, author, periodical title, keyword or according to a variety of other systems. An abstracting journal, on the other hand, contains more than bibliographic information about each item listed. According to the International Standards Organisation in its *Recommendation 105/R 214—1961*, '. . . an abstract is a brief indication of the content of an article or other work, is issued independently of it and includes the appropriate bibliographical reference. It is usually compiled by a person other than the author though it may be based on the (author's) synopsis which accompanies the article or work. An abstract should set out the essential features of the original article or work indicating new observations

Olivine composition determination with small-diameter X-ray powder cameras[1]

By J. L. JAMBOR and CHARLES H. SMITH

Geological Survey of Canada, Ottawa

(1)

[Read 30 January 1964]

Summary. Olivine composition can be rapidly and accurately obtained by using standard (Straumanis-mounting) 57·3 mm X-ray powder cameras. A determinative curve based on analysed natural olivines has been constructed by plotting the mol. % Fo against the spacing of back-reflection line 174. The regression equation for the composition range Fo_{100} to Fo_{30} is

$$Fo \text{ (mol. \%)} = 4151 \cdot 46 - 3976 \cdot 45 \, d_{174}$$

using Fe-$K\alpha_1$ radiation. The fayalite end (Fo_{30} to Fo_0) of the determinative curve shows a relatively sharp break from a straight-line relationship of d-spacing versus composition. Practical applications of the determinative curve are briefly presented.

JAMBOR (J. L.) & SMITH (C. H.). *Olivine composition determination with small-diameter X-ray powder cameras.* Min. Mag., 1964, **33**, 730–741, 3 figs.

(2)

A determinative curve based on 22 analysed natural olivines has been constructed by plotting the mol.% of forsterite against the X-ray spacing of back reflection line 174, at d 1·0189Å or 2θ 143·61° for Fo_{100} and Fe$K\alpha_1$ radiation, using a standard Straumanis-mounting 57·3 mm powder camera. The regression equation for the range Fo_{100} to Fo_{30} is Fo (mol.%)=4151·46 − 3976·45 d_{174}. The fayalite end of the curve (Fo_{30}–Fo_0) shows a relatively sharp break from the straight-line relationship, possibly due to the non-uniform variation of b and perhaps related to magnetic interactions between ferrous ions. Nine new analyses of olivines are tabulated : these include six from serpentinized peridotite, Mt. Albert ultramafic intrusion, Gaspé, Quebec, and one of fayalite ($Fo_{0·7}$) from hornblende-biotite granite, Rockport, Massachusetts. [M.A. **14**–196, 351, **15**–219]

R. A. H.

(3)

Olivine composition determination with small-diameter x-ray powder cameras. J. L. Jambor and Charles H. Smith (Geol. Surv. Canada, Ottawa). *Mineral. Mag.* **33**(264), 730–41(1964). Olivine compn. can be rapidly and accurately obtained by using standard (Straumanis mounting) 57.3-mm.-x-ray powder cameras.

Oscar Guire

(4)

3151 **Jambor, J. L.; Smith, Charles H.** Olivine composition determination with small-diameter X-ray powder cameras: Mineralog. Mag., v. 33, no. 264, p. 730–741, illus., tables, 1964.

Jambor, J. L. (and C. H. Smith). Olivine composition determination with small-diameter x-ray powder cameras: Miner. Mag. v. 33, no. 264, p. 730–741, illus., 1964. *"Olivine composition can be rapidly and accurately obtained by using standard (Straumanis-mounting) 57.3 mm x-ray powder cameras. A determinative curve based on analysed natural olivines has been constructed by plotting the mol. % Fo against the spacing of back-reflection line 174."* (5)

25-10-3723. *JAMBOR (J. L.), SMITH (C. H.)* [Geol. Surv. Canada, Ottawa]. **Olivine composition determination with small-diameter X-ray powder cameras.** (Détermination de la composition de l'olivine avec des chambres à RX pour diagrammes de poudre de petit diamètre). *Mineral. Mag., G. B.* (1964), *33,* n° 264, 730-41. — Utilisation de la raie (174) avec petite chambre de 57.3 mm et FeK$_\alpha$. (6)

2301. Jambor, J. L. & Smith, Ch. H.: Olivine composition determination with small-diameter X-ray powder cameras. — Miner. Mag. 33, 730—741, 1964; 3 Fig., 2 Tab.

Der Forsterit-Anteil natürlicher Olivine (9 neue chemische Analysen) kann aus der Lage des Reflexes 174 bestimmt werden. Es handelt sich um denselben Reflex, der von HECKROODT (Trans. Geol. Soc. South Africa 61, 1958, 377) als 0.10.0 indiziert und zur Bestimmung benutzt wurde. Der Reflex ist ziemlich stark, so daß wenig Pulver nötig ist; die Genauigkeit entspricht der von Zählrohrgoniometer-Verfahren, die niedriger indizierte Reflexe benutzen. K. v. GEHLEN (7)

1 B142　　　　　　УДК 549.621.14:539.26
Определение состава оливина в рентгеновских камерах малого диаметра. Jambor J. L., Smith Charles H. Olivine composition determination with smalldiameter X—ray powder cameras. «Mineral. Mag.», 1964, 33, № 264, 730—741 (англ.)

Состав оливина может быть быстро и точно определен в рентгеновской порошковой камере диаметром 57,3 мм. Определяющий график, основанный на результатах анализа природных оливинов, строится нанесением молекулярных процентов Fo (форстерита) против расстояния между обратно отраженными линиями 174. Уравнение для состава колеблющегося от Fo$_{100}$ до Fo$_{30}$:Fo (мол%) = 4541,46—3976,45·d$_{174}$. Фаялитовый конец (от Fo$_{30}$ до Fo$_0$) определяющего графика показывает сравнительно короткий перегиб от прямолинейной зависимости межплоскостного расстояния от состава. И. Николаев (8)

(1) *Summary*
(2) *Mineralogical Abstracts*
(3) *Chemical Abstracts*
(4) *Bibliography of N. American Geology*
(5) *Bibliography and Index of Geology Exclusive of N. America*
(6) *Bulletin Signalétique*
(7) *Zentralblatt für Mineralogie*
(8) *Referativnyi Zhurnal*

Figure 5.1

and any conclusions drawn from them so that the reader can decide whether or not he need consult the original. . . . If designed for a particular group of readers an abstract may be selective and emphasize certain features of particular interest to them . . .'

Attempts have been made to classify abstracts into 'indicative' and 'informative'. The former is typically 'a short abstract written with the intention of enabling the reader to decide whether he should refer to the original article', while the informative abstract is defined as one which 'summarizes the principal arguments and gives the main data and therefore to some extent replaces the original article'. In actual fact these two types of abstract are the end members of a series and between them lie the majority of published abstracts. Even within one journal the quality varies considerably. These two points are illustrated by *Figures 5.1* and *5.2*. *Figure 5.1* presents seven different abstracts of the same paper. A summary which appeared with the original paper is also shown. The most detailed entry is from *Mineralogical Abstracts*, while that from *Chemical Abstracts* is only slightly more useful than the title. In *Figure 5.2*, however, the abstract from *Chemical Abstracts* is highly informative, while that from *Mineralogical Abstracts* is merely indicative.

History

Although they are more necessary today than ever before, abstracting and indexing journals are not new phenomena. According to D. A. Kronick in *A History of Scientific and Technical Periodicals* (Scarecrow Press, 1962), the first journal devoted to abstracts seems to have been *Aufrichtige und unpartheyische Gedancken über die Journale, Extracte und Monaths-Schriften, Worrinnen dieselben extrahiret, wann es nutzlich suppliret oder wo es nothig, emediret werden; nebst einer Vorrede von der Anneemlichkeit, Nutzen und Fehlern gedachter Schrifften*. Edited by C. Gottfried Hoffmann, only two volumes appeared between its birth in 1714 and its death in 1717. Each contained 12 issues and nearly 40 journal titles were scanned, including *Journal des Sçavans*.

From 1714 to 1790 over 40 abstracting journals were published (60 per cent in Germany) but only 14 lasted longer than 10 years and 24 did not continue for more than a year. By 1830, the date often incorrectly quoted for the birth of the abstracting journal, the concept of 'guides to the literature' was well established and the first of the world's major scientific abstracting services made its appearance. This was *Pharmaceutisches Centralblatt*, later to become *Chemisches Zentralblatt*, which continued publication (with a short break after the Second World War) until the end of 1969. The nineteenth and twentieth centuries have seen the proliferation of abstracting and indexing services to the extent that there are currently over 2000 such publications.

The first regularly produced guide devoted exclusively to the

(1)

53605g Comments on serpentinization and related metasomatism. Cerny, P. (Ceskoslov. Akad. Ved, Prague, Czech.). *Amer. Mineral.* 1968, 53(7–8), 1377–85 (Eng). In highly metamorphosed regions, alpine peridotites are serpentinized after consolidation of the enclosing siliceous rocks, under much lower pressure-temp. (p-T) conditions than existed during the preceding regional metamorphism. Mg outflow during an essentially equal vol. serpentinization often causes chloritization and zeolitization of surrounding rocks. On the other hand, in lowgrade metamorphic and sedimentary environment serpentinization takes place during the tectonic emplacement of ultramafic bodies and yields rodingite and jadeite zones that were produced by Ca outflow from completely serpentinized ultramafites. Some contact reactions of cold-intrusive ultramafites with adjacent siliceous rocks take place above the serpentinization p-T range and some within its limits. Above the p-T range, the products and extent of the reaction are controlled by the overall activity of the surrounding siliceous rocks, e.g., granulites, migmatites, plutonic masses; the reaction is essentially that of contact equilibration. Within the p-T limits of serpentinization this process combines with metasomatism caused by components released upon concomitant serpentinization. Resulting contact assemblages are extremely variable with regard to varying metamorphic grade, tectonic mobility, and other factors. Specific serpentine minerals seem to be formed by certain serpentinization processes, depending on static vs. shear pressures and on generally lower vs. higher p-T conditions. Recent investigation shows, however, that these relations are much more complicated than previously supposed. Breakdown of chrome spinel to Cr chlorite + magnetite procedes dominantly in preserpentinization stages. If contemporaneous with serpentinization, Cr chlorite is regarded as a metastable transient phase.   RCQD

(2)

69-796. Černý, P. *Comments on serpentinization and related metasomatism.* Amer. Min., **53**, 1377–1385, 1968.

Alpine peridotites are serpentinized under much lower P-T conditions than existed during the preceding regional metamorphism. Magnesium outflow during equal volume serpentinization often causes chloritization and zeolitization of surrounding rocks.   A. P.

Figure 5.2 (1) Chemical Abstracts; (2) Mineralogical Abstracts

world's earth sciences literature appears to have been the *List of Geological Literature Added to the Geological Society's Library during the Year . . .* This started publication in 1895 and was produced annually until 1936 (covering the literature up to 1934). It was an index arranged alphabetically by author and title, of books, articles and other geological material acquired by the Geological Society of London. Subject indexes were provided.

The US Geological Survey 'Bibliography of North American Geology' commenced publication slightly earlier than the *List*, in 1886, but was much more restricted in coverage. It has continued to the present day and is an index to material concerning the geology of the North American continent; Greenland; the West Indies; and Hawaii, Guam and other island possessions of the United States. Articles by American authors published in foreign journals are cited if they deal with North America or are of a general nature. Articles on North America by foreign authors are included regardless of place of publication, while those of a general nature are included if they appeared in North American journals. The 'Bibliography' has always been issued in the *Bulletin* series of the US Geological Survey. Publication has been more or less annual and cumulations are available up to 1960. Details of these cumulative volumes and the subsequent annuals are given in *Table 5.1*.

Table 5.1. PUBLICATION DETAILS OF 'BIBLIOGRAPHY OF NORTH AMERICAN GEOLOGY'

Years covered	Title	Bulletin number	Year of publication
1732–1891	Catalog and Index of Contributions to North American Geology	127	1896
1785–1918	Geologic Literature on North America		
	Part I Bibliography	746	1923
	Part II Index	747	1924
1919–1928		823	1931
1929–1939		937	1944
1940–1949		1049 (2 vols)	1957
1950–1959		1195 (4 vols)	1965
1960	Bibliography of North American Geology	1196	1964
1961		1197	1965
1962		1232	1966
1963		1233	1968
1964		1234	1966
1965		1235	1969
1966		1266	1970

Complementing the 'Bibliography of North American Geology' and superseding the London Geological Society's *List of Geological Literature* . . . was *Bibliography and Index of Geology Exclusive of North America*. Publication, by the Geological Society of America, started in 1934 (covering the literature for 1933) and continued annually until Vol. 30 (1968, covering the literature for 1965 as well as some earlier material). Vols. 31 (1967) and 32 (1968) were issued in monthly parts but annual cumulative indexes are available. These indexes are issued in two parts—a 'Cumulative Bibliography', which is a compilation of the citations arranged in author order, and a 'Cumulative Index', which is a subject index. The *Bibliography and Index* . . . covers 'the literature on the geology of South America, Europe, Asia, Africa, Australia, Antarctica, Iceland and islands of the eastern Atlantic and western Pacific Ocean basins'. Short abstracts accompany the majority of entries. Although the volume sequence is continued, the *Bibliography and Index of Geology Exclusive of North America* broadened its scope in 1969 and at the same time shortened its title to *Bibliography and Index of Geology*. A description of this publication is given later (see p. 90).

Another early guide in the geological field was *Geologisches Zentralblatt* (1901–42). From 1931 onwards it was issued in two parts: Abteilung A. *Geologie*, which was issued fortnightly, and Abteilung B. *Palaeontologie* (also called *Palaeontologisches Zentralblatt*), which was issued monthly. Each volume possesses an author and subject index.

Between 1925 and 1942 a second German abstracting journal covering geology was produced. Having its origins in the primary journal *Taschenbuch für die Gesammte Mineralogie mit hinsicht auf die Neusten Entdeckungen* (1807–27), it came into existence with the title *Neues Jahrbuch für Mineralogie, Geologie und Paläontologie: Abteilung A. Mineralogie und Petrographie, Referate*; Abteilung B. *Geologie und Paläontologie, Referate*. Between 1827 and 1925 its predecessors appeared under a variety of titles; for a detailed account of the publication history, reference should be made to BUCOP. From 1926 the title was modified slightly to *Neues Jahrbuch für Mineralogie, Geologie und Paläontologie, Referate*. It was issued in the following two parts between 1926 and 1927: Abteilung A. *Mineralogie, Petrographie* and Abteilung B. *Geologie, Paläontologie*. Between 1927 and 1942 it was issued in three parts (six issues per volume): Teil 1. *Kristallographie, Mineralogie*; Teil 2. *Allgemeine, Geologie, Petrographie, Lagerstattenlehre*; Teil 3. *Historische und Regionale Geologie, Paläontologie*.

In 1943 both the *Neues Jahrbuch für Mineralogie, Geologie und Paläontologie, Referate* and *Geologisches Zentralblatt* were incorporated in *Zentralblatt für Mineralogie, Geologie und Paläontologie*, which from that date changed from a primary to a secondary publication. Between 1943 and 1949 it was issued in three parts: Teil I, dealing with mineralogy and crystallography; Teil II, with petrography and geochemistry; and Teil III, with palaeontology and general geology. In 1950 it split into two separate titles *Zentralblatt für Geologie und Paläontologie* and *Zentralblatt für Mineralogie* (see p. 311).

Bibliographie des Sciences Géologiques was published during approximately the same period as the *Neues Jahrbuch für Mineralogie, Geologie und Paläontologie, Referate* and served the French-speaking community. Produced by the Societé Géologique de France, it started in 1923 and until 1930 was issued quarterly. No abstracts were given and entries were arranged first by country and then by periodical title. In 1930 it became an annual publication with entries arranged in 18 subject sections, each of which was broken down into subsections. Each volume was accompanied by an author index. From 1947 until it ceased in 1960 the publication was produced by extracting the earth sciences references first from *Bulletin Analytique* and after 1955 from *Bulletin Signalétique*.

Bulletin Analytique commenced publication in 1940. Produced by the Centre de Documentation du Centre National de la Recherche Scientifique (CNRS), it was a monthly journal containing abstracts of the world's scientific and technical literature. Between 1940 and 1945 the only earth sciences material covered was in the mineralogical and geochemical fields, but in 1946 a 'Sciences de la Terre' section was introduced. For the first year this was divided into three parts—'Minéralogie', 'Pétrographie' and 'Géologie'—but in 1947 and 1952, respectively, subsections on 'Paléontologie' and 'Physique du Globe' were added. In 1956 the journal changed its title and became *Bulletin Signalétique* (see p. 86).

Current abstracting and indexing services

Over 100 current services are concerned wholly or in part with earth sciences literature. They are produced by a wide range of organisations—learned societies, government bodies, industrial firms, trade associations and organisations which have in some cases been specifically set up to publish them. Not surprisingly, the publications vary considerably in size, format, method of indexing, frequency and subject coverage. In some cases the services are available on cards, in others on microform and in others on magnetic tape. In most instances, however, these are usually alternative formats to the more normal printed version.

Current abstracting and indexing journals can be classified into three groups. In the first are those that concern themselves with the literature of a particular subject or subjects. Within this category one can distinguish between multidisciplinary services and those that deal with relatively specialised fields. Only rarely does either type attempt to cover much more than the periodical literature and a few of the more important books, conference proceedings and reports. The second group comprises those that index or abstract particular types of literature, e.g. books, dissertations and reports, irrespective of the subject. The third includes those that concern themselves with the literature (usually the periodical literature) of a particular country or group of countries.

The guides in the second group have been discussed in Chapters 3 and 4, and some of those in the third group are mentioned in Chapter 6. A list of current subject-oriented guides is given in the Appendix to this chapter. This list has been confined mainly to services in the English language, although a few of what are considered the major non-English guides are included. Further details of these listed publications are to be found either in this chapter or, in the case of specialised services, in the later subject-oriented chapters. Additional information can be obtained from a number of published guides to abstracting and indexing services. These include:

A Guide to the World's Abstracting and Indexing Services in Science and Technology (National Federation of Science Abstracting and Indexing Services, Report 102, 1963)
Abstracting Services. Vol. 1. *Science, Technology, Medicine, Agriculture* (2nd edn., International Federation for Documentation, 1969)
A KWIC Index to the English Language Abstracting and Indexing Publications Currently being Received by the National Lending Library (4th edn., NLL, 1972)

Multidisciplinary services. Applied Science and Technology Index (H. W. Wilson and Co.) has been published under that title since 1958 and is a companion tool to *Business Periodicals Index.* From 1913 to 1957 both constituted the *Industrial Arts Index. Applied Science and Technology Index* appears monthly (except July) and cumulates quarterly and annually. Entries consist of bibliographical information arranged under alphabetical subject headings. The periodicals covered (only US, British and Canadian) are determined by a subscriber rating system and currently include 13 earth sciences titles, among which are the major North American geological journals but none from the UK.

Bulletin Signalétique came into existence in 1956 as a continuation of *Bulletin Analytique* (see p. 84). Produced monthly by the Centre de Documentation of CNRS, it is issued in 31 separately available sections. Four sections are of potential interest to earth scientists: Section 161, 'Cristallographie'; Section 210, 'Minéralogie. Géochimie. Géologie Extraterrestre. Pétrographie'; Section 214, 'Géologie Appliquée. Formations Superficielles'; and Section 216, 'Géologie Paléontologie'. The complete service covers some 7000 journals as well as reports and dissertations. Abstracts (about 50 words long on average) are in French but titles are given in their original language. Entries are arranged according to a special classification scheme and an English-language booklet describing the scheme is available from CNRS. Each section is provided with annual author and subject indexes and, with the exception of the 'Cristallographie' section, all earth sciences sections also have monthly subject indexes. In 1970 the four sections mentioned above contained 29 400 abstracts from the world's literature. Section 161 contained 8102; Section 210, 8197; Section 214, 4550; and Section 216, 8551. It has been announced that in 1972 the geology parts of *Bulletin Signalétique* are to be combined with *Bibliographie des Sciences de la Terre* (see pp. 92, 258).

Internationale Bibliographie der Zeitschriftenliteratur aus allen Gebieten des Wissens (*International Bibliography of Periodical Literature Covering All Fields of Knowledge*) is often referred to as *IBZ*. It started in 1911 but a new series of volumes began publication in 1965. It consists of half-yearly volumes, each of which comprises a list of periodicals scanned, an alphabetically arranged subject index which gives bibliographic details of the articles covered and an author index. The subject index contains German, French and English keywords but bibliographic information appears only under the German entry. Cross-references are given for the French and English equivalents. In 1970 *IBZ* covered articles from nearly 10 000 periodicals but its value is severely impaired by the fact that delays in coverage are at least 12 months and can be as much as 4 years.

Pandex: Current Index to Scientific and Technical Literature is a twice-monthly service covering some 2000 journals, 6000 books, 5000 patents and 35 000 reports per year. Entries are arranged

under journal title within broad subject groups. Author and subject indexes are provided in each issue and an unusual feature of
the service is that quarterly and annual cumulative indexes are
published on microfilm and microfiche. The service is also available on magnetic tape.

Referativnyi Zhurnal is sponsored by the USSR Academy of
Sciences and is one of the most comprehensive abstracting services
in the world. It started in 1954 and over 21 000 journals are
covered as well as monographs, dissertations, patents, standards
and maps. The journal is currently issued in 61 subject sections, of
which the following seven are likely to be of use to earth scientists:
Geodeziya i Aeros'emka (Geodesy and Aerial Surveying), 3500
abstracts per year; *Geofizika* (Geophysics), 21 000 abstracts per
year; *Geografiya* (Geography), 40 000 abstracts per year; *Geologiya* (Geology), 40 000 abstracts per year; *Gornoe Delo* (Mining),
18 000 abstracts per year; *Khimiya* (Chemistry), over 100 000
abstracts per year; *Pochvovedenie i Agrokhimiya* (Soil Science and
Agricultural Chemistry), 3600 abstracts per year. All the sections
are issued monthly and are provided with monthly author and
annual author (avtorskii) and subject (predmetnyi) indexes. The
Khimiya section also has formula and patent indexes, and the
Geografiya section has a geographical index. Author indexes are
arranged in two alphabetical sequences, Cyrillic and Roman,
according to the language in which the original material was written. The Roman author index allows abstracts to be located without any knowledge of Russian. Furthermore, although abstracts
are in Cyrillic, full bibliographic details are given in the language
of the paper. The subject indexes are unfortunately only in Cyrillic but it is possible to use them with only a slight knowledge of
Russian and a dictionary. Actually the journal can be used without
reference to the indexes, since in some of the series (e.g. *Geologiya*) there are English-language contents lists. Also, the subject
sequence of the abstracts is the same from issue to issue and it is
therefore possible to scan the same section of each issue in search
of references on a particular topic. As an aid to identifying which
of the various series contain information on a certain subject, a
useful English-language index to the subject divisions has been
published in *A Guide to Referativnyi Zhurnal*, by E. J. Copley
(National Reference Library of Science and Invention, Occasional
Publications, 1970).

With the exception of *Geodeziya . . .* and *Pochvovedenie . . .*
sections, all the sections of interest to earth scientists are available
in separate subseries. For example, the *Geologiya* section is divi-

ded into the following subject units: (1) General Geology; (2) Stratigraphy and Palaeontology; (3) Geochemistry, Mineralogy, Petrography; (4) Quaternary Period, Geomorphology of Land and Ocean Bottom; (5) Geological and Geochemical Prospecting; (6) Hydrogeology, Engineering Geology, Geocryology; (7) Ore Deposits; (8) Non-Metallic Useful Minerals; (9) Deposits of Useful Combustible Minerals; (10) Technique of Geological Prospecting. Details of the various subseries in other sections can be found in the FID publication *Abstracting Services* (see p. 85).

Science Citation Index (*SCI*) is published by the Institute for Scientific Information, a commercial organisation with its head-quarters in Philadelphia. It is based on the principle that current papers on a topic cite older literature deemed by the author to be relevant to his own work. *SCI* enables one to identify those articles which cite particular papers, and provided that one has a know-ledge of the older literature the system thus permits the retrieval of up to date material on the same subject. *SCI* consists of two basic units (both issued quarterly with annual cumulations): the *Citation Index* and the *Source Index*. The latter consists of a list of items, with full bibliographic details, which have appeared in the journals scanned by the service. These items, which are arranged in alphabetical order of author, may be abstracts, discussions, edi-torials, letters, reviews or (more commonly) technical papers. The service currently covers 2200 journals, of which 886 are American, 364 are British, 196 are German, 97 are French and 57 are Russian. The earth sciences account for 119 titles, of which 34 are in the field of oceanography and 17 in the field of meteorology. British geological journals are very poorly represented. In 1969, 341 430 items appeared in the *Source Index*, more than double the figure for 1964, when the service started.

The *Citation Index* consists of references cited in source items and is arranged in alphabetical order of first author. Under each entry appears a list of the items which cited it during the year covered by the *Index*. Only abbreviated journal title, year, volume and page are given for each source item. Titles of articles have to be obtained by referring to the *Source Index*.

When using *Science Citation Index*, a typical searcher would enter the *Citation Index* knowing details of one or more key papers in his field, written some time ago. The *Index* will lead him forward from that item or items to the current citing articles listed beneath it. If one wants to go further, the citing articles can be obtained and selected references from them may be used as new *Citation Index* entry points. This process, known as 'cycling', again brings the searcher forward in time to a crop of current citing articles. A network of articles interconnected by references is thus built up.

Inevitably a system which relies on citations produces much irrelevant material but it is a simple tool to use and, assuming that papers are currently being published on a given subject, rapidly produces results.

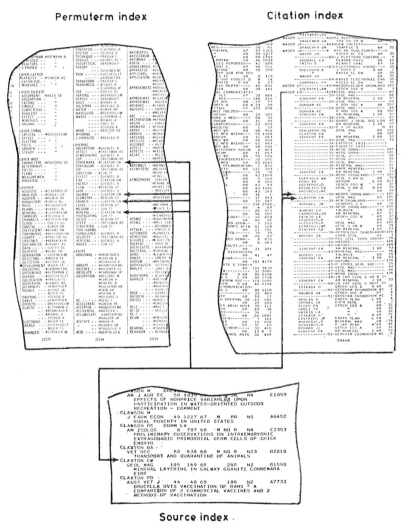

Permuterm index

Citation index

Source index

Figure 5.3 Science Citation Index

In 1967 a third unit of the *Science Citation Index* made its appearance. This is the *Permuterm Subject Index*, which is an index using the words contained in the titles of the articles listed in the *Source Index*. All the words in the titles are arranged in alphabetical order and under each there appears a list of the words which occur with it in any of the titles. Thus, in *Figure 5.3*, an article on mineral layering in granite can be located by looking under 'layering'. In the list both 'mineral' and 'granite' appear and alongside each is the name of the author (C. W. Claxton) of a paper on that subject. Bibliographic details can then be located in the *Source Index*. The same paper could have been found by using the *Citation Index* and looking under Wager, L. R. (1939). *Meddr Grønland*, **105**. This is a classic paper on mineral layering in igneous intrusions and is one of the references cited by Claxton.

The publishers offer a search service to those organisations which are unable to acquire *Science Citation Index*. These searches may be by subject or citation.

Much has been written about the use and value of *Science Citation Index*, and a bibliography appears regularly in the publication itself. From a descriptive point of view, two of the best articles are those by Cawkell (1968) and Malin (1968).

General earth sciences services. The most comprehensive English-language index to earth sciences publications is the *Bibliography and Index of Geology*, produced jointly by the Geological Society of America and the American Geological Institute. It started in 1969 as a continuation of the *Bibliography and Index of Geology Exclusive of North America* (see p. 83), and is issued monthly. Material is presented in 21 subject groups and within each group entries are listed in accession number order and consist of bibliographic information, descriptions or descriptive phrases and a UDC number. Each issue contains an author index and a three-level subject index. A 'Cumulative Bibliography' consisting of references in author order and a 'Cumulative Index' are provided annually. In 1970 the *Bibliography and Index of Geology* covered over 1400 serial titles from 66 countries and indexed approximately 30 000 items. Most of these were periodical articles but a few books, reports, conference proceedings and dissertations were included. The distribution among the various subject sections is shown in *Table 5.2*.

Although it is restricted in geographical coverage, *Abstracts of North American Geology* lists over 8000 items per year and is therefore sufficiently important to be mentioned in this section. A successor to *Geoscience Abstracts* (1959–65), it was started in 1966

Table 5.2. SUBJECT DISTRIBUTION OF REFERENCES IN *Bibliography and Index of Geology*

Subject field	Approximate number of items in 1970
Areal geology	890
Economic geology	3 580
Engineering geology	440
Extraterrestrial geology	740
Geochemistry	1 130
Geochronology	530
Geohydrology	740
Geomorphology	1 280
Igneous and metamorphic petrology	3 430
Marine geology	890
Mineralogy and crystallography	2 960
Miscellaneous	560
Paleobotany	770
Paleontology, general	380
Paleontology, invertebrate	1 460
Paleontology, vertebrate	710
Sedimentary petrology	1 970
Soils	260
Solid-earth geophysics	2 650
Stratigraphic and historical geology	2 950
Structural geology	1 670

by the US Geological Survey in an effort to keep earth scientists currently informed about geological developments in North America. The need for it arose because of the long delays in the production of the *Bibliography of North American Geology* (see p. 82). Although coverage is the same, it does not replace the latter but rather supplements it. The *Bibliography* continues to list citations for a calendar year. *Abstracts* . . . is produced monthly and does not restrict itself to material published in a particular year. Furthermore, each reference is accompanied by a summary of up to 150 words in length. Entries are arranged in alphabetical order of senior author and each issue contains a three-level subject index. Surprisingly, no annual indexes are provided. The publication covers papers, books and maps but not such things as reports and dissertations, which are considered to be difficult to obtain. The continued existence of *Abstracts of North American Geology* is in question. The US Geological Survey have indicated that they intend to end its publication after 1971.

Performing a similar function to the *Bibliography and Index of*

Geology but serving the French-speaking community is *Bibliographie des Sciences de la Terre*, produced by the French Bureau de Recherches Géologiques et Minières. Formerly a card service, it started publication in journal form in January 1968. It is issued monthly in eight subject sections, most of which are further subdivided. Each division is provided with at least an author, subject and geographic index, and some have additional indexes—for example, the 'Minéralogie' division has an index of minerals and the 'Paléontologie' division has a stratigraphic and fossil index. References are listed in subject groups and besides bibliographic data the entry contains information on which of 14 major libraries in France hold the item. A list of the libraries with addresses appears on the inside back cover of each issue. In 1969 *Bibliographie des Sciences de la Terre* listed approximately 30 000 items from the world earth sciences literature, which made it one of the most comprehensive guides to geological and related publications. These 30 000 items were split between the subject sections as indicated in *Table 5.3*.

Coverage and coverage overlap

Faced with all the abstracting and indexing publications listed in

Table 5.3. SUBJECT DISTRIBUTION OF REFERENCES IN *Bibliographie des Sciences de la Terre*

Subject sections	Subdivisions	Approximate number of references in 1969
Cahier A	Minéralogie	1 600
	Géochemie	1 720
	Géochemie isotopique	760
	Géologie extra-terrestre	460
Cahier B	Economie minière	980
	Gitologie	1 520
Cahier C	Roches cristallines	2 660
Cahier D	Roches sedimentaires	3 420
Cahier E	Géologie régionale	1 300
	Stratigraphie	2 440
√ Cahier F	Tectonique	1 220
	Physique du globe	1 400
	Géophysique appliquée	750
Cahier G	Hydrogéologie	1 740
	Géologie de l'ingenieur	880
	Formations superficielles	2 935
Cahier H	Paléobotanique	910
	Paléozoologie	2 580

the Appendix to this chapter, the reader might be forgiven for enquiring why there are so many covering the same or overlapping fields and whether it is really useful to look regularly at more than one or two. The answer to the first question is that each service supposedly caters for a different market, i.e. general, specialised, English-speaking, Russian-speaking, American-oriented and so on. The answer to the second question is that although many services overlap, none of the guides offers complete coverage of its stated field, and to be kept well informed the reader may find it necess-

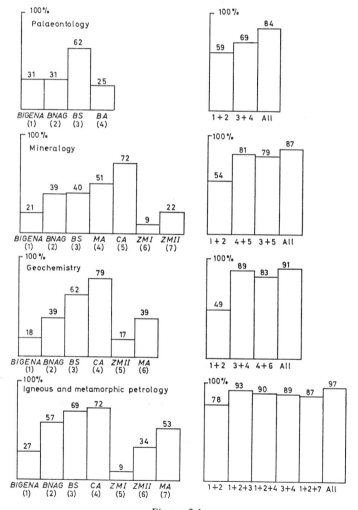

Figure 5.4

ary to consult several guides. This latter point was proved by a recent investigation carried out by the author. The investigation aimed at discovering the extent of coverage and coverage overlap of the services in each of the following fields: Palaeontology; Mineralogy; Geochemistry; Igneous and metamorphic petrology; Economic geology; Geomorphology; Geophysics; and Sedimentary petrology. In each case a basic bibliography of 100 references was compiled by scanning the references at the end of articles in the 1968 issues of relevant journals and selecting not more than four of those dated 1964 from any one article. In this way the references from at least 25 articles contributed to each bibliography.

Each of the references in each of the eight bibliographies thus compiled was then checked against several years' issues of appropriate abstracting journals. The check was made with the help of annual author indexes. No attempt was made to determine whether

Table 5.4

Subject field	Source journals	Abstracting journals
Palaeontology	*Palaeontology* *Journal of Paleontology*	*BIGENA*; *BNAG*; *BS*; *BA*
Mineralogy	*Mineralogical Magazine* *American Mineralogist*	*BIGENA*; *BNAG*; *BS*; *MA*; *CA*; *ZMI*; *ZMII*
Geochemistry	*Geochimica et Cosmochimica Acta*	*BIGENA*; *BNAG*; *BS*; *CA*; *MA*; *ZMII*
Igneous and metamorphic petrology	*Journal of Petrology*	*BIGENA*; *BNAG*; *BS*; *CA*; *MA*; *ZMI*; *ZMII*
Economic geology	*Economic Geology*	*BIGENA*; *BNAG*; *BS*; *ABEG*; *CA*; *ZGP*; *ZMI*; *ZMII*; *MA*
Geomorphology	Citations taken from papers abstracted in *Geomorphological Abstracts*	*BIGENA*; *BNAG*; *BS*; *GA*; *BGI*; *ZGP*
Geophysics	*Journal of Geophysical Research*	*BIGENA*; *BNAG*; *BS*; *GPA*; *MGA*; *PA*; *NSA*
Sedimentary petrology	*Journal of Sedimentary Petrology*	*BIGENA*; *BNAG*; *BS*; *ZMII*; *MA*

Key to abstracting journals. *ABEG—Annotated Bibliography of Economic Geology; BGI—Bibliographie Géographique Internationale; BIGENA—Bibliography and Index of Geology Exclusive of North America; BNAG—Bibliography of North American Geology; BS—Bulletin Signalétique; BA—Biological Abstracts; CA—Chemical Abstracts; GA—Geomorphological Abstracts; GPA—Geophysical Abstracts; MA—Mineralogical Abstracts; MGA—Meteorological and Geoastrophysical Abstracts; NSA—Nuclear Science Abstracts; PA—Physics Abstracts; ZGP—Zentralblatt für Geologie und Paläontologie; ZMI—Zentralblatt für Mineralogie,* Teil I: *ZMII—Zentralblatt für Mineralogie,* Teil II.

the references could be located through the subject indexes. The journals used for the compilation of the bibliographies and the abstracting journals checked are listed in *Table 5.4*.

The results of these tests are presented graphically in *Figures 5.4, 5.5* and *5.6*. *Figures 5.4* and *5.5* show (a) the coverage of individual abstracting journals for each of the fields considered and (b) the coverage that can be obtained by using various combinations of services.

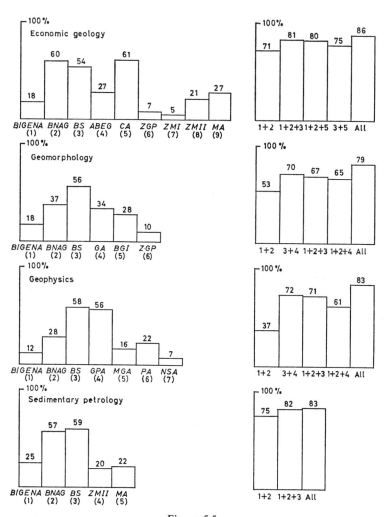

Figure 5.5

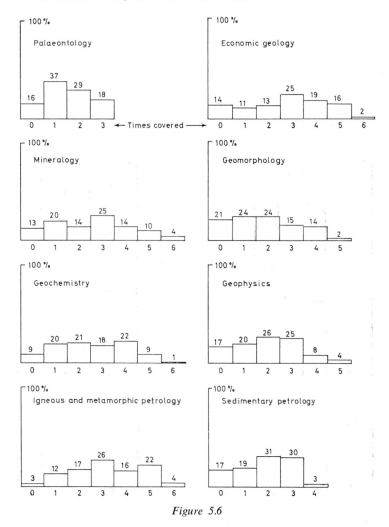

Figure 5.6

The best coverage of any service is by *Chemical Abstracts* in the field of geochemistry but even in this case only 79 per cent of the references can be found. In each case the combined use of the services results in a retrieval rate of between 80 and 90 per cent. For igneous and metamorphic petrology the use of six abstracting journals enables all but three references to be found.

The degree of overlap in coverage is quite surprising. Depending on the field, between 47 per cent (palaeontology) and 85 per cent (igneous and metamorphic petrology) of the references appear in two or more journals and some references are covered as many as six times.

It is interesting to consider for a moment the references which could not be located in the abstracting journals. Of the 110 items concerned, 49 were journal articles; but of these, 17 were non-English and the majority of the remainder were from journals outside the field of earth sciences, e.g. *Journal of Scientific Instruments* and *IEEE Transactions*. Books accounted for 20 references, reports for 16, conference proceedings for 14, theses for 7, maps for 2, government publications for 1 and standard specifications for 1. These latter figures emphasise the need to consult the guides mentioned in Chapter 3 in addition to the subject-oriented guides, when one searches for information.

Speed of coverage. Because of the time taken to acquire and abstract the primary literature and then assemble, print and distribute the abstracts, it is often many months, and sometimes years, before the research worker discovers the existence of a paper which is of interest to him. *Table 5.5* illustrates the delays involved in the preparation of some of the major earth sciences abstracting services. It will be seen that for most services between 80 and 90 per cent of the abstracts are available in libraries within 18 months of the appearance of the primary literature. Only three services, however, cover more than 30 per cent of the literature within 6 months of its appearance.

CURRENT AWARENESS PUBLICATIONS

Because of the time delays referred to above, there has developed a need for services to keep scientists currently informed. This need has been partially satisfied by the appearance of what have been termed 'current awareness' journals. Unlike abstracting journals, which contain summaries of the papers, these current awareness services are merely title announcement lists. Moreover, they usually appear more frequently than the abstracting publications.

There are two current awareness publications which serve the earth sciences community in general. These are *Current Contents* and *Geotitles Weekly*.

The former is produced by the Institute of Scientific Information in Philadelphia and started in 1959. It is currently issued in several subject editions one of which—*Physical and Chemical Sciences*—is of interest to earth scientists. *Current Contents* is

Table 5.5. SPEED OF COVERAGE OF VARIOUS ABSTRACTING JOURNALS

Title	Month on cover	Month of receipt at NLL	Percentage of references published in different years, where x is the year of receipt of the abstracting journal at NLL						
			x	$x-1$	$x-2$	$x-3$	$x-4$	$x-5$	$x-6$
Abstracts of North American Geology	March	June	0	100	0	0	0	0	0
Bibliography and Index of Geology	April	June	31	59	8	2	0	0	0
Biological Abstracts	April	June	1	75	18	3	3	0	0
Bulletin Signalétique (Géologie, Paléontologie)	June	June	3	87	10	0	0	0	0
Chemical Abstracts	May	June	45	43	10	2	0	0	0
Geophysical Abstracts	May	June	0	84	13	2	1	0	0
Geographical Abstracts. A. Geomorphology	June	June	0	57	29	8	4	2	0
Meteorological and Geoastrophysical Abstracts	March	June	0	79	17	2	2	0	0
Mineralogical Abstracts	June	July	6	59	15	8	5	1	6
Physics Abstracts	June	June	47	52	1	0	0	0	0
Referativnyi Zhurnal (Geologiya)	May	June	12	80	6	1	0	1	0

issued weekly and consists of contents tables reproduced from journals that have appeared in recent weeks. About 50 earth sciences titles are covered in the *Physical and Chemical Sciences* section. The tables are arranged in broad subject groups, e.g. 'Earth Sciences', 'Crystallography', 'Multidisciplinary', and a list of journals covered in each issue appears at the front. Speed of preparation and distribution is the essence of the publication, and to this end ISI have arrangements with some publishers to acquire contents lists at proof stage. Furthermore, copies for distribution in Europe are sent first of all in bulk and by air freight to Holland, from where they are despatched to individual addresses. These measures sometimes result in material being notified to subscribers within a week or two of its publication and in geographically remote places even before the primary publications are available in local libraries. At the back of each issue is an author index and address directory to facilitate the acquisition of reprints. This activity is encouraged by the Institute, who will provide personalised 'Request-A-Print' cards at $50 per thousand. If reprints are not available from authors, ISI itself will provide copies under its OATS (Original Article Tear Sheet) service.

Geotitles Weekly (1969–) is a British publication produced by Geosystems (Lea Associates Ltd.). Each weekly issue consists of three parts: yellow pages which contain news of conferences, organisations, equipment and services; white pages which contain a subject-classified list giving details of author, title and source of around 50 000 items per year (much of it ephemeral); and green pages which contain an author index and a key to the coded sources in the previous sections. Throughout the publication titles are given in English. The subject classification used (Geosystems Decimal Classification) has been specially developed for Geosystems and is outlined on the back of each publication.

Besides these two current awareness publications there are others covering mainly other disciplines but impinging on the earth sciences. These include *Chemical Titles, Current Papers in Physics* and *Marine Science Contents Tables*.

Current awareness services properly used have no lasting use, for ultimately the papers they index are covered by the abstracting services. Libraries should discourage their use as retrospective searching tools by throwing them away after not more than 2 years.

MECHANISED INFORMATION SERVICES

During the last 10 years many abstracting and indexing services have turned to the computer for help in coping with the ever-increasing flood of literature with which they have to deal. Bibliographic information relating to the documents covered by the services still has to be fed into the computer by human indexers but, once in, the computer with its sophisticated manipulative powers can arrange this material, produce a wide range of indexes, and control the typesetting and printing operations. This has allowed abstracting organisations to produce publications more quickly, with more varied and up to date indexes and with less labour than previously. Among the tools available to earth scientists which are currently produced with the aid of a computer are: *Bibliography and Index of Geology, Bibliographie des Sciences de la Terre, Geophysical Abstracts, Abstracts of North American Geology, Chemical Titles, Chemical Abstracts, Biological Abstracts, Physics Abstracts, Current Papers in Physics* and *Science Citation Index.*

One type of index which has become particularly common with the advent of the computer is the permuted title index. In such indexes each of the key words in a title serves as a subject heading with the title being printed as many times as there are keywords.

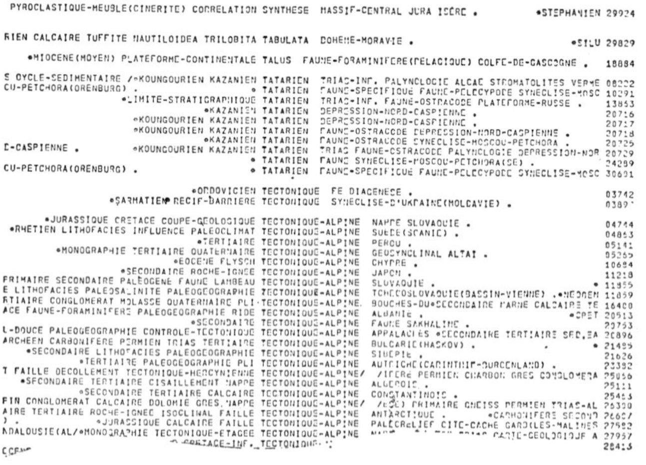
... SYNCLINAL ... 00899
04758
...ACIES CORREL 05998
...FAUNE SYNCLINAL CARTE-GEOLOGIQUE ROUMANIE(VALACHIE-MOLDAVIE) . 08774
•CRETACE-INF. SYNCLINAL DISCORDANCE PORTUGAL(TORRES-VEDRAS) . 10065
•ORDOVICIEN SILURIEN GRES QUARTZITE SYNCLINAL VENDEE(POIROUX) . 11619
•CORRELATION GNEISS MICASCHISTE SYNCLINAL ANALYSE-MINEURS BOUCLIER-D'UKRAINE(KRIVOI-ROG) 12850
•FAMENNIEN(NEHDEN) SYNCLINAL FAUNE-AMMONITE RHENANIE-PALATINAT . 15022
THOFACIES ECOLOGIE SPONGIA PALEORELIEF RIDE SYNCLINAL BOHEME-MORAVIE . •CENOMANIEN TURONIEN LI 15817
•KIMMERIDGIEN(SUP.) EROSION SYNCLINAL PHOTOGEOLOGIE POLOGNE(SAINTE-CROIX) . 18491
•COUPE-GEOLOGIQUE GIVETIEN SYNCLINORIUM BELGIQUE . 21720
AISSEUR CARBONIFERE CONTROLE-TECTONIQUE PLI SYNCLINORIUM DNIEPR-DONETS(DONBASS) . •EP 23239
•CAMBRIEN SYNCLINORIUM METAMORPHISME SAIAN-OUEST . 25779
•ALGONKIEN INFRACAMBRIEN SYNECLISE GRABEN(AULACOGENE) GHANA TOGO DAHOMEY . 06051
VONIEN-SUP. PALEOGEOGRAPHIE BA SO(BARYTINE) SYNGENESE SARRE . •DE 24919
PYROCLASTIQUE-MEUBLE(CINERITE) CORRELATION SYNTHESE MASSIF-CENTRAL JURA ISERE . •STEPHANIEN 29924
RIEN CALCAIRE TUFFITE NAUTILOIDEA TRILOBITA TABULATA BOHEME-MORAVIE . •SILU 29829
•MIOCENE(MOYEN) PLATEFORME-CONTINENTALE TALUS FAUNE-FORAMINIFERE(PELAGIQUE) GOLFE-DE-GASCOGNE . 18884
S CYCLE-SEDIMENTAIRE /•KOUNGOURIEN KAZANIEN TATARIEN TRIAS-INF. PALYNOLOGIE ALGAE STROMATOLITES VERME 08222
CU-PETCHORA(ORENBURG) . • TATARIEN FAUNE-SPECIFIQUE FAUNE-PELECYPODE SYNECLISE-MOSC 10291
•LIMITE-STRATIGRAPHIQUE TATARIEN TRIAS-INF. FAUNE-OSTRACODE PLATEFORME-RUSSE . 13853
•KAZANIEN TATARIEN DEPRESSION-NORD-CASPIENNE . 20716
•KOUNGOURIEN KAZANIEN TATARIEN DEPRESSION-NORD-CASPIENNE . 20717
•KOUNGOURIEN KAZANIEN TATARIEN FAUNE-OSTRACODE DEPRESSION-NORD-CASPIENNE . 20718
•KAZANIEN TATARIEN FAUNE-OSTRACODE SYNECLISE-MOSCOU-PETCHORA . 20725
D-CASPIENNE . •KOUNGOURIEN KAZANIEN TATARIEN TRIAS FAUNE-OSTRACODE PALYNOLOGIE DEPRESSION-NOR 20729
• TATARIEN FAUNE SYNECLISE-MOSCOU-PETCHORA(SE) . 24289
CU-PETCHORA(ORENBURG) . • TATARIEN FAUNE-SPECIFIQUE FAUNE-PELECYPODE SYNECLISE-MOSC 30691
•ORDOVICIEN TECTONIQUE FE DIAGENESE . 03742
•SARMATIEN• RECIF-BARRIERE TECTONIQUE SYNECLISE-D'UKRAINE(MOLDAVIE) . 0389[illegible]
•JURASSIQUE CRETACE COUPE-GEOLOGIQUE TECTONIQUE-ALPINE NAPPE SLOVAQUIE . 04744
•RHETIEN LITHOFACIES INFLUENCE PALEOCLIMAT TECTONIQUE-ALPINE SUEDE(SCANIE) . 04853
•TERTIAIRE TECTONIQUE-ALPINE PEROU . 05141
•MONOGRAPHIE TERTIAIRE QUATERNAIRE TECTONIQUE-ALPINE GEOSYNCLINAL ALTAI . 05265
•EOCENE FLYSCH TECTONIQUE-ALPINE CHYPRE . 10684
•SECONDAIRE ROCHE-IGNEE TECTONIQUE-ALPINE JAPON . 11218
PRIMAIRE SECONDAIRE PALEOGENE FAUNE LAMBEAU TECTONIQUE-ALPINE SLOVAQUIE . • 11855
E LITHOFACIES PALEOSALINITE PALEOGEOGRAPHIE TECTONIQUE-ALPINE TCHECOSLOVAQUIE(BASSIN-VIENNE) .•NEOGEN 11859
RTIAIRE CONGLOMERAT MOLASSE QUATERNAIRE PLI TECTONIQUE-ALPINE. BOUCHES-DU•SECONDAIRE MARNE CALCAIRE TE 16400
ACE FAUNE-FORAMINIFERE PALEOGEOGRAPHIE RIDE TECTONIQUE-ALPINE ALBANIE . •CRET 20513
•SECONDAIRE TECTONIQUE-ALPINE FAUNE SAKHALINE . 20753
L-DOUCE PALEOGEOGRAPHIE CONTROLE-TECTONIQUE TECTONIQUE-ALPINE APPALACHES •SECONDAIRE TERTIAIRE SED.EA 20896
ARCHEEN CARBONIFERE PERMIEN TRIAS TERTIAIRE TECTONIQUE-ALPINE BULGARIE(HASKOV) . • 21495
•SECONDAIRE LITHOFACIES PALEOGEOGRAPHIE TECTONIQUE-ALPINE SIBERIE . 21626
•TERTIAIRE PALEOGEOGRAPHIE PLI TECTONIQUE-ALPINE AUTRICHE(CARINTHIE-BURGENLAND) . 23392
T FAILLE DECOLLEMENT TECTONIQUE-HERCYNIENNE TECTONIQUE-ALPINE/ /IFERE PERMIEN CHARBON GRES CONGLOMERA 25056
•SECONDAIRE TERTIAIRE CISAILLEMENT NAPPE TECTONIQUE-ALPINE ALGEROIS . 25111
•SECONDAIRE TERTIAIRE CALCAIRE TECTONIQUE-ALPINE CONSTANTINOIS . 25463
FIN CONGLOMERAT CALCAIRE DOLOMIE GRES NAPPE TECTONIQUE-ALPINE/ /ESE) PRIMAIRE GNEISS PERMIEN TRIAS-AL 26330
AIRE TERTIAIRE ROCHE-IGNEE ISOCLINAL FAILLE TECTONIQUE-ALPINE ANTARCTIQUE . •CARBONIFERE SECOND 26607
) . •JURASSIQUE CALCAIRE FAILLE TECTONIQUE-ALPINE PALEORELIEF CITE-CACHE GARD(LES-MALINES 27592
ANDALOUSIE(AL/•MONOGRAPHIE TECTONIQUE-ETAGEE TECTONIQUE-ALPINE ... CARTE-GEOLOGIQUE A 27957
...CRETACE-INF. TECTONIQUE... 28415

Figure 5.7

Commonly the keywords are arranged in alphabetical order down the middle of a page with the rest of the title (or as much of the rest as the layout will allow) wrapped around. These indexes have become known as KWIC (keyword in context) indexes and an example is given in *Figure 5.7*.

The mechanisation of the production of abstracting and indexing services is an expensive undertaking. However, the fact that the resultant data base can be manipulated quickly and relatively cheaply has meant that additional services can be offered to help offset the initial cost. Thus subscribers can obtain one-off retrospective bibliographies on the subject of their choice or a regular notification of papers in their specialised field, i.e. an SDI (Selective Dissemination of Information) service. In some cases the whole data base is available for purchase in machine-readable form, thus allowing subscribers to manipulate it as they wish on their own computers. Current developments are moving towards remote access to the data base. Thus a subscriber can sit at a computer terminal in his own laboratory and, via the normal telephone line, search the machine record held in the service headquarters.

In the earth sciences field the most advanced mechanised service is operated by the American Geological Institute. The Institute's Geological Reference File (GEO-REF) was initiated in 1967 and contains over 100 000 citations together with content description, annotation and in-depth indexing using both natural language and the Universal Decimal Classification. This file forms the basis of *Bibliography and Index of Geology* and is also used to produce : one-off bibliographies; indexes for many primary journals, including *American Mineralogist* and *Bulletin of the Geological Society of America*; and retrospective bibliographies for subscribers. The entire file is available for browsing via remote terminal.

Mechanised information services are also offered by Geoservices (the publishers of *Geotitles Weekly*); Chemical Abstracts Service (UKCIS in the United Kingdom); the Institute of Electrical Engineers, who produce *Physics Abstracts*; *Biological Abstracts*; and *Engineering Index*. Since all these services are developing rapidly, users are recommended to write for up to date information.

REFERENCES

Cawkell, A. E. (1968). 'Search Strategies Using the *Science Citation Index*', in *Computer Based Information Retrieval Systems.* Clive Bingley
Malin, M. V. (1968). 'The *Science Citation Index*: A New Concept in Indexing'. *Library Trends*, **16** (3), 374–387

APPENDIX: A CHECK-LIST OF CURRENT SUBJECT-ORIENTED ABSTRACTING AND INDEXING SERVICES COVERING THE EARTH SCIENCES

MULTIDISCIPLINARY SERVICES

Applied Science and Technology Index (1913–)
Bulletin Signalétique (1946–)
Current Contents—Physical and Chemical Sciences (1971–)
Internationale Bibliographie der Zeitschriftenliteratur aus allen Gebieten des Wissens (IBZ) (1911–)
Pandex (1967–)
Referativnyi Zhurnal (1953–)
Science Citation Index (1961, 1964–)

EARTH SCIENCES AND RELATED SERVICES

General

Abstracts of North American Geology (1966–)
Bibliographie des Sciences de la Terre (1968–)
Bibliography and Index of Geology (1969–)
Bibliography of North American Geology (1886–)
Geotitles Weekly (1967–)
Zentralblatt für Geologie und Paläontologie (1950–)
Zentralblatt für Mineralogie (1950–)

Specialised*

Aluminium Abstracts (1963–)
Antarctic Bibliography (1965–)
Arctic Bibliography (1953–)
Articles in Civil Engineering (1969–)
Asbestos Bulletin (1960–)
Bibliography of Geological Literature on Atomic Energy Raw Material (1964–)
Bibliography of Seismology—New Series (1965–)
Bibliography of Snow, Ice and Frozen Ground (1951–)
Bibliography and Subject Index of South African Geology (1959–)
Biological Abstracts (1926–)
British Ceramic Abstracts (1942–)
Building Science Abstracts (1928–)
Chemical Abstracts (1907–)
Chemical Titles (1961–)
Current Geographical Publications (1938–)
Current Papers in Physics (1966–)
Deep Sea Research and Oceanographic Abstracts (1953–)
Engineering Index (1885–)
Excerpta Botanica—Sectio A (1959–)
Fuel Abstracts and Current Titles (1960–)
Gas Abstracts (1945–)
Geocom Bulletin (1968–)

* Some of these services are only partially devoted to earth sciences material.

Geodetic Abstracts (1956–)
Geographical Abstracts—Section A, Section B, Section D (1965–)
Geophysical Abstracts (1929–)
Geotechnical Abstracts (1970–)
Highway Research Abstracts (1931–)
Highways, Current Literature (1922–)
HRIS Abstracts (1968–)
Hydata (1965–)
IMM Abstracts (1950–)
Journal of the American Ceramic Society and Ceramic Abstracts—section
    Ceramic Abstracts (1918–)
Journal of Glaciology—section *Glaciological Literature* (1947–)
Marine Science Contents Tables (1967–)
Meteorological and Geoastrophysical Abstracts (1961–)
Meteorological Office Library—Monthly Bibliography (1919–)
Mineralogical Abstracts (1959–)
NCB Abstracts, C, Coal and Mining Geology (1965–)
New Geographical Literature and Maps (1918–)
Nuclear Science Abstracts (1948–)
Oceanic Abstracts (1971–)
Oceanic Index (1964–)
Petroleum Abstracts (1961–)
Physics Abstracts (1898–)
Rock Mechanics Quarterly Abstract Bulletin (1970–)
Selected Water Resources Abstracts (1968–)
Soils and Fertilizers (1938–)
Soviet Bloc Research in Geophysics (1961–)
Speleological Abstracts (1962–)
Structure Reports (1940–)
Zoological Record (1864–)

6

Translations and foreign literature

D. N. Wood

THE LANGUAGE PROBLEM

Hawkes (1967) reports the results of a study into the size, source and language of primary earth sciences literature. The investigation was based on a sample of the 1961 literature indexed in 22 of the world's leading documentation services in the earth sciences. The sample was restricted to scientific and technical periodicals; regular publications of government agencies, research laboratories and universities; symposia, guidebooks and conference proceedings; and textbooks or reference books. Excluded were semi-published reports, theses, translations, patents and abstracts of papers given at meetings. Oriental literature was excluded from the survey and earth science was defined so as to exclude meteorology, oceanography and certain papers in the fields of hydrology, soil science, fuels and mineral deposits.

The results of the investigation indicated that in 1961 slightly more than 30 000 articles and books, etc., were published in the earth sciences and that of these only 27 per cent were in English. A detailed breakdown by language is given in *Table 6.1*.

Despite the volume of publications in languages other than English, the use made of them is not high. A study of the citations in the 1961 issues of leading Anglo-American geological journals reveals that only 12.3 per cent of the documents cited by British and North American authors are foreign-language publications. Of these the most commonly cited are German publications, followed by Spanish, French and Russian. The full results are presented in *Table 6.2*.

Table 6.1. LANGUAGE BREAKDOWN OF THE 1961 PRIMARY LITERATURE OF EARTH SCIENCE (AFTER HAWKES, 1967)

Language	Estimated number of titles	Percentage in each language
Russian	9 370	30
English	8 350	27
French	3 300	11
German	3 340	11
Non-Russian Slavic	2 770	9
Other European	3 270	11
Non-European	180	1
Total	30 580	100

Table 6.2. LANGUAGE BREAKDOWN OF THE REFERENCES CITED BY BRITISH AND NORTH AMERICAN AUTHORS IN THE 1961 ISSUES OF 11 ENGLISH-LANGUAGE GEOLOGICAL JOURNALS*

Language	North American authors		British authors		Total	
English	1 775	(89,2%)	451	(84.3%)	2 226	(87.7%)
German	74	(3.4%)	46	(8.6%)	120	(4.7%)
Spanish	46	(2.3%)	1	(0.2%)	47	(1.9%)
French	27	(1.4%)	16	(3.0%)	43	(1.7%)
Russian	21	(1.1%)	7	(1.3%)	28	(1.1%)
Others	59	(2.9%)	14	(2.6%)	73	(2.9%)
(none of these individually accounted for more than 1% of the citations)						
Total	2 002	(100.3%)	535	(100.0%)	2 537	(100.0%)

* The journals scanned were:
American Journal of Science
Bulletin of the American Association of Petroleum Geologists
Bulletin of the Geological Society of America
Economic Geology
Geological Magazine
Journal of Geology
Journal of Paleontology
Journal of Petrology
Journal of Sedimentary Petrology
Palaeontology
Quarterly Journal of the Geological Society of London

The low level of use of non-English-language material can be attributed to three main causes. In the first place, much geological information is of purely regional interest; secondly, many earth scientists are unfamiliar with the guides to the literature and, hence, fail to discover the existence of foreign-language articles and books which might be of interest to them; and thirdly, there is the language problem involved (probably the main cause) (Wood, 1967).

LOCATING EXISTING TRANSLATIONS

In an effort to help scientists overcome the language barrier, many thousands of translations are produced each year. The annual output of individual article translations in all subject fields into English is about 15 000 and, in addition, over 250 foreign-language journals are translated in their entirety. To help scientists to locate and obtain these existing translations, a number of national and international services have been developed over the last 15 years.

In the UK the National Lending Library and Aslib are the main sources of information on the location of translations. The NLL maintains a card index of translations into English. Until 1967 this was exclusively concerned with translations from the so-called difficult languages, i.e. Russian, Japanese, Chinese, Czech, Serbo-Croat, etc., but since then it has been broadened in scope and now includes translations from all languages into English. The index contains information only on the translations which the NLL possesses. These include translations produced by the NLL as well as many others which the Library has collected from countries all over the world. The translations indexed, many of which are in microfiche form, now number over 250 000 and, in addition, the Library has a complete collection of cover-to-cover translated journals. All the translations can be loaned to the Library's registered borrowers or can be consulted in the public reading room. Photocopies of the majority can be supplied on request.

Requests for information on translations can be sent in by any individual (by letter or telex, or on an NLL loan request form) and the Library is currently dealing with over 15 000 such requests per year. Enquiries are generally dealt with by return of post, and the required translations are available in about 25 per cent of the cases. Any requests which are unsatisfied at the NLL are passed on automatically to the Commonwealth Index of Unpublished Scientific and Technical Translations housed at Aslib.

The Commonwealth Index, copies of which are also available in New Zealand, India, Canada and Australia, contains information on the whereabouts of translations from all languages into English. The Index, which was established in 1951, currently contains records of over 200 000 translations and a further 12 000 are being added each year. It also includes information on translations in progress. A language breakdown of the translations recently indexed shows that 34 per cent are from German, 23 per cent from Russian, 8 per cent from Japanese, 7 per cent from French, 7 per cent from Italian and the remainder from some 14 other languages. Aslib is currently dealing with 18 000 translation enquiries per year, including those which are forwarded by the NLL. It should be noted that the function of the Aslib index is merely to refer enquirers to the place where the translation is located. The service, unlike that at the NLL, is not supported by a collection of the translations themselves.

Bearing in mind the slightly different functions of the two organisations and the fact that the NLL's terms of reference have changed with time, some guidance to potential users of the Commonwealth and NLL indexes is possibly required. In general, enquirers should in the first instance:

1. apply to Aslib for information on translations of all material less than 6 months old;
2. apply to Aslib for information on translations of all non-Russian material published before 1968;
3. apply to the NLL for all Russian translations more than 6 months old and for translations of all material more than 6 months old and published after 1967.

Besides the translations available at the NLL, a considerable collection of translated literature is maintained at the National Reference Library of Science and Invention. The Library subscribes to over 100 series of cover-to-cover translations of foreign periodicals and holds about 20 000 article translations. The latter are recorded in a card index and the majority are from Russian (42 per cent) and German (36 per cent).

In addition to the facilities available in the UK, scientists can consult by telex or letter a number of translation indexes in Europe. The largest of these is at the European Translations Centre (ETC), 101 Doelenstraat, Delft, Netherlands. The ETC is financed by a number of West European countries, Canada and Japan, and is attempting to collect and index translations from the difficult languages into any of the languages of Western Europe. It receives information on translations from co-operating centres all over the world and in many cases holds copies of the actual translations. Besides individual article translations the Centre has a collection of cover-to-cover translated journals.

ETC deals with about 14 000 requests annually but the majority of these come from Holland itself. For information on translations into English, it is doubtful whether the Centre can provide any better service than that available from the NLL and Aslib. However, the ETC service may be of value for the English scientist who can read German or French.

Further information on translations into French and German is available at the national collecting centres in France and West Germany, i.e. at the Centre National de la Recherche Scientifique, Centre de Documentation, 28 rue Saint-Dominique, Paris VIIe, and the Technische Informationsbibliothek, Am Welfengarten 1, Hanover. Most of the translations recorded at these centres eventually find their way into the ETC index but this may take several months, and for really up to date information or for actual copies of the translations, the national centres themselves should be approached.

In the United States, the National Translations Center is at the John Crerar Library, 35 West 33rd Street, Chicago. Until late 1968 the Center was operated by the American Special Libraries Association (SLA) but is now administered by the Library itself with a grant-in-aid from the National Science Foundation. The function of the Center is to index and, where possible, collect unpublished translations into English from all over the world. It operates an information service and can supply full-size or microfiche copies of the translations it holds.

The organisations so far mentioned deal with translations in all scientific and technical subject areas. Other organisations exist which concern themselves with translations of literature within particular subject fields. Examples within the earth sciences field include the American Geological Institute and the American Geophysical Union. One of the largest subject-oriented translation programmes is that of Euratom. This organisation collects and indexes translations from the difficult languages into all the languages of Western Europe and covers all subjects related to atomic energy. It has many in the earth sciences, particularly in the geochemical and mineralogical fields.

In addition to operating information services regarding the availability of translations, most of the organisations mentioned above publish lists of translations. Until the end of 1970 the NLL, for instance, produced each month the *NLL Translations Bulletin*. Available through HMSO, this gave details in subject categories of the various translated articles which had been added to the Library's collection. From 1971 British translations have been listed in the *NLL Announcement Bulletin*. Information on the translations collected in France and West Germany are recorded, respectively, in *Bulletin des Traductions—Centre National de la Recherche Scientifique* (formerly *Catalogue Mensuel des Traductions*) and *Neu eingegangene bzw. gemeldete*, published by the Technische Informationsbibliothek der Technischen Universitaet in Hanover.

The most comprehensive published index of scientific and technical translations was *Technical Translations*, produced by the Clearinghouse for Federal Scientific and Technical Information, now called the National Technical Information Service (NTIS), in the USA. This ceased publication at the end of 1967 but is still a valuable guide to older translations. It was a twice-monthly publication in which translations were listed in broad subject groups. It included translations produced in North America, translations collected by North American organisations (including SLA), and translations of which CFSTI was notified by such bodies as the NLL and ETC. For most of its publication history, *Technical Translations* announced translations from the difficult languages into all the languages of Western Europe, but from 1966 this policy was changed and only translations into English were listed.

Since *Technical Translations* ceased publication, the function of acting as a comprehensive published index to the world's scientific and technical translations has been taken over by two relatively new publications: *World Index of Scientific Translations* and *Translations Register-Index*. Although originally a quarterly publication, the former is now produced monthly by ETC and announces translations from the difficult languages into the languages of Western Europe. It cumulates annually. *Translations Register-Index*, started by SLA, is now published twice monthly by the American National Translations Center. It announces in its *Register* section new accessions to the translations centre at the John Crerar Library, and in the *Index* section records these accessions as well as other items available from commercial translators, government and other sources. The publication has regular cumulations, thus allowing retrospective searches to be undertaken relatively easily,

To simplify the location of translations issued before the appearance of *Translations Register-Index*, the SLA has published a *Consolidated Index of Translations into English (CITE)*, which covers 140 000 translations issued between 1953 and 1966.

In addition to the national and international lists of translations referred to above, there are a number of other announcement bulletins produced which are designed to publicise the existence of translations prepared by particular organisations or within particular subject fields. For example, details of the Euratom translations are given in *Transatom Bulletin*, while US Government sponsored translations are announced in *Government Reports Announcements*. The American Geophysical Union announces translations in its *Transactions*, where they are listed in a section entitled 'Translations of Russian and other Foreign Publications'. The American Geological Institute (AGI) publishes details of new translations in its journal *International Geology Review*. In 1962, as Part 2 of the October issue of *International Geology Review*, the Institute published a 'Catalogue of Translations of Russian Papers in Geology, Solid Earth Geophysics and Related Sciences through 1961'. This catalogue, compiled by J. W. Squire and A. Eustus, lists translations recorded by the AGI translations office from 1953 to 1961 and includes cover-to-cover translations as well as 1650 individual article translations. Many commercial firms also produce their own lists of available translations.

Most of the information contained in these relatively specialised journals is eventually incorporated in the major international indexes and they are therefore of use primarily as current-awareness tools.

This chapter has dealt so far with the location of *ad hoc* article translations, and only brief mention has been made of cover-to-cover translations. The systematic translation of each and every issue of certain foreign journals was started in the 1950s in an attempt to make western scientists more aware of scientific and technological developments in countries such as the Soviet Union. Particularly in the early years, the majority of cover-to-cover translations were produced under government sponsorship in the USA and the UK. The NLL has been responsible for the UK programme, and since 1959/60 has supported the regular translation and publication of 16 Russian journals. In the USA the National Science Foundation has been responsible for government support. Since the cover-to-cover programme began, the number of journals being translated has grown steadily and there are currently over 250 scientific and technical periodicals being translated. Although a large number of these are still produced with government aid, many commercial firms have also entered the field. Over 90 per cent of the translated journals are Russian, the remainder being Polish, Yugoslavian, Japanese and Chinese. Most are available on an annual subscription, but a few (issued by NTIS) are obtainable as single issues only. The average time lag between the publication of the original and the appearance of the translation is about 9 months.

A list of cover-to-cover translations in the earth sciences is presented as Appendix 1 to this chapter. Further details of these and many other translated journals can be obtained from *Transatom Bulletin* (every month); *Translations Journals. List of Periodicals Translated Cover-to-Cover, Abstracted Publications, Periodicals Containing Selected Articles and Multilingual Publications* (ETC, (1972); *A Guide to Scientific and Technical Journals in Translation*, by C. J. Himmelsbach and G. E. Boyd (Special Libraries Association, 1968); and *English Language Equivalent Editions of Foreign Language Serials*, by R. C. Gremling (Literature Service Associated, 1966).

Since, according to the 1967 edition of the *Directory of Japanese Scientific Periodicals* (National Diet Library, Tokyo), there are 4929 current Japanese periodicals in the field of science and technology, it is somewhat surprising that only two are translated from cover to cover. It is particularly surprising when one also considers that less than 1 per cent of the English-speaking community can cope with Japanese. The reason for this is that compared with other non-English-speaking countries Japan publishes much of its literature in English and other European languages. Ten per cent of the journals, for instance, are published entirely in languages other than Japanese, and in addition many of the Japanese journals contain occasional articles in English, English summaries, or at least an English contents list. A list of Japanese earth sciences journals published in English is given in Appendix 2 to this chapter. Many of the English-language articles in Japanese journals are original articles, but some are translations of papers which originally appeared in Japanese in other publications.

Besides cover-to-cover translations a number of journals exist which publish translations of selected articles from the non-English literature. In the earth sciences field the best-known is *International Geology Review*, a monthly publication of the American Geological Institute. The journal started in 1959 and contains translations of papers selected from many Russian sources and sometimes from Chinese, Japanese and other foreign serials. Non-serial works are reviewed and on occasion large parts are translated. Another journal containing selected translations is *Geochemistry International*, published every 2 months since 1964. It carries translations of selected articles from the Russian journal *Geokhimiya* and of other geochemical papers from Russian, German, French and Japanese journals.

Periodical articles are, of course, not the only sources of useful information. Many thousands of scientific and technical books are also being published each year in foreign languages. In 1968, for instance, the NLL purchased nearly 5000 Russian books, of which 18 per cent were in the earth sciences field.

Unlike English scientific books, many of which are essentially review publications, Russian books and monographs in particular often contain original scientific information, and for this reason many of them are translated. As with periodicals, government support for book translations has been forthcoming in both the UK and the USA. In the UK the NLL has had the responsibility for organising the rather limited programme, and about 10 translations are issued per year. During the last 5 years seven earth sciences titles, mainly palaeontological, have been published. Commercial publishing houses have also been active in this field, and in 1968 the NLL, which attempts to obtain copies of all scientific book translations into English, acquired almost 500. The Library's collection of translated scientific books now numbers nearly 12 000, of which about 250 are earth sciences publications. Most book translations are announced in the national bibliographies of the countries in which they are produced but there are also a number of special guides. Most of the translation bulletins mentioned above give bibliographic details of selected translated books, but the most complete guide is *Index Translationum*, published annually by UNESCO. Another useful guide is the NLL's monthly publication *List of Books Received from the USSR and Translated Books*.

How to get a translation prepared

If it has not been possible to locate a translation of a scientific paper by consulting the relevant sources of information mentioned above, it may be thought necessary to have a translation prepared. Some scientists, particularly those employed by large commercial organisations and government departments, are fortunate in so far as translation facilities are often provided locally. In the UK, for instance, almost 40 per cent of the organisations employing more than 100 people provide a translation service for German and French literature, and 30 per cent do so for Russian.

For UK scientists, help in obtaining a translation of Russian and, to a more limited extent, Japanese articles or books is available through the NLL's translation programme. The Library will arrange for an article to be translated free of charge on the following conditions:

1. The article should be in the field of science, technology (including agriculture and medicine) or social science.
2. The article should preferably be not more than 2 years old.

3. The article should not be from a journal which is scheduled for cover-to-cover translation.
4. The translation must be for private use and not for publication.
5. The requester must agree to edit the translation from a technical point of view.

The translations produced under this scheme are prepared by extramural translators who are selected for their subject knowledge as well as their linguistic ability. About 1200 articles, as well as the books already mentioned, are translated each year and are announced monthly in the *NLL Announcement Bulletin*.

Scientists requiring translations of material in other languages and who are employed by organisations which neither operate a local service nor have access to one are forced to find either individual translators or commercial translation bureaux which will undertake the work for them. In universities, for instance, it has been reported that translation work is done very much on a 'get together over beer basis' with lecturers from the language departments. This is a convenient arrangement, but often results in translations of doubtful scientific merit.

Those scientists who do not have colleagues willing to provide translations can be put in touch with an appropriate translator by either the Institute of Linguists, Aslib or ETC. The Institute publishes an *Index of Members of the Translator's Guild* (with quarterly supplements) giving a list of registered translators together with their subjects and languages, and Aslib and ETC both maintain a card index of translators. Additional help may be obtained by consulting the *Directory of Technical and Scientific Translators and Services*, by P. Millard (Crosby Lockwood, 1968); the *International Directory of Translators and Interpreters*, by B. Pond (Pond Press, 1967); *Translators and Translations: Services and Sources in Science and Technology*, by F. E. Kaiser (2nd edn., Special Libraries Association, 1965); or *Who's Who in Translating and Interpreting*, by A. Flegon (Flegon Press, 1967). Making use of these guides to translators should help to ensure that the translator is technically as well as linguistically competent, but the system is by no means foolproof, since many translators over-generalise when stating their areas of subject competence.

Apart from the difficulty of locating a technically suitable translator, the cost may prove prohibitive to the individual who wishes to have a translation produced. Prices vary considerably from one language to another, but in no case is translating cheap. If an arrangement is made directly with the translator, the cost in the UK will range from £4·50 per 1000 words for French and German, £6·00 or more for Russian, to upwards of £10·00 per 1000 characters for Chinese, Japanese and other ideographic languages. If a commercial translating bureau is used, prices may well be higher.

GUIDES TO FOREIGN SCIENTIFIC LITERATURE

Before the need for a translation of a particular article or book is apparent, it is necessary to discover not only that it exists but also something about its contents. To conclude this chapter, attention will therefore be drawn to some of the published guides to foreign literature. Some of these are indexes and merely give details of the titles of publications; others give lengthy abstracts which may on occasion obviate the need for a full translation.

For those scientists who are capable of coping moderately well with foreign languages there is an abundance of foreign national bibliographies, abstracting services and publications lists of foreign surveys and institutions. Many of these have been referred to in Chapters 3, 4 and 5. Probably the most comprehensive such service is the Russian abstracting publication *Referativnyi Zhurnal* (see p. 87). Dealing solely with the Soviet literature although somewhat out of date is the publication of the Soviet Ministry of Geology *Geologicheskaya Literatura SSSR, Bibliograficheskii Ezhegodnik* (Geological Literature of the USSR, Bibliographical Yearbook). The latest volume covers the literature of 1960 and was published in 1967. Russian publications can also be traced in *Knizhnaya Letopis* (a weekly list of books which have just been published), *Novye Knigi* (a weekly list of books about to be published) and *Letopis Zhurnalnykh Statei* (a weekly index to the contents of Soviet journals). *Novye Knigi* is translated from cover to cover and, in addition, translations of selected entries appear monthly in *International Geology Review*.

For the English-speaking scientist the most common way of discovering the existence and contents of a foreign language book or paper is by using the English-language abstracting and indexing journals, particularly the larger ones such as *Bibliography and Index of Geology*, *Geophysical Abstracts* and *Chemical Abstracts*. In addition, there are a number of publications in English which deal with the literature of particular countries. Details of some of these are given in *Table 6.3*.

One of the most ambitious programmes for drawing the attention of English-speaking scientists to foreign literature (particularly the literature of communist countries) is conducted by NTIS. This organisation produces three bibliographies, *USSR Scientific Abstracts*, *East European Scientific Abstracts* and *Communist*

Title	Publisher	Year of commencement	Number of issues per year	Number of items indexed/abstracted per year
Abstracts of Bulgarian Scientific[*] Literature—Geology and Geography	Bulgarian Academy of Sciences, Scientific Information Centre for Natural, Mathematical and Social Sciences	1958	2	250–300
Abstracts of Bulgarian Scientific Literature—Mathematics, Physics, Astronomy, Geophysics, Geodesy	do.	1958	2	200–350
Abstracts of Romanian Technical Literature	Institutul Central de Documentare Tehnica	1965	4	1 000–1 500
Indian Science Abstracts	Indian National Scientific Documentation Centre	1965	12	12 000
Korean Scientific Abstracts	Korea Scientific and Technological Information Center	1969	4	500
Pakistan Science Abstracts (formerly Pakistan Scientific Literature—Current Bibliography)	Pakistan National Scientific and Technical Documentation Centre	1961	4	1 000
Philippine Abstracts	Division of Documentation, National Institute of Science and Technology—Manila	1960	4	600
Polish Scientific Periodicals—Contents	Polish Academy of Sciences, Documentation and Scientific Information Centre	1962	10	10 000
Polish Technical and Economic Abstracts	Centralny Instytut Informacji Naukowo-Technicznej i Ekonomicznej	1951	4	1 000
Romanian Scientific Abstracts	Academy of the Socialist Republic of Romania, Scientific Documentation Centre	1964	12	5 000

Chinese Scientific Abstracts. All are issued in subject sections, including one entitled 'Earth Sciences'.

Another organisation in the USA which has paid particular attention to the problem of foreign literature is the Library of Congress. Each month between 1949 and May 1969 it published, in co-operation with over 300 other North American libraries, the *Monthly Index of Russian Accessions.* This four-part publication consisted of a list of Russian monographs arranged in alphabetical order of author, a list of Russian periodicals together with translated contents pages, a subject index to periodical articles and monographs, and an index to monographs on the Soviet Union or by Soviet authors in English or Western European languages.

For Russian books, another useful guide is the NLL's monthly publication *List of Books Received from the USSR and Translated Books.* This lists in subject order the Library's Russian book accessions (about 5000 per year).

There are, of course, many other abstracting and indexing publications that are intended to lead scientists to foreign literature. It is impossible to mention them all here, but details can be found in the various guides to abstracting and indexing publications mentioned on p. 85.

APPENDIX 1

(a) COVER-TO-COVER ENGLISH TRANSLATIONS OF EARTH SCIENCE JOURNALS

Title of original journal	Years available in translation	Title of translation	Availability
Antarktika—Doklady Komissii	1960–61	*Antarctica—Commission Reports*	NTIS
Byulleten Moskovskogo Obshchestvo Ispytatelei Prirody, Otdel. Geologicheskii	1956–58	*Bulletin of the Moscow Society for Natural Research, Geological Section*	NTIS
Doklady AN SSSR Geokhimiya, Geologiya, Geofizika, Petrografiya, Mineralogiya, Gidrogeologiya, Paleontologiya	1959–61	*Proceedings of the Academy of Sciences of the USSR, Earth Science Sections: Geochemistry, Geology, Geophysics, Petrography, Mineralogy, Hydrogeology and Paleontology*	Kraus
do.	1962–	do.	American Geological Institute

Title of original journal	Years available in translation	Title of translation	Availability
Doklady AN SSSR Geokhimiya	1956–58	Proceedings of the Academy of Sciences of the USSR, Geochemistry Section	Plenum
Doklady AN SSSR Geologiya	1957–58	Proceedings of the Academy of Sciences of the USSR, Geological Sciences Section	Plenum
Doklady AN SSSR Okeanologiya	1961–64	Soviet Oceanography	American Geophysical Union
Fizikotekhnicheskie Problemy Razrabotki Poleznykh Iskopaemykh	1965–	Soviet Mining Science	Plenum
Geodeziya i Kartografiya	1959–61	Geodesy and Cartography	American Geophysical Union
Geofizicheskii Byulleten	1966–	Geophysical Bulletin	NTIS
Geokhimiya	1956–63	Geochemistry (after 1963 selected articles only published in Geochemistry International)	Scripta Technica Inc.
Geologiya i Geofizika	1965–	Geology and Geophysics	Aztec School of Languages
Geologiya Nefti i Gaza	1958–	Petroleum Geology	Petroleum Geology
Geologiya Rudnykh Mestorozhdenii	1960–	Economic Geology, USSR	Economic Geology Publishing Company
Geomagnetizm i Aeronomiya	1961–	Geomagnetism and Aeronomy	American Geophysical Union
Geomorfologiya	1970–	Geomorphology	Plenum
Geotektonika	1967–	Geotectonics	American Union Geophysical

Title of original journal	Years available in translation	Title of translation	Availability
Informatsionnyi Byulleten Sovetskoi Antarkticheskoi	1958–61	Soviet Antarctic Expedition	American Elsevier Publishing Co.
do.	1961–	Information Bulletin of the Soviet Antarctic Expedition	American Geophysical Union
Izvestiya AN SSSR, Fizika Atmosfery i Okeana	1965–	Izvestiya. Atmospheric and Oceanic Physics	American Geophysical Union
Izvestiya AN SSSR, Fizika Zemli	1965–	Izvestiya. Physics of the Solid Earth	American Geophysical Union
Izvestiya AN SSSR, Seriya Geofizicheskaya	1957–64	Bulletin of the Academy of Sciences of the USSR, Geophysics Series	American Geophysical Union
Izvestiya AN SSSR, Seriya Geologicheskaya	1958–61	Bulletin of the Academy of Sciences of the USSR, Geologic Series	Kraus
do.	1962–	The whole journal is translated but selected translations only are published in International Geology Review	American Geological Institute
Izvestiya VUZ Geodeziya i Aerofotosemka	1962–	Geodesy and Aerophotography	American Geophysical Union
Kristallografiya	1956–	Soviet Physics—Crystallography	American Institute of Physics
Litalogiya i Poleznye Iskopaemye	1966–	Lithology and Mineral Resources	Plenum
Meteoritika	1963–	Meteoritica	Spectrum Translation and Research Inc.
Meteorologiya i Gidrologiya	1966–	Meteorology and Hydrology (selected articles only)	NTIS
Novye Knigi SSSR	1964–	Forthcoming Russian Books. Technical Sciences	Scientific Information Consultants

Title of original journal	Years available in translation	Title of translation	Availability
Novye Knigi SSSR	1965–	Index to Forthcoming Russian Books. Technical Sciences	Scientific Information Consultants
Okeanologiya	1965–	Oceanology	American Geophysical Union
Osnovaniya Fundamenty i Mekhanika Gruntov	1964–	Soil Mechanics and Foundation Engineering	Plenum
Paleontologicheskii Zhurnal	1962–	Until 1967 the whole journal was translated but selected translations only were published in International Geology Review. Since 1967 the whole journal has been published in translation under the title Paleontological Journal	American Geological Institute
Pochvovedenie	1958–61	Soviet Soil Science	NTIS
do.	1962–68	do.	Soil Science Society of America
Prikladnaya Geofizika	1966–	Exploration Geophysics	Plenum
Problemy Severa	1958–	Problems of the North	National Research Council of Canada
Rudarsko-Metallurski Zbornik	1962–	Mining and Metallurgy Quarterly	NTIS
Sovetskaya Geologiya	1960–	The whole journal is translated but selected translations only are published in International Geology Review	American Geological Institute
Trudy Geofizicheskogo Instituta, AN SSSR	1957 only	Soviet Research in Geophysics	Plenum
Trudy Morskogo Gidro-fizicheskogo Instituta, AN SSSR	1961–64	Soviet Oceanography	American Geophysical Union
Vestnik Zavod za Geo-losko i Geofizicko Istrazivanje	1962–	Bulletin of the Institute for Geological and Geophysical Research —issued in three series	NTIS

(b) JOURNALS CONSISTING OF TRANSLATIONS OF ARTICLES SELECTED FROM
VARIOUS FOREIGN JOURNALS

Title of journal	Originating date	Availability
Geochemistry International (selected articles from *Geokhimiya* and other journals)	1964–	American Geological Institute
International Geology Review (selected articles from Russian journals)	1959–	American Geological Institute
Soviet-Bloc Research in Geophysics, Astronomy and Space (selected articles from Russian and Chinese journals)	1961–	NTIS
Soviet Geography: Review and Translation	1960–	American Geographical Society
Soviet Hydrology: Review and Translation	1962–	American Geophysical Union
Soviet Oceanography (selected articles from *Doklady AN SSSR*, and *Trudy Morskogo Gidrofizicheskogo Instituta*)	1961–64	American Geophysical Union
Soviet Soil Science	1969–	Soil Science Society of America

APPENDIX 2: JAPANESE EARTH SCIENCES JOURNALS PUBLISHED IN ENGLISH

Aerological Data of Japan
Annual Report of the International Polar Motion Service
Annual Report of the Meteorological and the Seismological Observations Made at the International Latitude Observatory of Mizusawa
Bulletin of the Geographical Survey Institute
Bulletin of the International Institute of Seismology and Earthquake Engineering
Bulletin Time Service, Mizusawa Observatory
Contributions, Marine Research Laboratory, Hydrographic Office of Japan
Data Report of Hydrographic Observations, Series Geomagnetism
Geochemical Journal
Geographical Reports, Tokyo Metropolitan University
Geology and Mineral Resources of Japan
Geophysical Magazine, Tokyo
Japanese Journal of Geology and Geography
JARE Scientific Reports, Series A, Aeronomy

JARE Scientific Reports, Series B, Meteorology
JARE Scientific Reports, Series C, Geology, Geography, Glaciology,
  Seismology, Geodesy and Geochemistry
JARE Scientific Reports, Series D, Oceanography
Journal of Earth Sciences, Nagoya University
Journal of the Faculty of Science, Hokkaido University, Series 4, Geology
  and Mineralogy
Journal of the Faculty of Science, Hokkaido University, Series 7, Geophysics
Journal of the Faculty of Science, University of Tokyo, Section 2, Geology,
  Mineralogy, Geography, Geophysics
Journal of Geomagnetism and Geoelectricity, Kyoto
Journal of Geosciences, Osaka City University
Journal of the Japanese Association of Mineralogists, Petrologists and
  Economic Geologists
Journal of Physics of the Earth
Journal of Science, Hiroshima University, Series C, Geology and Mineralogy
Kumamoto Journal of Science, Series B, Section 1, Geology
Memoirs of the Faculty of Science, Kyushu University, Series D, Geology
Mineralogical Journal, Sapporo
Monthly Notes, International Polar Motion Service
National Report of the Japanese Antarctic Research Expedition
Oceanographical Magazine
Proceedings of the Research Institute of Atmospherics, Nagoya University
Publications of the International Latitude Observatory of Mizusawa
RAAG Memoirs of the Unifying Study of Basic Problems in Engineering
  and Physical Sciences by Means of Geometry
Records of Oceanographic Work in Japan
Report, Kakioka Magnetic Observatory
Report, Simosato Magnetic Observatory
Report of Ionosphere and Space Research in Japan
Report of Radiation Observation
Science Reports, Niigata University, Series E, Geology and Mineralogy
Science Reports, Tohoku University, Second Series, Geology
Science Reports, Tohoku University, Third Series, Mineralogy, Petrology
  and Economic Geology
Science Reports, Tohoku University, Fifth Series, Geophysics
Science Reports, Tohoku University, Seventh Series, Geography
Science Reports, Tokyo Kyoiku Daigaku, Section C, Geology, Mineralogy
  and Geography
Seismological Bulletin, Japan Meteorological Agency, Tokyo
Special Contributions, Geophysical Institute, Kyoto University
Tokyo Journal of Climatology
Transactions and Proceedings, Palaeontological Society of Japan

REFERENCES

Hawkes, H. E. (1967). 'Geology', in R. B. Downs and F. B. Jenkins (eds.),
  'Bibliography: Current State and Future Trends. Part 2'. *Library Trends*,
  15 (4), 816–828
Wood, D. N. (1967). 'The Foreign Language Problem Facing Scientists and
  Technologists in the United Kingdom—Report of a Recent Survey'.
  Journal of Documentation, **23** (2), 117–130

7

Geological maps

E. L. Martin

THE HISTORY AND DEVELOPMENT OF GEOLOGICAL MAPS

Although in 1683 Martin Lister read a proposal to the Royal Society for 'a new sort of maps of countrys, together with sands and clays' (Lister, 1684) the earliest true geological maps date from around the beginning of the nineteenth century. There are a few earlier mineralogical, lithological and soil maps, such as that of France by L. Coulon (*Rivières de France*, 1664) and Christopher Packe's *A New Philosophico-Chorographical Chart of East Kent* published in 1743.

One of the earliest true geological maps, a manuscript map of the City of Bath, was prepared by William Smith in 1799 and is now in the library of the Geological Society of London. Smith produced numerous maps, including the first geological map of England and Wales, published in 1815. Details of Smith's manuscript and published maps have been given by several authors (Judd, 1897; Sheppard, 1917; Eyles and Eyles, 1938; Cox, 1942; and Eyles, 1969).

The Geological Survey of Great Britain was established in 1835 under the direction of H. T. de la Beche and by 1840 he had prepared eight 1 : 63 360 geological sheets of SW England. By 1850, B. Cotta was able to publish a catalogue of 571 geological maps of various parts of the world (*Geognostische Karten unseres Jahrhunderts*. Englehardt, Freiberg).

Geological surveys were quickly established in many parts of the world, and by the turn of the century small-scale geological maps of various degrees of perfection were available for most countries, and for some areas large-scale mapping had begun. During the period 1752–1881, 924 maps of the Americas were published (Marcou and Marcou, 1884). The first of a series of international maps, issued under the aegis of the International Geological Congresses, was initiated in 1893 with the publication of the first sheet of the *International Geological Map of Europe* (1 : 1.5M).

During the twentieth century there has been a steady increase in geological map production. The now widespread use of aerial photography has revolutionised the production both of topographic base maps and of geological maps produced largely by techniques of photo interpretation. In recent years there has been increasing use of colour photography (true and false colour), infra-red photography and various types of radar imagery. Satellite photographs sometimes show gross structural features not evident from aerial photographs of more limited coverage. The US Geological Survey have prepared geological maps of the moon based on the interpretation of telescope photographs (this series of maps at a scale of 1 : 1M bear *USGS Miscellaneous Geologic Investigation Map* (I) and *Lunar Atlas Chart* (LAC) numbers). A 1 : 5M map of the near side of the moon has also been published and 1 : 100 000, 1 : 25 000 and 1 : 5000 maps of lunar landing sites were prepared in connection with the Apollo missions; the large-scale maps were prepared from Lunar Orbiter photographs. J. Harbour has given a general account of the USGS lunar mapping programme in *Geotimes*, **14** (7) (1969).

The last two decades have also witnessed a great upsurge in the production of small-scale international maps produced under the aegis of the Commission for the Geological Map of the World (CGMW).

Accounts of the history and development of geological maps are given by Jules Marcou in *Sur les Cartes Géologiques à l'Occasion du 'Mapoteca Geologica Americana'* (Dodivers, Besançon, 1888); by T. Sheppard in *Report of the British Association for the Advancement of Science* 1920; in *Methods in Geological Surveying*, by E. Greenly and H. Williams (Murby, 1930); by H. A. Ireland in *Bulletin of the Geological Society of America*, **54** (9) (1943); and in *A Source-Book of Geological, Geomorphological and Soil Maps for Wales and the Welsh Borders (1800–1966)*, by D. A. Bassett (National Museum of Wales, 1967).

A brief survey of the current state of geological mapping has been given by K. C. Dunham in *Quarterly Journal of the Geological Society of London*, **123** (1) (1967).

TYPES OF GEOLOGICAL MAPS

The conventional geological map represents in pictorial form the distribution of rocks at the earth's surface. It results from the geologist's interpretation of his field observations, often supplemented by earlier records and drilling information. In areas of simple geological structure with ample surface and subsurface data the interpretation may be unequivocal and the resultant map may be extremely accurate. In more complex and poorly exposed areas there may be more than one possible interpretation and many of the geological boundaries shown on the map may be conjectural. Conjectural boundaries are often represented as broken lines on geological maps. The presentation and reliability of geological maps are discussed by Robertson (1956), Hageman (1968) and Harrison (1963).

Useful sourcebooks on mapping and the construction of geological maps are: *Manual of Field Geology*, by R. R. Compton (Wiley, 1961); *Die Geologische Karte. Ihre Anfertigung und Ausdeutung*, by H. Falke (De Gruyter, 1967); *Methods in Geological Surveying*, by E. Greenly and H. Williams (Murby, 1930); and *Field Geology*, by F. J. Lahee (6th edn., McGraw-Hill, 1961).

Elementary books intended for students, which are primarily devoted to map interpretation and the construction of geological sections, include: *Introduction to Geological Structures and Maps*, by G. M. Bennison (Arnold, 1964); *Geological Maps and their Interpretation*, by F. G. H. Blyth (Arnold, 1965); *Coupes et Cartes Géologiques*, by A. Foucault and J. F. Raoult (SEDES, Paris, 1966); *Simple Geological Structures*, by J. I. Platt and J. Challinor (4th edn., Murby, 1968); *Geological Structure and Maps*, by A. Roberts (2nd edn., Cleaver-Hume, 1958); and *Geological Maps*, by B. Simpson (Pergamon, 1968).

The most detailed geological maps normally published are at scales of from 1:10 000 to 1:25 000; they include the coalfield maps of Germany (1:10 000) and Great Britain (1:10 560) and the 1:25 000 maps of Germany and Switzerland and the 1:24 000 maps of the USA.

Maps at smaller scales must normally be more or less generalised, either because all the mapped detail cannot be shown at the scale chosen or because the mapping is of a reconnaissance nature and only the broad outlines of the geological succession can be determined. For many areas the largest-scale maps issued in regular series are at scales of from 1:40 000 to 1:60 000; most European countries have series at scales within this range. Maps at still smaller scales (1 : 200 000 to 1 : 500 000) are chiefly produced from photogeological mapping or to supplement series at larger scales. National maps are frequently at scales of 1 : 1M and smaller and continental maps at 1 : 5M and smaller.

Sometimes only part of the rock succession may be recorded on a geological map, either to isolate and clarify the distribution of a particular rock unit or system or to differentiate the bedrock from the superficial deposits. In the latter case two editions of the same map are produced, one showing the 'solid', bedrock or pre-Quaternary geology, and the other the 'drift', superficial or Quaternary deposits. Drift maps commonly show the distribution of the pre-Quaternary rocks over drift-free areas. This is usually the case with the 1 : 63 360 series of England and Wales, although on some sheets the 'solid' geology has been simplified in comparison with the 'solid' maps. On some Scottish 1 : 63 360 sheets the drift is shown as a stipple overprint, and on other drift maps the 'solid' rocks may be grouped as 'bedrock'.

In addition to conventional geological maps there are many other types of map which may be loosely grouped as geological maps. In some cases these do not form clear-cut types, but may comprise maps which vary in their method of construction. Further information on the various types of geological maps will be found in *Subsurface Mapping*, by M. S. Bishop (Wiley, 1960); and in *A Source-Book of Geological, Geomorphological and Soil Maps for Wales and the Welsh Borders (1800–1966)* mentioned earlier. A general account and principles of classification are given (in Hungarian, with a German summary) by Radócz (1968).

Photogeological maps are normally geological maps prepared largely by techniques of photointerpretation; in some cases they may merely indicate structural lineaments visible from the air. Two useful textbooks are *Photogeology and Regional Mapping*, by J. A. E. Allum (Pergamon, 1966), and *Photogeology*, by V. C. Miller and C. F. Miller (McGraw-Hill, 1961). The American Society of Photogrammetry publishes three basic handbooks: *Manual of Photogrammetry* (3rd edn., 1966), *Manual of Photographic Interpretation* (1960), and *Manual of Color Aerial Photography* (1968); and many useful articles are published in its journal *Photogrammetric Engineering*.

Structural and tectonic maps vary widely in their presentation. Some indicate small-scale features such as dips of bedding, cleavage, schistosity and foliation. Small-scale tectonic maps indicate gross regional structures, depths to basement and major sedimentary basins. Structure contour maps indicate the subsurface position of specified rock units.

Isopach maps indicate the variations in thickness of rock units or formations.

Lithofacies maps show the limits and variations of facies of rock units. Many examples of these maps are given in *Lithofacies Maps: An Atlas of the United States and Southern Canada*, by L. L. Sloss, E. C. Dapples and W. C. Krumbein (Wiley, 1960).

Palaeogeographical maps show the geography of past geological times. The principles used in their construction are given in *Paleogeologic Maps*, by A. I. Levorsen (Freeman, 1960). An early palaeogeographical map is included in Charles Lyell's *Principles of Geology* (3rd edn., Murray, 1834). More modern examples are included in the following books: *The Geologic Development of the Japanese Islands*, by M. Minato, M. Gorai and M. Hunahashi (Tsukiji Shokan, Tokyo, 1965); *Atlas of Paleogeographic Maps of North America*, by C. Schuchert (Wiley, 1955); and *A Palaeogeographical Atlas of the British Isles and Adjacent Parts of Europe*, by L. J. Wills (Blackie, 1951). Palaeogeographical maps which stress dynamic aspects are sometimes known as palaeotectonic maps; examples are the 'Paleotectonic Maps of the Jurassic, Triassic and Permian Systems' (*US Geological Survey Miscellaneous Geologic Investigations Maps* I-175, I-300, I-515; 1956, 1959, 1967). A refinement of the palaeogeographic map is the palinspastic map (Kay, 1945), in which an attempt is made to show the original geographic position of the units mapped by the removal of the effects of faulting, folding and other factors.

Palaeoclimatic maps depict temperature zones, wind directions and other climatic phenomena of past geological ages. Examples are given in *Descriptive Palaeoclimatology*, by A. E. M. Nairn (Interscience, 1961); and *Climates of the Past: An Introduction to Paleoclimatology*, by M. Schwarzback (Van Nostrand, 1963).

Geophysical maps show the variations of the earth's gravity, the intensity of the magnetic field, the electrical or seismic properties of the subsurface rocks or radioactivity measurements. Geophysical techniques are outlined in *Introduction to Geophysical Prospecting*, by M. B. Dobrin (2nd edn., McGraw-Hill, 1960).

Geochemical maps show the relative abundance of chemical elements as determined from soil, water or plant material. They are used in the location of ore deposits and to delimit areas unsuitable for various types of agricultural purposes (e.g. with a high selenium or low boron or copper content).

Hydrogeological maps indicate the availability and potential yield of groundwater resources, their chemical composition or the nature of the piezometric surface. Various types of information may be combined on a single map or may be distributed between several complementary maps of the same area. An international legend for hydrogeological maps is published by the International Association of Scientific Hydrology/International Association of Hydrologists.

Geotechnical maps are used for civil engineering and military purposes. They are frequently lithological maps on which the physical properties of each unit are tabulated. Some maps delimit areas of earthquake or tsunami risks (e.g. *New Zealand Industrial Maps*). A bibliography on geotechnical mapping was published in *Zeitschrift für Angewandte Geologie*, **7** (3) (1961). Terrain classification maps (1 : 250 000) have been prepared by the CSIRO Division of Soil Mechanics and are described in their *Technical Paper No. 2* (1968). There is no clear-cut distinction between terrain classification and geomorphological maps.

Geomorphological maps delineate land forms such as fault scarps, terraces and volcanic domes. Glacial maps are a special type of geomorphological map in which features of glacial accumulation, erosion, transport and deposition are indicated.

Soil maps delimit soil series or associations and are based on the study of soil profiles. Although they are not usually regarded as geological maps, they frequently reflect the nature of the underlying rocks.

Electronic data processing has allowed of the greater use of statistical techniques in the interpretation of geological data, particularly in the field of trend surface analysis. *An Introduction to Statistical Models in Geology*, by W. D. Krumbein and F. A. Graybill (McGraw-Hill, 1965) contains a chapter on map analysis. Multivariate facies maps are discussed by J. M. Parkes in *Computer Contributions, Kansas State Geological Survey*, No. 40 (1969).

GUIDES TO GEOLOGICAL MAPS

A very useful list of national and international maps prepared by the Commission for the Geological Map of the World (CGMW), was published in the *Geological Newsletter of the International Union of Geological Sciences*, No. 3 (1969); it contains a list of the addresses of publishing bodies and prices of maps are shown in most cases. Since 1963 the Commission has also published a *Bulletin* which gives details of new maps and bibliographies.

Most official geological maps are produced by national and state geological surveys or equivalent organisations, from whom lists of available maps can be obtained; the libraries of such organisations frequently have good reference collections of foreign maps in addition to maps of their own territories. A series of lists of maps for various countries, which usually include indexes to map series, is published in the *Circular Letter* (*Circ. Lett.*) now *Geological Newsletter* (*Geol. Newsl.*) of the IUGS. The title of the series is 'Inventory of Available Geological Maps'. Lists for the following countries have been published (up to *Geol. Newsl.*, (2), 1972):

EUROPE
Austria   *Geol. Newsl.*, (4), 1968
Belgium   *Geol. Newsl.*, (4), 1969
Cyprus   *Geol. Newsl.*, (1), 1967
Czechoslovakia
                 Geol. Newsl., (2), 1968
Denmark   *Geol. Newsl.*, (4), 1967
Finland   *Geol. Newsl.*, (1), 1970
France   *Geol. Newsl.*, (3), 1968
Germany (GFR)
                 Geol. Newsl., (1), 1971
Great Britain
                 Geol. Newsl., (3), 1969
Greece   *Geol. Newsl.*, (2), 1969
Hungary   *Geol. Newsl.*, (1), 1969
Italy   *Geol. Newsl.*, (1), 1967
The Netherlands
                 Geol. Newsl., (2), 1968
Northern Ireland
                 Geol. Newsl., (3), 1969
Norway   *Geol. Newsl.*, (4), 1968
Poland   *Geol. Newsl.*, (4), 1970
Portugal   *Geol. Newsl.*, (1), 1967
Romania   *Geol. Newsl.*, (4), 1967
Sweden   *Geol. Newsl.*, (3), 1968
Switzerland   *Geol. Newsl.*, (1), 1968

ASIA
Ceylon   *Geol. Newsl.*, (4), 1971
Israel   *Geol. Newsl.*, (2), 1969
Thailand   *Geol. Newsl.*, (1), 1972

AFRICA
Burundi   *Geol. Newsl.*, (3), 1969
Congo (Kinshasa)
                 Geol. Newsl., (3), 1969
Kenya   *Geol. Newsl.*, (3), 1971
Morocco   *Geol. Newsl.*, (2), 1970
Nigeria   *Circ. Lett.*,   18

Rwanda   *Geol. Newsl.*, (3), 1969
South Africa
                 Geol. Newsl., (4), 1969
Swaziland   *Geol. Newsl.*, (1), 1969
Tanzania   *Circ. Lett.*,   18
Tunisia   *Geol. Newsl.*, (1), 1968
Zambia   *Circ. Lett.*,   18

AUSTRALASIA
New Zealand
                 Geol. Newsl., (3), 1970
Western Australia
                 Circ. Lett.,   18
Exclusive of
  Western Australia
                 Geol. Newsl., (1), 1967

NORTH AMERICA
Canada   *Geol. Newsl.*, (2), 1972
Mexico   *Geol. Newsl.*, (3), 1971
United States
                 Geol. Newsl., (3), 1970

SOUTH AMERICA
Argentina   *Geol. Newsl.*, (3), 1967
Chile   *Geol. Newsl.*, (3), 1967
Colombia   *Geol. Newsl.*, (2), 1967
Ecuador   *Geol. Newsl.*, (2), 1967
French Guiana
                 Geol. Newsl., (2), 1967
Guyana   *Geol. Newsl.*, (2), 1967
Surinam   *Geol. Newsl.*, (2), 1967
Trinidad and Tobago
                 Geol. Newsl., (2), 1967

ANTARCTICA *Geol. Newsl.*, (2), 1972

ARCTIC REGIONS
Greenland   *Geol. Newsl.*, (2), 1971
Iceland   *Geol. Newsl.*, (4), 1971

In addition to official mapping a very large number of geological maps are the results of individual research. The majority of such maps are published in the journals of local and national geological societies but some may appear only in university theses. Many maps are produced by petroleum and mining exploration companies. A few of these are published (e.g. the excellent colour-printed maps of parts of Iran produced by the British Petroleum Company); others may be available to genuine enquirers at the offices of the companies if the information they contain is no longer confidential, and some may be deposited at the offices of the appropriate official geological survey organisation.

A number of abstracting and indexing publications (see also Chapter 5) contain details of geological maps. None of these is completely adequate but the following will be found useful.

Bibliographie Cartographique Internationale. An annual publication published since 1946/47. Separately published maps are listed under regional headings but the reference numbers of all geological maps are given in the subject index.

New Geographical Literature and Maps, published semi-annually since 1918 by the Royal Geographical Society. This lists maps and atlases under regional headings.

Referativnyi Zhurnal—Geografiya. Published monthly, it contains a *Kartografiya* sub-section which can also be purchased separately. Many items are cross-referenced to entries in the *Geologiya* section.

Bibliography and Index of Geology Exclusive of North America. Volumes 1–30 covering the years 1933–65 include lists of separately published geological maps at the end of the bibliography sections. Selected maps appearing in other publications are indexed under 'geological maps' in the subject index. Subsequent volumes of this bibliography do not list separately published geological maps and there is no primary subject index heading for 'geologic maps'. The word 'map' is used as a secondary or tertiary heading and entries must be looked for either under a regional heading or under another subject head such as 'engineering geology'. The index term 'map' is used in a broad sense and may refer to an actual map or merely to details of its preparation, or some comment concerning it. Articles of relevance to geological maps or mapping may also be found under the heading 'cartography' and 'geological exploration'. From Vol. 33, articles on North American geology were also indexed and the title was changed to *Bibliography and Index of Geology.* Prior to the publication of the *Bibliography,* maps were indexed (under the heading 'maps, geological') in *List of Geological Literature Added to the Geological*

Society's Library During the Year . . . , published by the Geological Society of London.

Bulletin Signalétique. Includes a few maps indexed under 'cartes' and 'cartographie'. From 1969 the relevant section is '216 *Géologie. Paléontologie*'. Earlier volumes are differently arranged. The monthly issues are indexed and there is an annual index.

Bibliographie des Sciences de la Terre. A small number of maps are included, and can be traced under 'cartes' in the KWIC indexes.

Geographical Abstracts Section D. A section on 'Cartography' is included from 1968. Atlases and articles on maps are listed.

Bulletin of the Geography and Map Division, Special Libraries Association.

A useful article by R. W. Stephenson (1970) lists published sources of information about maps and atlases. It contains details of geographical journals which regularly carry lists and reviews of maps and atlases; cartographic accession lists; national bibliographies containing map and atlas citations; catalogues; and lists of publishers and dealers.

Concerned only with marine atlases is 'Bibliography of Marine Atlases', compiled by P. A. Keehn (*American Meteorological Society, Special Bibliographies on Oceanography*, Contribution 6, 1968).

Among publishers who issue catalogues containing details of geological maps are Zumstein (Munich), Edward Stanford (London) and the Telberg Book Corporation (Sag Harbor, New York). The *Stanford Reference Catalogue* (1969–70) is a particularly valuable guide.

Additional information can be obtained from *Modern Maps and Atlases. An Outline Guide to Twentieth Century Map Production*, by C. B. M. Locke (Bingley, 1969).

The most recent guide to geological maps is a valuable contribution by M. W. Pangborn in *Geologic Reference Sources*, by D. C. Ward and M. W. Wheeler (revised edn., Scarecrow Press, 1972).

SELECT REGIONAL LIST OF GEOLOGICAL MAPS AND MAP SERIES

The principal geological maps and map series are listed below under regional headings. The list is not intended to be complete, nor does it contain details of older maps which are mainly of historical interest. A provisional list of small-scale national and international maps, prepared by the CGMW, is published in the *Geological Newsletter* (3) (1969). It also contains a list of addresses from which the maps listed can be obtained. Unless otherwise stated, the maps referred to in the list are publications of the various official geological surveys.

WORLD

Geological (1:10M in 20 sheets) and tectonic (1:15M in nine sheets) maps of the world are in preparation under the auspices of the CGMW. A series of 14 mineral maps (1:20M, 1963) is published by the Bureau de Recherches Géologiques et Minières (BRGM); each sheet shows the distribution of one or more minerals. A 1:27M structural map (*Skhema Struktury Zemnoi Kory*) is published by the Ukrainian Academy of Sciences, Kiev.

The great expansion of geological mapping in recent years has rendered earlier world maps largely obsolete; among these, however, may be mentioned Franz Beyschlag's *Geologische Karte der Erde* (Borntraeger, 1929–32), issued in 12 sheets at a scale of 1:15M.

A 'Descriptive Catalog of Selected Aerial Photographs of Geological Features in Areas Outside the United States' was published in 1969 as *USGS Professional Paper*, No. 591; the photographs are of areas in Central and South America, south-west Asia, China, Japan, the south-west Pacific and Antarctica.

EUROPE

The *International Geological Map of Europe* (1:1.5M) was the first of the maps to be published under the auspices of the International Geological Congress; some sheets are now in their third edition. A hydrogeological map at the same scale is in preparation. Quaternary, tectonic, metallogenic, iron ore, coal, natural gas, oil and soils maps (1:2.5M) are either published or in course of preparation. Geological and seismo-tectonic maps (1:5M, 1966) are published. The Bureau Gravimetrique International publishes a series of 1:1M Bouguer anomaly maps; four European sheets are currently available.

AUSTRIA

A 1:1M geological and tectonic map was published in 1964 and a hydrogeological map on the same scale in 1969. A 1:500 000 geological map was issued in 1968.

BELGIUM

A 1:1M geological map was issued in 1950. Map series are published at scales of 1:160 000 and 1:40 000. A few sheets of a new 1:25 000 series have been issued. (*Geol. Newsl.*, (2), 1968.)

BULGARIA

Geological, metallogenic, structural and oil and gas prospect maps (1:1M) were published in 1965.

CYPRUS

Geological and hydrogeological maps (1:250 000) were published in 1963 and 1970, respectively. (*Geol. Newsl.*, (1), 1967.)

CZECHOSLOVAKIA

Geological, tectonic, mineral resource, metallogenic, hydrogeological, Quaternary and residual deposit maps (1:1M) were published in 1966. A radiometric map at the same scale was published in 1968. Geological and aeromagnetic map series are issued at a scale of 1:200 000. (*Geol. Newsl.*, (2), 1968.)

DENMARK

A few sheets of the 1:100 000 and 1:160 000 series are available (*Geol.*

Newsl., (4), 1967). A geological map of the whole of Scandinavia (*Geologisk Översiktskarta over Norden, 1:1M*), compiled by the geological surveys of Denmark, Norway, Sweden and Finland, was published in 1933.

FINLAND

There is a 1:1M geological map, and map series at scales of 1:400 000 and 1:100 000. The map series are issued in two editions: 'bedrock' and 'Quaternary geology'. (*Geol. Newsl.*, (1), 1970.)

FRANCE

The Bureau de Recherches Géologiques et Minières (BRGM), which now incorporates the Service de la Carte Géologique de la France, is the chief publishing body. The BRGM catalogue, published annually, lists available geological, geophysical and other maps, and includes indexes to map series. A list of available maps is also published in *Geol. Newsl.*, (3) (1968). Two articles by Bodelle (1967, 1968) deal with the preparation and sale of French maps. The BRGM also publish maps for certain French or ex-French overseas territories; these are mentioned below under the appropriate regional heading.

A geological map (1:1M, 1969) covering the whole of France, including Corsica, is now in its fifth edition. There are three geological map series— 1:320 000, 1:80 000 and 1:50 000. Coverage is complete at the two smaller scales, although not all sheets are currently in print. Coverage at a scale of 1:50 000 is still very incomplete and no sheets have been published for Corsica. Sheets 7 and 12 of the 1:320 000 series include the Channel Islands and sheets 16 and 27 of the 1:80 000 series include part of the Channel islands. The 1:50 000 sheet of Chambery (XXXIII-32) is also available in the form of a plastic relief model obtainable from the Institut Géographique National, Paris. Gravity, magnetic, hydrogeological and metallogenic maps are also published. A series of gravity maps (1:200 000) covers most of northern and western France. Geological maps are also issued by the Service de la Carte Géologique d'Alsace et de Lorraine.

GERMANY

Prior to the partition of Germany much of the country had been covered by 1:25 000 maps. Some of these covered areas now in Poland. Maps were issued by the Prussian and other state surveys and, although various systems of sheet numbering were used, almost all maps were on the same sheet lines. Other map series were issued at scales of 1:50 000, 1:100 000 and 1:200 000. A few coalfields and other areas were covered at a scale of 1:10 000.

Post-war maps include a 1:600 000 map of the German Democratic Republic (1961) and a geological map of Thuringia (1:300 000, 1955); for West Germany the Bundesanstalt für Bodenforschung has published a tectonic map (1:2.5M, 1968) and a soil map (1:1M, 1962). A 1:1M geological map is in preparation. A groundwater map (1:2.5M, 1958) is issued by the Bundesanstalt für Länderkunde. Most of the states (Länder) have recent geological maps at scales of from 1:300 000 to 1:500 000. Map series at scales of 1:25 000 (now with a uniform sheet numbering system) and 1:100 000 are continued. There are 1:25 000 soil maps for many areas, and hydrogeological maps at various scales. (*Geol. Newsl.*, (1), 1971.)

GREECE

A geological map (1:500 000) was published in 1954. A few areas, including parts of Crete, are covered by the 1:50 000 series. A geological map of

Euboea and other maps at a scale of 1:200 000 are published in the *Annales Géologiques des Pays Helleniques*; these and other maps are listed in *Geol. Newsl.*, (2) (1969). A geological map of the Epirus (1:100 000) with accompanying text is issued by the Institut Français du Pétrole.

HUNGARY

Geological (1:300 000) and mineral resource (1:500 000, 1967) maps are available. The principal map series is published at a scale of 1:200 000 but 1:25 000, 1:10 000 and various special sheets are also issued (*Geol. Newsl.*, (1), 1969).

ITALY

A geological map (1:1M, 1969) includes Sicily and Sardinia. Most of Italy is covered by 1:100 000 sheets and gravity overlays are issued with some recent sheets. The list of available maps (*Geol. Newsl.*, (1), 1967) omits special sheets and does not mention the 1:25 000 series, sheets of which are available for a few areas. The 1:100 000 maps for parts of north-east Italy are entitled *Carta Geologica delle Tre Venezie* and are issued by the geological section of the hydrographic bureau.

LUXEMBURG

A 1:100 000 geological map (1966) and a 1:25 000 series are published.

THE NETHERLANDS

A geological map (1:600 000, 1958) and map series at scales of 1:200 000 and 1:500 000 are published (*Geol. Newsl.*, (2), 1968).

NORWAY

A geological map (1:1M, 1960), a map of mines and ore deposits (1:1M, 1958), a glacial map (1:2M, 1960) and map series at scales of 1:100 000 and 1:250 000 are published (*Geol. Newsl.*, (4), 1968).

POLAND

Several geological atlases which include geological, structural, facies, formation and other maps are published. These include a *Geological Atlas of Poland* (1:2M, 1968), a *Geological Atlas of the Polish-Carpathian Foreland* (1:500 000, 1968) and a *Geological Atlas of the Lower Silesian Coal Basin* (1:100 000, 1970). A 1:1M atlas of eight sheets was issued in 1957 and a 1:1M atlas of 13 sheets in 1960–65. Various map series have been published, at scales ranging from 1:10 000 to 1:300 000. Many 1:25 000 sheets issued by the former Prussian geological survey cover areas now included in Poland. (*Geol. Newsl.*, (4), 1970.)

PORTUGAL

A mineral map (1:500 000, 1965) was issued in two sheets and a geological map (1:1M) was published in 1968. There are a few published sheets of the 1:50 000 series, and there are also some 1:50 000 and 1:25 000 sheets for the Azores (*Geol. Newsl.*, (1), 1967). The Direcçaõ Geral dos Servicos Agricolas has issued a soil map (1:1M, 1967).

ROMANIA

Geological, Quaternary, tectonic and soil maps (1:1M) are published. A 1:500 000 geological map is issued in 12 sheets. There are geological and soil map series at a scale of 1:200 000 and geological and hydrogeological 1:100 000 sheets. (*Geol. Newsl.*, (4), 1967).

SPAIN

In addition to the 1 : 1M map of Iberia mentioned below, metallogenic (1 : 2.5M, 1964) and mineral (1 : 1M, 1962) maps are available. There are 1 : 400 000 and 1 : 50 000 series which also cover the Balearic and Canary Islands. A geological map (1 : 1M, 1966) published by the Instituto Geológico y Minero de España includes the whole of the Iberian peninsula, the Balearic and Canary Islands.

SWEDEN

All maps are issued in map series which are designated by letters: Aa 1 : 50 000 series; Ab 1 : 200 000 series; Ac 1 : 100 000 series; Ad agrogeological maps, 1 : 20 000; Ae 1 : 50 000 new series; Af 1 : 50 000 petrological maps; Ba various small-scale maps; Bb special maps; Ca maps with memoirs; D 1 : 100 000 maps of peat deposits. Bedrock and Quaternary geological maps of Sweden (1 : 1M, 1958) were issued as Ba 16 and Ba 17, respectively.

SWITZERLAND

A geological map is published at a scale of 1 : 500 000. Geotechnical and geological maps (1 : 200 000) cover the country in eight sheets. The whole of Switzerland was covered by the 1 : 100 000 series, now superseded for many areas by the Geological Atlas (1 : 25 000) sheets and special sheets at various scales. (*Geol. Newsl.*, (1), 1968.)

TURKEY

Eight sheets of a 1 : 800 000 geological map were published in 1942–46. There are tectonic (1966) and mineral maps (1969) at a scale of 1 : 2.5M, and a 1 : 500 000 geological map series.

UNITED KINGDOM

A great number of maps has been produced by the Geological Survey of Great Britain during the last 130 years. These maps are normally based upon the corresponding topographic sheets of the Ordnance Survey. The last catalogue of published maps is *List of Memoirs, Maps, Sections, etc., published by the Geological Survey of Great Britain and the Museum of Practical Geology to 31 December, 1936* (HMSO, 1937) and is now out of print. Current lists of maps only include available sheets. Index maps showing the coverage of 1-inch drift geological and soil maps for the British Isles are given by J. A. Taylor in *Geography*, **45** (1–2), 1960).

Many useful maps are published in the *Quarterly Journal of the Geological Society of London*, the *Proceedings of the Geologists' Association*, other local geological journals and the *Transactions of the Royal Society of Edinburgh*. Many of these journals are provided with cumulative indexes. A *Classified Index to the Maps in the Publications of the Geological Society of London, 1811–1885* (Bibliography of Special Subjects No. 4, Boston (Mass.) Public Library, 1887) was compiled by Richard Bliss.

Recent Geological Survey maps include: a *Geological Map of the British Islands* (5th edn., 1969) and a *Tectonic Map of Great Britain and Northern Ireland* (1966), both at a scale of 1 : 1 584 000; a geological map of Great Britain in two sheets at a scale of 1 : 625 000 (this map is also available overprinted with the sheet lines of the current 1-inch series and in an outline edition); and an aeromagnetic map (southern sheet only) at the same scale. A series of 16 aeromagnetic sheets covers Great Britain and Northern Ireland at a scale of 1 : 250 000; these are available as dyeline copies and show only magnetic contours and the national grid.

Many atlases contain small-scale geological maps. The most useful is the *Atlas of Britain and Northern Ireland* (Clarendon Press, 1963), which contains 1:2M maps of solid and drift geology, structural and lithological maps and distribution maps (with bore-hole details) for Precambrian, Lower Palaeozoic, Devonian, Carboniferous, Permo-Triassic, Jurassic, Cretaceous and Tertiary–Quaternary rocks. This atlas also contains a series of 1:1M solid and drift geological maps.

The latest edition of *Stanford's Geological Atlas of Great Britain* (Stanford, 1964) was compiled by T. Eastwood. The maps are based on the Geological Survey 10-mile uncoloured map and are reduced to a scale of 12 miles to 1 inch. The previous edition (1914) contained coloured geological maps and included Ireland and the Channel Islands.

England and Wales

A series of 19 1:253 440 geological sheets was published prior to World War II; four of these sheets (8, 12, 16, and 20/24, covering England east of a line from Bridlington to Worthing) were also published in a drift edition. Only a few of these sheets have been reprinted. A special sheet for the Bristol district was published in 1955. Gravity survey overlays are available for some sheets.

One-inch (1:63 360) New Series sheets are currently available for about half of England and Wales. Many of these, particularly for areas in the Midlands and northern England, are available in separate solid and drift editions; most sheets include one or more geological cross-sections. Special sheets were published for Anglesey, the Bristol district, Isle of Man, Isle of Wight, London, Nottingham and Oxford, but only the first 4 sheets are currently available. For areas where there are no New Series maps, recourse must be had to the Old Series, which covers the whole of England and Wales and is out of print. Some Old Series maps were issued as 'whole sheets', but the majority were published as 'quarter-sheets'. The sheet-lines for the Old Series quarter-sheets for northern England are the same as those of the New Series, the equivalent sheet numbers being as listed in *Table 7.1*:

Table 7.1

N.S.	O.S.	N.S.	O.S.	N.S.	O.S.	N.S.	O.S.
1	110 NW	18	106 SW	35	104 SE	55	95 SE
2	110 NE	19	106 SE	37	99 NE	58	91 NW
3	110 SW	20	105 SW	38	98 NW	59	91 NE
4	110 SE	21	105 SE	39	98 NE	60	92 NW
5	108 NE	22	101 NW	40	97 NW	61	92 NE
6	109 NW	23	101 NE	41	97 NE	62	93 NW
7	108 SW	24	102 NW	42	96 NW	63	93 NE
8	108 SE	25	102 NE	43	96 NE	64	94 NW
9	109 SW	26	103 NW	44	95 NW	65	94 NE
10	109 SE	27	103 NE	47	99 SE	66	91 SW
11	109 NE	28	101 SW	48	98 SW	67	91 SE
12	106 NW	29	101 SE	49	98 SE	68	92 SW
13	106 NE	30	102 SW	50	97 SW	69	92 SE
14	105 NW	31	102 SE	51	97 SE	70	93 SW
15	107 NE	32	103 SW	52	96 SW	71	93 SE
16	107 SW	33	103 SE	53	96 SE	72	94 SW
17	107 SE	34	104 SW	54	95 SW	73	94 SE

New Series sheets 36, 45, 46, 56, 57 are combined in the Isle of Man special sheet.

A series of 1:25 000 maps of classic geological areas has recently been commenced. Sheets already published include Cheddar (ST 45), Clevedon and Portishead (ST 47), Church Stretton (SO 49) and Craven Arms (SO 48).

Basic geological mapping has normally been carried out using county or national grid 6-inch (1:10 560) maps, except for the earlier surveys made before 6-inch topographic maps were available. Six-inch maps are only published for certain coalfield and other mineral areas, and colour-printed sheets are available for the London area. Several hydrogeological maps have been issued at scales of 1:63 360 and 1:126 720. A series of Horizontal Sections (the lines of section are engraved on the 1-inch Old Series sheets) and Vertical Sections were published. The Soil Survey of England and Wales publish soil maps for certain areas at scales of 1:25 000 and 1:63 360.

Scotland

Quarter-inch (1:253 440) sheets are available for most of Scotland except for the central Highlands and the Outer Hebrides. There is only one series of 1:63 360 maps for Scotland and there are no maps for part of the central Highlands, Outer Hebrides and the southern Shetland Islands. Maps of the Outer Hebrides are contained in a series of five papers by T. J. Jehu and R. M. Craig, published in the *Transactions of the Royal Society of Edinburgh* from 1924 to 1934. Special 1-inch sheets are available for Arran, Assynt, Glasgow, northern Skye, northern Shetland and western Shetland. A special 1:25 000 map of the Edinburgh district has recently been published. Six-inch (1:10 560) sheets are published for a few areas, mainly for the Midland Valley coalfields. A series of Horizontal Sections (mainly coalfield areas) and Vertical Sections (Midland Valley coalfields) were published. The Soil Survey of Scotland publishes some 1:63 360 soil maps.

Ireland

Before partition the whole of Ireland had been geologically surveyed at a 6-inch scale and 1:63 360 hand-coloured maps had been published. These maps showed drift deposits as an overprint, but glacial deposits were not subdivided. Revisions of the drift geology were made before the publication of the colour-printed sheets for Belfast, Cork, Dublin, Killarney, Limerick and Londonderry. A few 1:253 440 sheets were also published. Since World War II the Geological Survey of Northern Ireland has issued new editions of 1-inch sheets 7, 8, 29, 35 and 36; ¼-inch sheets 2 and 5; and a special map of Belfast at a scale of 1:21 120. A few 6-inch maps are published for the Tyrone and Ballycastle coalfields. Aeromagnetic (1968) and gravity anomaly (1967) maps at a scale of 1:253 440 are also available. The Geological Survey of Ireland (Eire) issued a geological map of the whole of Ireland on a scale of 1:750 000 in 1962. A map of the surface geology of Ireland compiled by A. Geikie (1:633 600) was published by Bartholomew in 1910. A few soil maps have been published by the National Soil Survey of Ireland.

YUGOSLAVIA

Geological (1:500 000, 1953) and tectonic (1:2.5M, 1960) maps are available. A 1:200 000 map of Bosnia-Herzegovina was issued in 1954. Geological map sheets are issued at 1:50 000 and other scales.

USSR

Geological (1968) and tectonic (1966) maps are published at a scale of

1:2.5M, and geological, tectonic, geomorphological, hydrogeological, hydro-chemical, Quaternary and metallogenic maps (1:1.5M) were issued in 1967–68. An atlas of lithological–palaeogeographic maps entitled *Atlas Lithologo–Paleogeograficheskikh Kart SSSR* at a scale of 1:7.5M is published by the Ministry of Geology and Academy of Sciences in four volumes (1. Proterozoic; 2. Palaeozoic; 3. Mesozoic; and 4. Palaeogene, Neogene and Quaternary). Geological and other maps are also available for many of the republics. The principal map series are 1:1M and 1:200 000. An obsolete 1:420 000 series covered many parts of European Russia.

ASIA

Geological (1:6M, 1954) and tectonic (1:5M, 1966) maps of Eurasia are published by the USSR Academy of Sciences. A geological map of central Asia was issued in 1967, and the CGMW published a geological map (1:5M) of Asia and the Far East in 1959.

Mapping by the US Geological Survey resulted in 1963 in a 1:2M 'Geological Map of the Arabian Peninsula' (*USGS Miscellaneous Investigations Map* I-270A) and a series of 21 1:500 000 sheets of Saudi Arabia (I-200A–I-220A). *Mineral Investigations Maps* (1:100 000) are issued by the Saudi Arabian Ministry of Petroleum and Mineral Resources.

Geological (1:2M) and tectonic (1:10M) maps of the Himalayas are included in *Geology of the Himalayas*, by A. Gansser (Interscience, 1964).

A *Geological Map of South East Asia and the Far East* (1960), an *Oil and Natural Gas Map of Asia and the Far East* (1958–62) and a *Mineral Distribution Map of Asia and the Far East* (1963) are published by the United Nations at a scale of 1:5M.

A geological atlas of Eastern Asia, containing 1:2M maps, was published by the Tokyo Geographical Society in 1929. More recently a series of 1:250 000 geological maps covering parts of China, Manchuria, Korea and Japan has been issued by the US Army Map Service and by the Tokyo Geographical Society.

AFGHANISTAN

A 1:2M soil map was published in 1962 and a 1:2.5M geological map in 1969.

CEYLON

Geological (1:2M) and mineral (1:1.5M) maps are available, and there are a 1:253 440 tectonic and a 1:63 360 geological map series. (*Geol. Newsl.*, (4), 1971.)

CHINA

Geological maps were formerly published by the Geological Survey of China and by the geological surveys of certain provinces. Some maps were also produced by the various railway companies. Many sheets of a 1:1M series were issued, and there were also 1:235 000, 1:200 000, 1:100 000 and 1:50 000 series. A 1:1.5M map of Manchuria and adjacent areas, in 4 sheets, was published by the US Army Intelligence Division in 1953.

INDIA

Geological maps of India are available at scales of 1:2M and 1:5M (1962), and metallogenic and tectonic maps (1:2M) were published in 1963.

A 1:2M tectonic map was also published by the Oil and Gas Commission in 1968. Regional maps (including areas in Pakistan and Burma) published in the *Memoirs* and *Records of the Geological Survey of India* are indexed in No. 77 (1947) of the former. A list of available maps is included in the Geological Survey of India *List of Publications* (1963).

INDONESIA
A 1:2M geological map (1966) was issued as *USGS Miscellaneous Investigations Map* I-414. There are 1:100 000 and 1:200 000 map series for certain areas.

IRAN
Geological (1959) and metallogenic (1965) maps were published at a scale of 1:2.5M. There is a 1:250 000 series, and a number of 1:250 000 sheets of south-west Iran were issued by British Petroleum.

IRAQ
A 1:1M geological map was published in 1957. There is also a 1:50 000 series.

ISRAEL
A 1:250 000 geological map in two sheets was issued in 1965. There are also 1:100 000 and 1:50 000 series; some sheets are accompanied by structural maps and sections.

JAPAN
Geological maps are published at scales of 1:5M (1968) and 1:2M (1966). Tectonic, hydrogeological and other maps are also available at a scale of 1:2M. There are map series at scales of 1:500 000, 1:400 000, 1:200 000, 1:100 000, 1:75 000 and 1:50 000. Oil and gas fields are shown on a separate 1:25 000 series. A five-volume index to geological maps of Japan was published by the Geological Survey of Japan in 1964–65. Maps issued by other organisations and in journals are included in this index. A series of palaeogeographical maps is included in *The Geologic Development of the Japanese Islands*, by M. Minato, M. Gorai and M. Hunahashi (Tsukiji Shokan, Tokyo, 1965).

JORDAN
A geological map covering the whole of Jordan was published in 1939 on a scale of 1:1M. More detailed geological maps (1:250 000) were issued in 1954 and 1968.

KOREA
A 1:1M geological map (1956) and a series of 1:50 000 sheets are published.

LEBANON
Geological (1955) and gravimetric (1959) maps are issued at a scale of 1:200 000 and there is a 1:50 000 map series. A 1:1M map of Lebanon, Syria and neighbouring areas was published in *Notes et Mémoires sur le Moyen Orient*, **8** (1966).

MALAYSIA
Geological and tectonic maps of the Thai–Malay peninsula (1:5M, 1958)

and a number of geological and mineral maps at larger scales are available for Malaya, Sarawak and Sabah. Some of these are listed in *Geol. Newsl.*, (3) (1969).

PAKISTAN

A 1 : 2M geological map was published in 1964. A series of 1 : 253 440 sheets is available for part of West Pakistan.

PHILIPPINES

Geological (1 : 1M, 1962) and mineral distribution (1 : 2.5M, 1964) maps are published.

SYRIA

Geological (1 : 500 000, 1 : 1M) and tectonic (1 : 1M) maps were published in 1964.

TAIWAN

Geological maps are issued at scales of 1 : 300 000 (1953), 1 : 100 000 and 1 : 50 000.

THAILAND

A 1 : 750 000 geological map of north-eastern Thailand was issued in 1964. A 1 : 2.5M geological map of Thailand is included in *Bulletin of the United States Geological Survey*, No. 984 (1951). (*Geol. Newsl.*, (1), 1972.)

VIETNAM, CAMBODIA, LAOS

A 1 : 2M geological map of the area was published in 1952. Sheets of the 1 : 500 000 series are available for some areas. A 1 : 450 000 geological map of North Vietnam was published by the Telberg Book Corporation in 1962.

AFRICA

Geological (1964) and tectonic (1968) maps at a scale of 1 : 5M and a mineral map (1 : 10M, 1968) are published by the Association of African Geological Surveys/Unesco. A 1 : 15M tectonic map (1969) is published by CGMW/Unesco; a coal map (1 : 10M, 1966) by AAGS/ECA; and a soils map (1 : 5M, 1963) by the Commission for Technical Co-operation in Africa South of the Sahara (CCTA).

A geological map of north-west Africa (1 : 2M, 1952) was published in six sheets in connection with the XIX International Geological Congress. A second edition series commenced in 1962.

The former French west African territories are covered by the nine sheets of the *Carte Géologique de l'Afrique Occidentale* (1 : 2M, 1960) and by a number of 1 : 500 000 sheets. Most maps are listed in the BRGM catalogue. The former French central African territories are covered by the four sheets of the *Carte Géologique de l'Afrique Equatoriale* (1 : 2M, 1956-58) and by a number of 1 : 500 000 sheets. Again most maps are listed in the BRGM catalogue (see p. 26).

There are two general geological maps of eastern Africa: the 'Carta Geologica dell'Africa Orientale' published in *Geologia dell'Africa Orientale*, by G. Dainelli (4 vols., Reale Accademia d'Italia, 1943), which covers Ethiopia and neighbouring regions; and the *Geological Map of East Africa*

(Geological Survey of Tanganyika, 1952), which covers Kenya, Tanzania and Uganda.

A geological map of the former Federation of Rhodesia and Nyasaland (1:2.5M) was published in 1961 as *Federal Atlas Map*, No. 4.

ALGERIA
Geological map series are issued at scales of 1:500 000, 1:200 000 and 1:50 000.

ANGOLA
Geological (1956) and mineral (1966) maps are published at a scale of 1:2M. There is also a 1:250 000 series.

BOTSWANA
A 1:2M map was published in 1964. There is also a 1:125 000 series.

CAMEROUN
A geological map (1:1M) accompanies the survey's *Bulletin*, No. 2 (1956).

CHAD
A geological map (1:1.5M) was issued in 1964.

CONGO
Geological and other 1:5M sheets are published in the series *Atlas Général du Congo*. Geological (1:2M, 1951) and tectonic (1:3M, 1952) maps are also published, and there are 1:200 000, 1:100 000 and 1:50 000 map series.

DAHOMEY
A geological map (1:1M) was published by the French Bureau de Recherches Géologiques et Minières in 1960.

EGYPT (UAR)
A 1:5M geological map was published in 1939. There are also some 1:125 000 and 1:100 000 sheets.

ETHIOPIA
A geological map (1:2M) was published in 1955. There are also a few 1:500 000 sheets.

GABON
A geological map (1:1M) accompanies *Mémoire du Bureau de Recherches Géologiques et Minières*, **72** (1970).

GAMBIA
A 1:500 000 map was issued with *Bulletin of the Gold Coast Geological Survey*, No. 3 (1925).

GHANA
Geological maps (1:1M, 1955, and 1:2M, 1963) and 1:62 500 and 1:125 000 map series are published.

IVORY COAST
A geological map (1:1M) was published in 1965.

KENYA
A 1:3M geological map was published in 1969. There is also a series of 1:125 000 sheets. (*Geol. Newsl.*, (3), 1971.)

LESOTHO
A geological map of Basutoland (1:380 160) was published in 1939.

LIBYA
A geological map on a scale of 1:2M was issued by the US Geological Survey in 1964 as *Miscellaneous Investigations Map* I-350A.

MALAGASY
Geological (1:1M, 1952 and 1964) and tectonic (1:3M, 1961) maps are published. The island is also covered by 13 1:500 000 sheets and there are series at scales of 1:100 000 and 1:200 000.

MALAWI
A geological map (1:1M, 1966) and a 1:100 000 series are published.

MOROCCO
Geological (1952), metallogenic (1962) and mineral deposit (1960) maps are published at a scale of 1:2M. A geological map in 6 sheets and a gravimetric map are also available at a scale of 1:500 000. Larger-scale maps of certain areas are also published, principally at scales of 1:50 000, 1:100 000 and 1:200 000. These maps are usually allocated a number in the *Notes et Mémoires* series of the Service Géologique.

MOZAMBIQUE
Geological (1:2M, 1957) and tectonic (1:3M, 1957) sketch maps and a 1:250 000 series are published.

NIGER
A geological map (1:2M) and explanatory text was published in 1967.

NIGERIA
A geological map (1:2M) was issued in 1965 and another (1:3M) which also shows mineral deposits in 1967.

PORTUGUESE GUINEA
Geological maps at scales of 1:1M and 1:2M were published in 1947.

RHODESIA
Geological (1:1M, 1961) and mineral (1:2M, 1951) maps have been published. There are also 1:100 000 maps for certain areas.

SENEGAL
Geological and mineral deposit maps (1:500 000), both in 4 sheets, were issued in 1962 and 1966, respectively.

SIERRA LEONE
A 1:1M geological map was published in 1960.

SOMALIA
Eight 1:500 000 sheets were published in 1957–59 by AGIP Mineraria. A

1 : 400 000 map of French Somaliland was published in 1946 and there is an incomplete 1 : 25 000 series for the former British Somaliland.

SOUTH AFRICA
A geological map (1 : 1M, 1955) has been published and is also available overprinted with Bouguer gravity contours (1958). Map series are published at scales of 1 : 125 000 and 1 : 250 000 with a few 1 : 50 000 sheets for special areas. Older sheets were published at 1 : 148 750 and 1 : 238 000. (*Geol. Newsl.*, (4), 1969.)

SOUTH-WEST AFRICA
A 1 : 1M geological map showing mineral localities and gravity contours was published in 1963. There is also a 1 : 125 000 series.

SPANISH GUINEA
A geological sketch map (1 : 500 000, 1954) was published by the Instituto de Estudios Africanos.

SPANISH WEST AFRICA
A 1 : 500 000 geological map was published in 1958 by the Instituto Geológico y Minero de España. There is also a 1 : 50 000 series.

SUDAN
The third edition of the 1 : 4M geological map was published in 1963. There are also some 1 : 250 000 sheets.

SWAZILAND
There is a 1 : 250 000 geological map (1966) and a 1 : 50 000 series.

TANZANIA
A 1 : 3M map was published in 1947. There is a 1 : 125 000 series.

TUNISIA
Geological and mineral deposit maps (1 : 500 000, 1951 and 1966) are published in two sheets. There are also series at scales of 1 : 50 000 and 1 : 200 000.

UGANDA
A geological map (1 : 1.25M, 1961) and a series of 1 : 250 000 sheets are published and are also available overprinted with Bouguer gravity contours. There are also a 1 : 100 000 geological series and mineral map (1 : 1.25M, 1961).

ZAMBIA
Geological (1 : 1M, 1959–60) and mineral (1 : 3M, 1959) maps are published. There is a 1 : 500 000 map of the Copper Belt and 1 : 75 000 and 1 : 100 000 map series.

AUSTRALASIA
Most of the 13 sheets of the international map of Australia and Oceania (1 : 5M, 1965–   ) have now been published. The area covered lies between

latitudes 24° N. and 48° S. and longitudes 108° W. and 132° E., with the omission of a part of the north-west and south-east. Metallogenic and tectonic maps at the same scale are in preparation.

AUSTRALIA

Official mapping is carried out by the geological surveys of the various states, supplemented in recent years by the Bureau of Mineral Resources, Geology and Geophysics (BMR). Mapping by the BMR is largely confined to undeveloped areas of northern and central Australia and to the Territory of Papua and New Guinea.

Systematic mapping commenced in Victoria during the latter half of the nineteenth century, and a series of 1 : 31 680 sheets was produced covering the southern and central parts of the state. An attempt at systematic mapping was made in Tasmania in the 1930s and two 1 : 15 840 sheets were published in the *Bulletin of the Geological Survey of Tasmania*, No. 41 (1934). Apart from these early attempts, most systematic mapping dates from the early 1950s. The standard scales are 1 : 63 360 (Geological Atlas 1-mile series) and 1 : 250 000 (formerly 1 : 253 440 Geological Atlas 4-mile series). All sheets, whether produced by the state geological surveys or by the BMR, are on the same sheet lines. Most sheets of the 4-mile series are accompanied by explanatory notes. Maps at a scale of 1 : 50 000 are issued by the New South Wales and Western Australia Geological Surveys. In New South Wales the University of New England has issued several sheets of a geological map of New England at scales of 1 : 100 000 and 1 : 250 000. A number of 1 : 63 360 sheets covering 10 kiloyard squares compiled by students of the University of Tasmania were published in the *Papers and Proceedings of the Royal Society of Tasmania* during the 1950s and 1960s.

Various geophysical maps are produced by the BMR: radiometric maps are published for parts of all states except South Australia; gravity maps (1 : 500 000) are published for parts of the Northern Territory, Queensland and Western Australia; and aeromagnetic maps at various scales are published for parts of all states. The Geological Survey of South Australia also issues 1 : 63 360 and 1 : 250 000 aeromagnetic maps. The CSIRO Division of Soils has produced soil maps (frequently at a scale of 1 : 63 360) for some areas.

The Department of National Development issue an *Index to Australian Resources Maps of 1940–59* (1961); a supplement published in 1966 covers maps issued during the period 1960–64. Both these publications include index maps.

The first geological map of Australia at a reasonably large scale was compiled by T. W. E. David (1 : 2 990 000, 1932). This map was also included in *The Geology of the Commonwealth of Australia*, by T. W. E. David (3 vols., Arnold, 1950). More recent maps issued by the BMR are a geological map (1 : 6 336 000, 1952) and a tectonic map (1 : 2 534 400, 1960). Geological and mineral resources maps (1 : 6M) are included in the *Atlas of Australia*, issued by the Department of National Development.

An *Atlas of Australian Soils* (1 : 2M, 1965–68) has been compiled by CSIRO and consists of 10 sheets with explanatory booklets. A map of underground water (1 : 5M, 1965) is issued by the Australian Water Resources Council.

All the states issue maps covering their own territories, at scales ranging from 1 : 506 880 to 1 : 2 534 400. The BMR have published several 1 : 500 000 maps of parts of northern Australia (Western Australia, Northern Territory and Queensland), and a 1:2 534 400 map of the Territory of Papua and New Guinea.

NEW CALEDONIA
Several sheets of a 1:50 000 series have been published.

NEW ZEALAND
The New Zealand Geological Survey has issued a 1:1 013 760 geological map in two sheets (1947) and a 1:2 000 000 map (1958). Map series at scales of 1:250 000, 1:63 360 and 1:25 000 are also published. The 1:25 000 *Industrial Series* has an air-photo mosaic base and geological (lithological) boundaries. An explanatory text is printed on the back of these maps and gives details of foundation conditions and landslip, volcanic and tsunami risk.

NORTH AMERICA
Geological (1965) and tectonic (1969) maps have been published by the US Geological Survey (USGS), and metallogenic and metamorphic maps are in preparation; all these maps are at a scale of 1:5M. A catalogue of geological maps of America (North and South) 1752–1881 by J. and J. B. Marcou was published as *Bulletin of the United States Geological Survey*, No. 7 (1884). *A Catalogue of Small-Scale Geologic Maps Useful for Broader Regional Studies* was compiled by W. H. Bucher *et al.* (National Research Council, Washington, 1933).

CANADA
The Geological Survey of Canada is producing a new series of small-scale maps. These include geological (1969), tectonic (1969), mineral deposit (1969), glacial (1968), magnetic anomaly (1968), permafrost (1967), physiographic (1970) and geochronology (1970) maps at a scale of 1:5M. The Observatories Branch of the Department of Energy, Mines and Resources has a Bouguer gravity anomaly map at the same scale (Map GMS 69-1, 1969) and four maps covering Canada at 1:2.5M (Maps GMS 67-1 to 4, 1968). The principal map series contain coloured geological maps at various scales ranging from small-scale maps of provinces or groups of provinces to large-scale maps of mining areas. Many maps are at scales of 1:63 360 and 1:253 440. A preliminary series of uncoloured maps, and various aeromagnetic and geophysical maps are also issued. A series of 1:1M indexes to geological mapping are issued and are revised at frequent intervals.

Geological maps are also produced by the geological surveys or mines departments of the provinces and by the Research Council of Alberta. The Ontario Department of Mines publishes a series of index maps at a scale of 1:1 013 760, and map indexes for Quebec are included in *Special Publication of the Quebec Department of Natural Resources*, No. 96 (1968).

A large number of lithofacies, isopach and palaeogeological maps are included in *Geological History of Western Canada*, edited by R. G. McCrossan *et al.* (2nd edn., Alberta Society of Petroleum Geologists, 1965). (*Geol. Newsl.*, (3), 1971.)

MEXICO
A geological map (1:2M) was published in 1968. A tectonic map (1:2.5M,

1961) is available from the Geological Society of America. There is also a 1:100 000 map series. (*Geol. Newsl.*, (3), 1971.)

UNITED STATES OF AMERICA

Official mapping is carried out by the USGS and the geological surveys or equivalent organisations of most states. Rather less than one-quarter of the conterminous USA is covered by maps at scales of 1:24 000, 1:31 680 and 1:62 500. A map showing the coverage at these scales appeared as the cover picture of *Geotimes*, **15** (1) (1970). Certain areas, including most of California, are covered by the new 1:250 000 map sheets. An index map showing the extent of coverage at this scale by the USGS and state surveys is published in Chapter A of 'Geological Survey Research', an annual series issued as *Professional Papers of the United States Geological Survey*.

The USGS publishes state indexes to geological maps and national indexes to geological maps, air photographs and air-photo mosaics. There is also a series of booklets entitled *Geologic and Water-Supply Reports and Maps*, published for each state. Maps are also listed in the cumulative *Publications of the Geological Survey (1879–1961)* (USGPO, 1964) and in the annual and monthly supplementary lists of new publications.

The USGS has published geological (1932), tectonic (1962), basement rock (1968), oil and gas (1955) and coal (1942) maps at a scale of 1:2.5M. A glacial map of the USA east of the Rocky Mountains (1:1.75M, 1959) is issued by the Geological Society of America.

The USGS has also published the *National Atlas of the United States* (1970). This contains landform, earthquake, tectonic, geological, soil, water and mineral maps at a scale of 1:7.5M.

A 1:2.5M (1957) map of Alaska is published by the USGS, and geological (1:250 000) and tectonic (1:1M) maps of the Alaskan peninsula were published as Parts 2 and 3 of *Memoir of the Geological Society of America*, No. 99 (1966).

The chief USGS map series are as follows, the letters in parentheses being the series designations:

Geologic Quadrangle Maps (CQ). Geological maps of the USA, mainly at scales of 1:24 000 and 1:62 500, but including some 1:63 360 and 1:125 000 sheets. The series is a continuation of the *Folio Atlases* which were published between 1894 and 1946.
Miscellaneous Geologic Investigations Maps (I). Map sheets of this series are 1:24 000 photogeological maps; preliminary and reconnaissance maps, maps of foreign countries and the geological versions of the *Lunar Atlas Charts* are also included.
Oil and Gas Investigations Maps (OM).
Oil and Gas Investigations Charts (OC).
Coal Investigations Maps (C).
Geophysical Investigations Maps (GP). Aeromagnetic and radiometric maps.
Hydrologic Investigations Atlases (HA).
Mineral Investigations Field Studies Maps (MF). Mainly preliminary geological and tectonic maps.
Mineral Investigations Resource Maps (MR). Principally maps showing the distribution of mineral occurrences.

A series of palaeotectonic maps of various geological systems is published as *Miscellaneous Geologic Investigations Maps*. The numbers are as follows: I-450, Permian (1967); I-300, Triassic (1960); and I-175, Jurassic (1956).

'A Descriptive Catalog of Selected Aerial Photographs of Geologic Features in the United States', by Charles S. Denny *et al.*, is published as *USGS Professional Paper*, No. 590 (1968) and 'An Airphoto Index to Physical and Cultural Features in Eastern United States' has been compiled by D. M. Richter (*Photogrammetric Engineering*, **31** (5), 1965). A similar index covering the western USA is published in **33** (12) of the same journal.

Many geological maps and map series are produced by the geological surveys or equivalent organisations of the various states. State geological maps are usually issued by the state surveys, but some are issued by the USGS. Many geological maps are also to be found in *Bulletin of the Geological Society of America, Bulletin of the American Association of Petroleum Geologists* and other journals. Strip maps, showing the geology along highways, are often included in excursion guidebooks produced by various geological societies, and the American Association of Petroleum Geologists has recently initiated a *Geologic Highway Map Series*. Other map sources include: *Structural Geology of North America*, by A. J. Eardley (2nd edn., Harper & Row, 1962); *Lithofacies Maps—An Atlas of the United States and Southern Canada*, by L. L. Sloss, E. C. Dapples and W. C. Krumbein (Wiley, 1960); and *Geol. Newsl.*, (3) (1970).

CARIBBEAN REGION

Geological maps of this region are mostly scattered in a wide variety of books and journals. Among recently published maps the following may be mentioned.

CUBA

Geological (1 : 1M, 1962) and tectonic (1 : 1.25M, 1966) maps are published.

DOMINICAN REPUBLIC

A 1 : 250 000 geological map was published in 1967. (*Geol. Newsl.*, (3), 1969).

GUADELOUPE, MARTINIQUE

A series of 1 : 50 000 maps was published in 1962–65 by the Bureau de Recherches Géologiques et Minières.

HAITI

A 1 : 250 000 geological map is contained in *Mémoire de l'Institut des Hautes Etudes de l'Amérique Latine*, **6** (1960).

JAMAICA

A geological map (1 : 250 000) was published in 1969.

PUERTO RICO

A geological map of Puerto Rico and adjacent islands was issued in 1964 as *USGS Miscellaneous Geologic Investigations Map* I-392. There is a series of maps issued as *USGS Quadrangle Maps.*

TRINIDAD AND TOBAGO

Geological maps of Trinidad (1959) and Tobago (1948) at a scale of

1 : 100 000 are available. The latter was published in *Bulletin of the Geological Society of America*, **59**, (8).

CENTRAL AND SOUTH AMERICA

A 1 : 1M geological map of Central America was included in *Bulletin of the United States Geological Survey*, No. 1034 (1957). A series of useful volumes issued by the Pan American Union in the 1960s is entitled *Annotated Index of Aerial Photographic Coverage and Mapping of Topography and Natural Resources*. A geological map of South America (1 : 5M, 1964) is issued by the Geological Society of America and a metallogenic map at the same scale is in preparation. A catalogue of maps was compiled by Marcou and Marcou (1884), and other useful guides are *Catalogue of the Geologic Maps of South America*, by H. B. Sullivan (American Geographical Society, 1922); and *Sources of Geological Information in Latin America*, by H. W. Smith (Working Paper No. 4, Seminar on Acquisition of Latin American Library Materials, Pan American Union, 1969). A select bibliography was issued by the Commission for the Geological Map of the World in 1963.

ARGENTINA

Geological (1 : 2.5M, 1964, and 1 : 5M, 1963), hydrogeological (1 : 5M, 1963) and mineral deposit (1 : 4.5M) maps are published. There is also a 1 : 200 000 series. (*Geol. Newsl.*, (3), 1967.)

BOLIVIA

Geological maps have been compiled by V. Oppenheim (*Boletin de la Sociedad Geológica del Perú*, **17** (1944), 1 : 2M) and by F. Ahlfeld (*Revista del Museo de la Plata, N.S., Geología*, **3** (1946), 1 : 1 212 000).

BRAZIL

A geological map (1 : 5M, 1960) and a 1 : 250 000 series are published. There is a 1 : 1M geological map (1963) and a 1 : 100 000 series for the state of Sao Paulo; for Paraná there is a geological map and a 1 : 50 000 series; for the province of Rio de Janeiro a 1 : 50 000 series; and for Minas Gerais a 1 : 25 000 series. A geological map of Rio Grande do Sul (1 : 1.5M, 1966) was published in *Publicaçaõ Especial, Escola de Geológia, Universidade do Rio Grande do Sul*, **11** (1966). Much of north-east Brazil is covered by photogeological maps at a scale of 1 : 250 000 (*Reconhecimento Fotogeológico da Regiaõ Nordeste do Brasil*).

CHILE

Geological (1 : 1M, 1960) and geology, mineral and metalliferous deposits (1 : 500 000, 1965) are published. Several sheets of the 1 : 50 000 quadrangle series have been issued. (*Geol. Newsl.*, (3), 1967.)

COLOMBIA

A geological map (1 : 1.5M) was published in 1962. A few quadrangle maps at scales of 1 : 200 000 and 1 : 100 000 have been issued. (*Geol. Newsl.*, (2), 1970.)

COSTA RICA

A geological map (1 : 700 000) was published in 1968. Reference should also be made to *Maps of Costa Rica—An Annotated Cartobibliography*, by A. E. Palmerlee (University of Kansas Libraries, 1965).

ECUADOR
A geological, mineral and metallogenic map (1 : 500 000) was published in 1969. (*Geol. Newsl.*, (2), 1967.)

THE GUIANAS
A geological map (1 : 2M) of the three territories is published by the Geological and Mining Service of Surinam. There is a 1 : 500 000 geological map (1960), a mineral exploration map (1 : 1M, 1966) and a 1 : 100 000 series for French Guiana; a 1 : 1M map (provisional, 1962) and 1 : 200 000 series for Guyana; and a geological map (1 : 1M), mineral exploration map (1 : 1M) and 1 : 100 000 and 1 : 200 000 series for Surinam.

PARAGUAY
A 1 : 1M geological map and a 1 : 2M soil map are included in *USGS Professional Papers*, No. 327 (1959).

PERU
A 1 : 4M geological map was published in 1964. A 1 : 2M geological map was issued by the Sociedad Geologico del Perú in 1956.

URUGUAY
A geological map (1 : 750 000) was published in 1946.

VENEZUELA
A 1 : 1M geological map was published in 1955.

THE ANTARCTIC
A comprehensive series of geological, tectonic and other maps (at scales of 1 : 10M, 1 : 20M and 1 : 40M) is contained in the *Atlas Antarktiki I* (Glavnoe Upravlenie Geodezii i Kartografii, Moscow, 1966). A 1 : 10M map was published by the Comité National Français des Recherches Antarctiques in 1961. Many maps are to be found in the reports of the various Antarctic expeditions and in the 'Special Antarctic Issues' of the *New Zealand Journal of Geology and Geophysics*. Magnetic and gravity maps on a scale of 1 : 15M are contained in *Antarctic Map Folio*, Vol. 9 (American Geographical Society, 1968). A useful list of maps of this part of the world is contained in *Selected Maps and Charts of Antarctica. An Annotated List of the Maps of South Polar Regions published since 1945*, compiled by R. W. Stephenson (Library of Congress, 1959). (*Geol. Newsl.*, (2), 1971.)

ARCTIC REGIONS
A geological map of the Arctic (1 : 7.5M) was published by the University of Toronto Press in 1960. A map of the Soviet Arctic (1 : 2.5M) was included in *Trudy Nauchno-Issledovatel'skogo Instituta Geologii Arktiki*, **81** (1957). A 1 : 10M tectonic map was issued by the USSR Academy of Sciences in 1963.

ICELAND
A series of 1 : 250 000 geological sheets is published. (*Geol. Newsl.*, (4), 1971.)

GREENLAND
Geological maps (1 : 100 000) of various parts of Greenland are published by the Geological Survey of Greenland and in the journal *Meddelelser om Grønland*. (*Geol. Newsl.*, (2), 1971.)

OCEANS
Physiographic diagrams of the North Atlantic (1 : 5M, 1958), South Atlantic (1 : 11M, 1962) and Indian Oceans (1 : 11M, 1964) are issued by the Geological Society of America; these maps are also reproduced at a reduced scale in the *International Dictionary of Geophysics*, edited by S. K. Runcorn (2 vols. and map supplement, Pergamon, 1968). Four maps and an explanatory text of 'World Subsea Mineral Resources' are published as *USGS Miscellaneous Investigations Map* I-632 and a geological map of the Indian Ocean and bordering lands (1 : 13 655 000, 1963) was issued as I-380 in the same series. A tectonic map of the south-west Pacific (1 : 10M, 1970) was issued as *New Zealand Oceanographic Institute Chart, Miscellaneous Series*, **20**, and was also included in *New Zealand Journal of Geology and Geophysics*, **13** (1) (1970).

REFERENCES

Bodelle, J. (1967). 'La Vente des Cartes Géologiques en France'. *Annales des Mines* (4), 61–69
Bodelle, J. (1968). *Bulletin du Comité Français de Cartographie*, **2**, 66–70
Cox, L. R. (1942). 'New Light on William Smith and his Work'. *Proceedings of the Yorkshire Geological Society*, **25** (1), 1–99
Eyles, J. M. (1969). 'William Smith (1769–1839): A Bibliography of his Published Writings, Maps and Geological Sections, Printed and Lithographed'. *Journal of the Society for the Bibliography of Natural History*, **5** (2), 87–109
Eyles, V. A. and Eyles, J. M. (1938). 'On the Different Issues of the First Geological Map of England and Wales'. *Annals of Science*, **3** (2), 190–212
Hageman, B. P. (1968). 'The Reliability of Geological Maps'. *International Yearbook of Cartography*, **8**, 144–154
Harrison, J. M. (1963). 'Nature and Significance of Geological Maps'. In C. C. Albritton (ed.), *The Fabric of Geology* (Freeman)
Judd, J. W. (1897). 'William Smith's Manuscript Maps'. *Geological Magazine*, **4** (4), 439–447
Kay, M. (1945). 'Paleogeographic and Palinspastic Maps'. *Bulletin of the American Association of Petroleum Geologists*, **29** (4), 426–450
Lister, M. (1684). 'An Ingenious Proposal for a New Sort of Maps of Countrys with Tables of Sands and Clays such as are Found in the North Parts of England'. *Philosophical Transactions of the Royal Society,* **14** (164), 739–746
Marcou, J. and Marcou, J. B. (1884). 'Mapoteca Geologica Americana. A Catalogue of Geological Maps of America (North and South) 1752–1881 in Geographic and Chronologic Order'. *Bulletin of the United States Geological Survey*, No. 7
Radócz, G. (1968). 'A Földtani Vonatkozású Térképek Attekintése és a Rendszerezés Néhany Szempontja'. *Evi Jelentese, Magyar Allami Földtani Intézet 1966*, 335–358

Robertson, T. (1956). 'The Presentation of Geological Information in Maps'. *The Advancement of Science*, **13** (50), 31–41

Sheppard, T. (1917). *Proceedings of the Yorkshire Geological Society*, **19**, 75–253

Stephenson, R. W. (1970). 'Published Sources of Information about Maps and Atlases'. *Special Libraries*, **61** (2), 87–98, 110–112

8

Searching the literature, keeping records and writing reports

D. N. Wood and A. G. Myatt

THE TECHNIQUE OF LITERATURE SEARCHING

It is probably true to say that to make the most intelligent use of scientific literature one must appreciate its structure. That is to say, one must appreciate that the literature does not comprise a miscellaneous collection of scientific papers, books and reports, etc., but that it can be subdivided into basic and quite definite categories. Earlier chapters have already drawn attention to these categories—encyclopaedias, dictionaries, handbooks, textbooks, monographs, review journals, abstracting and indexing periodicals, current awareness tools and a whole host of primary publications such as theses, semi-published reports and of course the ubiquitous scientific periodicals.

These categories have come into being partially at least as a result of the information requirements of the scientist and consequently the structure of the literature can to some degree be related to these requirements. From this it follows that, provided the scientist can define his information requirements, he can restrict his literature searching to one or more of the appropriate categories. The first and most fundamental thing to do at the outset of any literature search therefore is to define as precisely as possible the requirements of the search.

For instance, let us consider the case of someone who wants what we can call, for convenience, everyday information: such information as the optical properties of a particular mineral; the formula of a certain compound; the UK annual output of, say, tin; the melting point of a substance at atmospheric pressure; the family to which a particular fossil belongs; or the annual rainfall in, for example, Lagos. This sort of information can be obtained from the standard reference works, i.e. the types of publication mentioned in Chapter 4. In another case a scientist who is starting to investigate a completely new topic may very well require a short introduction to help clarify the subject. He can get this most easily from an encyclopaedia or, if a more detailed introduction is required, from a textbook.

In none of the instances mentioned so far would reference to periodical literature really be appropriate. Now let us consider the case of a university lecturer who is preparing a series of lectures on different subjects. He hardly has the time to read everything that has been written on these subjects, but on the other hand he obviously wants to make his lectures as comprehensive as possible. The solution to his problem lies in a 'state-of-the-art' review and the most appropriate place to look for such an article would be in one of the journals devoted to the publication of such reviews, i.e. in one of the *Advances* or *Progress* series referred to in Chapter 4.

In the case of a very detailed literature search in which all available information on all aspects of a subject is required, attention would have to be given to abstracting and indexing publications.

Finally, for the worker who wishes to keep up to date with his subject even the abstracting journal is hardly an appropriate tool to use, since, as was pointed out in Chapter 5, many of these are considerably out of date by the time they are published. If the requirement, therefore, is to discover whether anything has been published on a particular subject in recent weeks, the literature search can be confined to the current awareness publications such as *Current Contents* and *Chemical Titles* or alternatively to the primary literature itself.

Very occasionally it may be necessary for a complete literature search to be undertaken. This might be the case prior to a student's starting a Ph.D. research project. Such a student would obviously wish to discover the majority of what had been published on his subject so that he could define the area in which to concentrate his own research effort.

In an attempt to illustrate the technique of literature searching a very detailed search will now be described. The subject of the search will be 'Palaeomagnetism and its bearing on the hypothesis of continental drift'. For the purposes of the exercise it will be assumed that the search is being undertaken by someone who knows little or nothing about either palaeomagnetism or continental drift and who wishes to acquire a detailed and up to date knowledge of the subject. Since few people would be in this position, it will be a rather more comprehensive search than is usual, but on the other hand it has the advantage of allowing the use and relationship of a wide range of bibliographic tools to be demonstrated.

At the beginning of any literature search the first step is to define the subject of the search as closely as possible so that its ramifications can be appreciated and its terminology understood. At this stage also an attempt should be made to establish the boundaries of the search in terms of the time period to be covered and the types of sources in which information is likely to be found. In the case of this particular subject, for instance, there would be little point in searching the patent literature or trade publications. The types of literature which should be consulted at the outset of the search are encyclopaedias, dictionaries and general textbooks. As well as providing a basic introduction to the subject and some suggestions for further reading, these tools enable one to compile a list of keywords which will prove useful when attention is subsequently turned to the subject indexes of, for example, abstracting and indexing publications.

At the start of this search the dictionaries consulted included the *Encyclopaedic Dictionary of Physics, Dictionary of Geology* and the AGI *Glossary of Geology and Related Sciences*. Each of these, using entries of variable length, defines both palaeomagnetism and continental drift. The *Dictionary of Geology* entry also includes five references to general works on palaeomagnetism and four on continental drift.

The next step was to seek a more detailed introduction to the subject. The general encyclopaedias can be useful at this stage, and in *Encyclopaedia Britannica* the entry under 'Continents' contains a section dealing with continental drift. It is interesting to note that there is no actual entry under 'Continental drift', and the only way one can locate the information is by using the subject index (published as a separate volume). This is a point well worth remembering when using any encyclopaedia in which entries are arranged in alphabetical order. In *Chambers's Encyclopaedia* a lengthy entry including several illustrations and references was located.

General subject-oriented reference works are also useful for locating introductory material, and in this case *Principles of Physical Geology* by A. Holmes (Nelson, 1965) was found to contain a chapter entitled 'Palaeomagnetism and Continental Drift'. A select list of references accompanies the chapter and these, like those contained in the encyclopaedias and dictionaries, were noted down but not actually consulted at this stage.

As a result of the introductory information gleaned from the publications referred to above, a list of appropriate keywords was compiled. This included such terms as geomagnetism, palaeo-magnetism, terrestrial magnetism, earth's crust, earth's magnetic field, rock magnetism, polar wandering, continental drift and continents.

For the reader who is unfamiliar with the subject of the search it might be useful at this juncture to summarise the information obtained during this initial investigation. Palaeomagnetism is the faint magnetic polarisation of rocks that may have been preserved since their formation. The magnetic particles of these rocks are oriented with respect to the earth's magnetic field as it existed at the time and place of formation. Continental drift is the concept that the continents can drift on the surface of the earth because of the weakness of the sub-oceanic crust, much as ice can drift through water. Since the late nineteenth century there have been developed a number of theories concerning continental drift. Briefly, it has been proposed that for much of geological time the earth's continents were united into a single, or at the most two, gigantic land masses. During the last 150 million years these land masses are thought to have split up and the individual continents to have gradually assumed their present positions. In support of this theory a great deal of physiographic, geophysical, geodetic, biological, palaeoclimatic and geological evidence was put forward, but the theory was largely abandoned in the 1940s and early 1950s. In recent years, however, palaeomagnetic evidence has been intro-duced into the argument. If the remanent magnetism of a rock is examined, the position of the magnetic pole at the time of its formation can be calculated. If it is taken for granted that the axis of rotation of the earth has remained fixed in space and that throughout geological time the magnetic and rotational axes have been virtually coincidental, then any apparent movement of the pole as deduced from palaeomagnetic results will indicate a move-ment either in part or whole of the earth's crust. If an examination of rocks of the same age in, say, North America and Europe reveals different polar positions, then this can be taken to indicate that since those rocks were formed there has been relative movement between the continents, i.e. continental drift must have occurred. By analysing palaeomagnetic data derived from all over the world and for rocks of all geological ages it is possible to build up a picture of the nature and speed of any drift that might have taken place.

Armed with this preliminary information and with the list of keywords, to enable subject indexes to be searched effectively, in the next stage of the literature search one looked for a more de-tailed but still fairly general work on the subject. Such an account is usually to be found in a textbook or monograph, and enables one to develop an understanding of the subject under investigation as opposed to supplying detailed information on a limited aspect of the topic.

References to a number of books had already been obtained from the publications consulted earlier. For additional titles attention was turned to the various published guides to books. One of the more useful tools in this field is *Subject Guide to Books in Print* (see Chapter 4). Both 'Paleomagnetism' and 'Continental drift' were found to be subject headings, and three relevant books were listed: *Continental Drift*, by G. D. Garland (University of Toronto Press, 1966); *Continental Drift*, by S. K. Runcorn (Academic Press, 1962); and *Paleomagnetism and its Application to Geological and Geophysical Problems*, by E. Irving (Wiley, 1964). Additional guides which were consulted were *British National Bibliography* and *Cumulative Book Index* (see Chapter 4). Although none of the books located dealt solely with the relationship between palaeomagnetism and continental drift, all devoted reasonably long sections to the subject, and most contained substantial lists of references. The latter were recorded for future use.

Before one passes on to the location of more detailed and up to date information in journal articles and reports it is advisable to try and locate an existing bibliography on the subject. Bibliographies, the existence of which may obviate the need to search for references prior to the date of their appearance, may be published separately or may be part of a review summarising work in a particular field. Those published separately are listed in a number of guides, including those used to discover the existence of books. In this case, however, such guides proved to be of little value and attention was therefore turned to the *World Bibliography of Bibliographies* (see p. 76) and *Bibliographic Index* (see p. 77). The former proved of no use, but the latter, which includes, as well as separate bibliographies, review articles and books which contain substantial lists of references, provided details of a number of fairly short bibliographies. It is worthy of note that the *Bibliographic Index*, which lists entries under alphabetised subject headings, changes these headings from issue to issue. Thus, literature on palaeomagnetism is to be found indexed under 'Paleomagnetism' in one issue and under 'Magnetism—terrestrial' in another, while references on continental drift appear in different volumes under 'Continental drift' and 'Continents'.

Review articles can be traced through abstracting and indexing publications, but the process of locating them may take a considerable time. It is a somewhat easier task to scan those periodicals devoted to the publication of reviews. A fairly comprehensive list of English-language review services has been made at the NLL (the list is available free) and a glance at this list revealed two titles in the field of geophysics, namely *Advances in Geophysics* and *Reviews of Geophysics*. By use of the cumulative subject index for Vols. 1–11 of the former title an article entitled 'Paleomagnetism' was located in Vol. 8, 1961. The article proved to be well written and comprehensive and was accompanied by a bibliography containing over 200 references. The discovery of this review and the earlier books on continental drift and palaeomagnetism meant that the remainder of the search could be confined to locating detailed information published in relatively recent years.

At this stage of the search it became essential to turn to the abstracting and indexing journals. The first problem was to discover which of more than 2000 current titles were most likely to be of use. The selection was made by making use of the various guides to abstracting literature, the most useful of which are *A Guide to the World's Abstracting and Indexing Services in Science and Technology* (see p. 85), *Abstracting Services. Vol. 1. Science, Technology, Medicine, Agriculture* (see p. 85) and the NLL publication *A KWIC Index to the English Language Abstracting and indexing Publications Currently Being Received by the NLL*. The first of these publications includes details of title, periodicity, coverage and form of indexes, and with the help of this information it was possible to select from more than 20 appropriate subject-oriented indexing journals six which merited detailed searching. These were:

> *Bibliography and Index of Geology Exclusive of North America*
> *Bibliography of North American Geology*
> *Meteorological and Geoastrophysical Abstracts*
> *Geophysical Abstracts*
> *Bulletin Signalétique*
> *Physics Abstracts*

The criteria used in selecting these six titles were coverage, delays in publication, availability of good indexes and language. Coverage includes subject scope, the number and place of origin of the periodicals covered and the total number of articles indexed per year. Some abstracting journals, e.g. *British Geological Literature*, cover only a portion of the world's literature, and although they might be useful in the case of a regionally oriented search, are hardly comprehensive enough to warrant searching where the subject is of world-wide importance. In any case, the entries in these relatively specialised tools are usually duplicated in the more comprehensive publications. On the other hand, other abstracting journals, although their titles give the impression of being comprehensive, are so selective in their coverage of the literature that they are virtually worthless as retrospective searching tools. Such is the case with, for example, *Geophysical Exploration*, which lists only 250 papers per year compared with nearly 2000 per year in *Geophysical Abstracts*.

Many abstracting and indexing publications are less useful than they might be because of the delays involved in their production. In this case it would have been inadequate to confine the search to the *Bibliography of North American Geology* and the *Bibliography and Index of Geology Exclusive of North America*, because at the time the search was carried out both publications were about 3 years out of date when published.

When carrying out a lengthy retrospective search of abstracting publications, one is helped enormously by the availability of good cumulative indexes. Some journals such as the *Quarterly Checklist of Geophysics* have no such indexes, while the indexes of others such as *Zentralblatt für Geologie und Paläontologie* are published somewhat belatedly. In the search being described only publications with good up to date indexes were used.

Language is obviously an important criterion when abstracting and indexing journals at which to look are being selected. In view of the linguistic limitations of the searcher, only English and French publications were consulted. A knowledge of Russian would have been useful, since this would have made it easier to search the appropriate sections of *Referativnyi Zhurnal*.

The selection having been made, each abstracting publication was searched in turn, the searcher commencing in each case with the latest cumulative index available and working back. Information was sought under a variety of subject headings, and found under such terms as 'poles', 'polar wandering', 'Wegener's theory', 'rock magnetism', 'terrestrial magnetism', 'magnetism—terrestrial', and the more obvious ones such as 'continental drift', 'continents' and 'palaeomagnetism'. Because of the existence of the review article and books located earlier, the search was confined to literature more recent than 1964. Having ascertained, by using the cumulative indexes, in what subject sections of the abstracting journals appropriate articles were listed, one was enabled to continue the search in the recent issues by looking in the relevant sections. For example, in *Geophysical Abstracts* articles on palaeomagnetism and continental drift are to be found in the section entitled

'Magnetic Properties and Palaeomagnetism'. As relevant references were located, they were noted down but not consulted at this stage.

The subject-oriented abstracting journals do not cover adequately certain types of literature such as reports, dissertations, conference proceedings and in some cases foreign-language literature. Because of this the search was extended to include such tools as *Dissertation Abstracts International, US Government Research and Development Reports,* the Aslib publication *Index to Theses . . .* , *Index of Conference Proceedings Received by the NLL* and the *Monthly Index of Russian Accessions.* From the latter a number of papers in Russian were located and the availability of translations was checked in such publications as *Translations Register-Index* and *World Index of Scientific Translations.* One translation was located in this way in *International Geology Review.*

From use of the abstracting journals mentioned above it was apparent that because of the delays involved in the production of such journals the references were up to 3 years out of date and in no case were the papers referred to less than 6 months old. To bring the search up to date, therefore, attention was turned to the current awareness literature. *Current Contents* was useful at this juncture, and a search of the previous 6 months' issues revealed a number of useful papers, one of which had been published only 3 weeks earlier.

Since the coverage of abstracting journals is by no means comprehensive, the search concluded with a look at the recent issues of those journals which in the course of the earlier search had revealed themselves as regularly carrying articles on palaeomagnetism. Two such journals were *Journal of Geophysical Research* and the *Bulletin of the Geological Society of America.*

All the references retrieved from the sources referred to above having been recorded, the individual papers were then consulted. These in their turn referred to additional works, and at the conclusion of the search many hundreds of references had come to light.

The literature search described illustrates a number of points which, since they apply to any comprehensive search, are worth discussing in further detail.

First, the order in which the search was carried out should be noted. Encyclopaedias and dictionaries were the first to be consulted, followed in turn by guides to books, bibliographies of bibliographies, guides to review journals, abstracting publications and, finally, current awareness tools and the primary literature itself. This search strategy allows for the maximum number of references to be located in the minimum amount of time. It should, of course, be borne in mind that the search described was intended to be comprehensive and assumed no knowledge at the outset. Depending on the nature of the enquiry and the previous knowledge of the searcher, the strategy used would have to be modified.

Apart from the reading carried out during the earlier part of the search in order to become subject-oriented, it will be noted that a complete list of references was made before the papers themselves were actually consulted. This tactic is a useful safeguard against the tendency to leave a literature search incomplete after having once obtained a substantial bulk of information following a partial survey.

During the search of abstracting literature it is important, if comprehensive coverage is required, that attention should not be confined to a single journal. No abstracting publication, not even *Chemical Abstracts*, is completely comprehensive, and for an exhaustive search, therefore, many abstracting and indexing services should be consulted. In this particular search the value of this strategy is indicated by the fact that of all the references obtained from abstracting journals 72 per cent were in *Bulletin Signalétique*, 60 per cent in *Meteorological and Geoastrophysical Abstracts*, 53 per cent in *Geophysical Abstracts*, 25 per cent in *Physics Abstracts*, and 40 per cent in the *Bibliography and Index of Geology Exclusive of North America* and the *Bibliography of North American Geology*. By confining the search to one journal only, the best retrieval rate that one could have achieved would have been 72 per cent and the worst 25 per cent. This bears out what was said about coverage of abstracting journals in Chapter 5.

When abstracting journals are used, it is advisable to locate the available cumulative indexes. Searching through individual issues of journals is an extremely tedious job and it is frustrating in the extreme to discover afterwards that a cumulative index is available. The various guides to abstracting journals will indicate whether or not such indexes exist but it should be noted that in certain cases they may be shelved some distance away from the issues to which they refer. If they cannot be located the librarian should be consulted.

Another important point to bear in mind when searching the indexes to abstracting journals is that relevant information will be found under a variety of subject headings. In the present search it has already been pointed out that useful references were located under many different headings. Those used in one journal were not always the same as those used in another and even within one journal relevant articles were indexed under different headings. The latter point is well illustrated by *Geophysical Abstracts*, where a paper entitled 'A Review of the Variation in Position of Land Masses during Geological Times' is to be found indexed under 'Paleomagnetism' but not under 'Continental drift'. Had the searcher confined his attention to the latter heading, this very relevant paper would have been missed. When subject indexes are used, therefore, it is essential to look under all possible keywords starting with the most specific ones and working to the more general as necessary. It is in this formulation of subject headings that preliminary subject orientation proves its value.

The problem of using subject indexes is admirably summarised by Smith (1951), who states: 'Indexers devote much thought and effort in placing subject matter where searchers are most likely to seek it. In a sense, each indexer projects his mind into the future hoping that searchers will project their minds back to meet his. The whole operation is essentially a guessing game of indexers and searchers playing on the same team against the invisible gremlins of error, mischance, mishap and false trails. The minds win when the searcher arrives at the wanted information.'

Additional points to bear in mind when English-language indexes are used are that the spelling of various words varies on each side of the Atlantic and that American terminology differs in some cases from that employed in the UK.

When one is undertaking a literature search it is often a matter of debate as to how far back in time the literature should be searched. A knowledge of the history of a subject can be useful and a brief history can usually be obtained at an early stage in the search when encyclopaedias and books are being consulted. In the case of the search outlined above it soon became apparent that although continental drift has been a subject of discussion since the middle of the nineteenth century, palaeomagnetic results played very little part in the argument until after 1950. There was little point, therefore, in searching for information prior to this date. The extent of the search in the abstracting literature is also determined by the information which one has unearthed prior to this stage of the search. Again, as far as the present search was concerned, good review articles covering the period up to 1964 had been found, so the search of the abstracting services was confined to the years after 1964. In general, it is probably true to say that a search should stop when further efforts fail to produce proportionate results. There are obviously exceptions to this rule, however; for example, in a patent search (which in any case earth scientists will rarely undertake) it may be necessary to retrieve every paper ever written on the subject irrespective of how long it takes.

During the search a complete record was kept of all the relevant references located. This is a particularly important point, since it is extremely frustrating to have to go back to the guides to the literature to discover additional details of incomplete references. Unless detailed records have been kept, this might be necessary when the results of a search are being written up for publication. There are many ways of keeping records of references and these will be discussed in the following section.

PERSONAL RECORD KEEPING

Various systems can be used for creating, maintaining and using an index of references and the following account shows how such systems can serve the individual research scientist (additional information is available in the references cited in Appendix 1 to this chapter). What will be attempted is a detailed examination of the alternative and economical methods of recording the details of references. These details, which will include author, title, subject, date and so on, are termed concepts.

It should be borne in mind that the nature of the information file may change with increasing size and it is possible to start with one system and subsequently change to another.

Systems available

The simplest method of recording references is to utilise ordinary $5in \times 3in$ or $6in \times 4in$ file cards. For each reference a single card is made out, giving the full bibliographic details of the document being recorded (*Figure 8.1*). The card can then be filed either in a single alphabetical sequence of authors or arranged into groups according to subject. If the file is a large one, or if a more detailed subject breakdown is required, the author file can be supplemented by an alphabetical subject file. In this case, in addition to an author card giving full bibliographic details and probably an abstract, one or more subject cards can be made out for each reference. Each of the subject cards (*Figure 8.2*) is annotated with the author's name and thus allows the bibliographic details to be retrieved from the other file.

Simple systems such as these are adequate where only a small number of references is recorded, but as the file grows, certain disadvantages become apparent. In the first instance, it becomes difficult to carry and to keep in order and more time is taken for

```
┌─────────────────────────────────────┐
│                                     │
│  DANIELSEN E.F.                     │
│                                     │
│  Stratospheric-Tropospheric exchange│
│  based on radioactivity, ozone and  │
│  potential vorticity                │
│                                     │
│  Jnl. of the Atmospheric Sciences. (Boston)│
│     25 (3) :502-518 May 1968.       │
│                                     │
└─────────────────────────────────────┘
```

Figure 8.1

| Stratosphere-Troposphere exchange. See   Danielsen, E.F. | Potential vorticity calculation. See   Danielsen, E.F. | Ozone transport Mechanisms. See   Danielsen, E.F. |

Figure 8.2

retrieval. Secondly, the system continues to work only so long as the same subject descriptors are used to index the documents, and as the file and terminology grow, information can be lost. For recording large numbers of references, therefore, a more sophisticated system is necessary. To be really effective such a system should have the following characteristics:

(a) speedy retrieval of information;
(b) the cards do not need to be in any order;
(c) cross-references should be on the same card;
(d) the file should be portable.

There are two card systems which go some way towards satisfying the above requirements. The first employs what are known as feature cards (sometimes called 'cards with visual punching', 'superimposed coincidentally punched cards', 'peek-a-boo' cards or 'sneek-a-peek' cards). Feature card systems have a numerical grid on each card and each card represents a different concept. The source of information, i.e. article, book, thesis, etc., is given a number, and where a concept applies to a particular document, a hole is made in the card in the appropriately numbered grid position (*Figure 8.3*).

From *Figure 8.3* we note that the documents numbered 12, 20 and 40 are related to concept A. Other concept cards are punched in the same way and the cards filed in concept order.

If one wishes to select documents which have been indexed with concepts A, B and D, it is necessary to remove the three cards from the file, align them and hold them up to the light. Where light

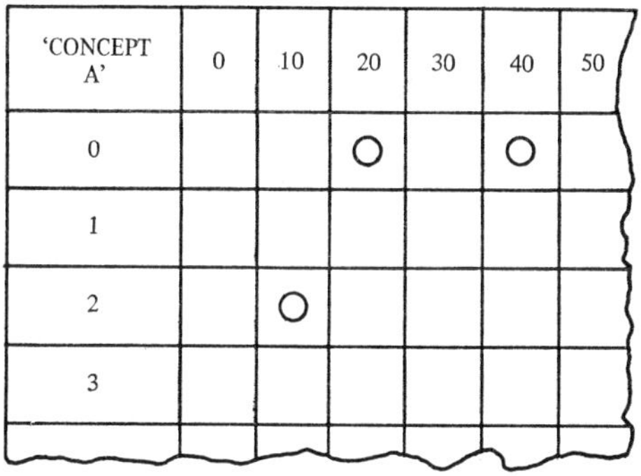

'CONCEPT A'	0	10	20	30	40	50
0			O		O	
1						
2		O				
3						

Figure 8.3

passes through the three cards, the relevant document number(s) are shown. To obtain the actual details of the documents it is then necessary to consult a separately constructed numerical file.

Cards 11in×6in can conveniently be used in such systems to record about 1000 documents. A standard 80-column punched card, on the other hand, can be used to record 800 documents. Such feature card systems are particularly appropriate for indexing numbered objects—for example, specimens or slides—although fine selection by concept is only possible if large numbers of cards are used.

Because accurate selection of cards is an essential first step in the searching process, feature cards must be kept in strict conceptual order in the file.

The second system having the characteristics referred to above makes use of edge-notched cards (for suppliers of these see Appendix 2 to this chapter). Such cards can be any size and are distinguished by the fact that around their margins they contain one or more rows of alphabetically or numerically coded holes. The centre of the card is either blank or printed so as to facilitate the recording of information. It may also contain an aperture in which can be mounted a micro-copy of a document. For each reference to be recorded it is possible to note full bibliographic details and summary of contents in the centre of the card. For future retrieval purposes this information is also recorded around the edge of the card by clipping out appropriate holes. A special pair of clippers is available for this purpose. Suppose, for example, one records an abstract on the face of the card which is concerned with the subject 'Basalt'. If the card has a margin of holes which are labelled with letters of the alphabet, it would be reasonable to clip hole 'B' on the margin of the card and then return the card to the pack.

To locate all records containing 'Basalt' as a key word at a later date, a long knitting needle is pushed through the 'B' hole of the whole pack (the pack of cards must be aligned to do this) and all cards with 'B' clipped will drop out when the pack is shaken gently (*Figure 8.4*).

Figure 8.4

Holes marked with letters are suitable for authors or subject keywords. With the author 'Brown' or the keyword 'Basalt' one does not require a dictionary in order to note that one must clip or needle hole 'B'. Similarly, holes on the card marked with numbers are useful for coding dates, document numbers or classification numbers. Where numbers are used to express letters or words, or letters are used to express numerical concepts, then it is necessary to have a dictionary before accurate coding or retrieval is possible. For this reason, cards have been designed which contain both numerical and alphabetical fields, thus allowing a variety of information to be recorded without the use of dictionaries.

An information file making use of edge-notched cards has a number of advantages:

(1) Because concepts are recorded round the edge of the card, the greater part of the card is available for written information.
(2) There is no need to keep the cards in any particular file order.
(3) Sorting is rapid and easy.
(4) Cross-references can be kept on one card.
(5) The file is portable.

Many edge-notched cards offer a variety of letters and numbers around the margin and are suitable for various types of record. With some cards designed for literature references, it is possible to code simultaneously all details of author, source (report, thesis, journal, paper, etc.), journal title, issuing body, number of publication, date of publication and several subject keywords. An example of a specially designed card is shown in *Figure 8.5*.

On the card illustrated, the numbering along the upper edge provides for the coded representation of 10 000 subjects on a decimal system (a dictionary of codes would, of course, be necessary in this case). The right-hand side of the bottom margin is used for recording the year of publication (only the last two digits of the year are recorded). The separate alphabetical panels 1, 2 and 3

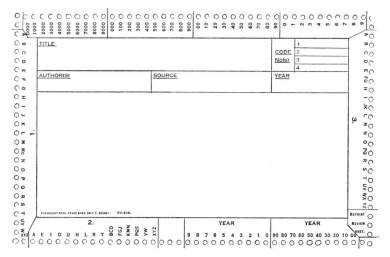

Figure 8.5

are used for filing according to first, second and third letters of the author's surname. The bottom of the right-hand margin indicates whether a reprint is available locally, whether the article is a general review of a major topic and whether only an abstract of the paper has been seen. The unmarked holes at the bottom are for special classifications according to individual preference. The centre of the card contains spaces for a lengthy summary or notes.

General notes on the coding of edge-notched cards

Cards with a single row of holes. So far we have been considering direct coding techniques, where the notching of a single hole conveys a particular piece of information—for example, the first, second or third letter of the author's name, the individual digits of a document number, the year of publication or the subject. In view of the limited number of holes on a card, the amount of information that can be recorded in this way is somewhat restricted. To conserve space and provide more fields, it is possible to use some type of indirect coding. For example, whereas ten holes would be required for the direct coding of digits 0 to 9, only four are needed if an indirect method is used. The four holes are labelled 1, 2, 4 and 7, respectively, and numbers 0 to 9 are recorded by notching a combination of 1, 2, 4 or 7, i.e.:

For — 0 notch 7 and 4   5 notch 4 and 1
1 notch 1   6 notch 4 and 2
2 notch 2   7 notch 7
3 notch 1 and 2   8 notch 7 and 1
4 notch 4   9 notch 7 and 2

With this system, false drops (unwanted cards) will occur when one wishes to select the numbers 1, 2, 4 or 7. Needling 1 will cause 3, 5 and 8 to drop as well. To overcome this, a parity code (P) can be added which thus enables every number to have a unique two-hole code, i.e.:

For — 0 notch 7 and 4   5 notch 4 and 1
1 notch 1 and P   6 notch 4 and 2
2 notch 2 and P   7 notch 7 and P
3 notch 1 and 2   8 notch 7 and 1
4 notch 4 and P   9 notch 7 and 2

Where a parity code is used, the edge of the card will appear as in *Figure 8.6.*

Figure 8.6

In alphabetical coding, the alphabet may be split up into groups, with extra notches to indicate the group. In each case, a two-hole unique code will identify separate letters. The methods of doing this are illustrated in *Figures 8.7* and *8.8.* Alternatively, the corner of the card will adapt well to punching two holes per letter, and

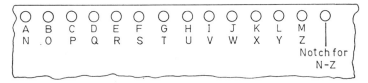

Figure 8.7

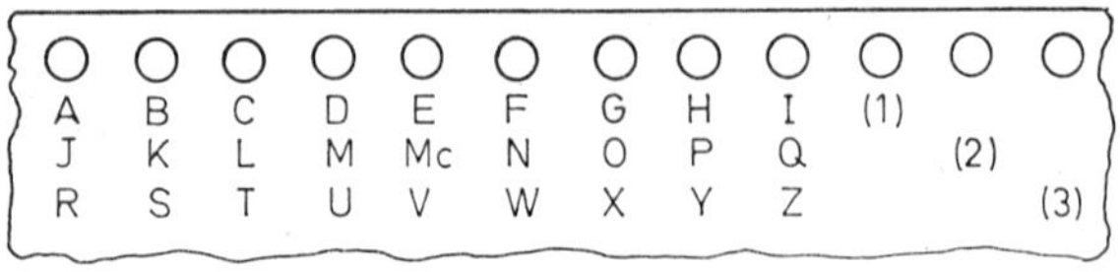

Figure 8.8

the alphabet can be condensed into only eleven holes of margin (*Figure 8.9*).

Figure 8.9

A variant of the corner sort is the pyramid code. Codes are spread along the edge of the card and again two holes are made for each digit or letter of the alphabet. The pyramid code is illustrated in *Figure 8.10*. The advantage of using the pyramid prin-

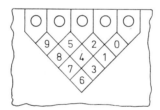

Figure 8.10

ciple is that the cross-referencing is self-explanatory and the number of holes used is reduced. The code does, however, extend further into the body of the card, which, in certain cases, may be a disadvantage.

It is possible to represent letters by numbers, in which case the 7, 4, 2, 1, system can be used for the alphabet, with the letter 'M' being punched for the second half of the alphabet. This system is illustrated in *Figure 8.11*.

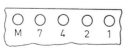

$$\begin{bmatrix} A-L &= 1-12 \\ Mc &= 13 \\ M &= M \\ N-Z &= M + (1-12) \\ Z &= M,7,4,2 \end{bmatrix}$$

Figure 8.11

Cards with double and multiple rows of holes. Where very large numbers of concepts are required, margin space can be saved by adopting a system which employs either two, three or four rows of holes. In *Figure 8.12* some examples are given of how a two-row

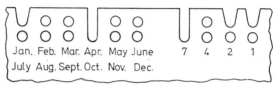

Figure 8.12

card may be used. In these examples the months January and October have been coded and the numbers 7 and 3. The parity code mentioned earlier can be eliminated in the latter case, since numbers requiring two punches are coded into the outer row, while 7, 4, 2 and 1 (requiring only one punch each) are coded into the inner row. When such a double row of holes is alphabetised, an effective name selection code can be condensed into only thirteen holes of margin. Surname letters are coded into the inner row, other letters into the outer row.

If these principles are extended, the capacity of an edge-notched card can be considerably increased. *Figures 8.13* and *8.14* illustrate how two systems may each be used to code a large number of concepts. By use of five codes per concept (one punch into each of three inner rows and two punches into the outer row) 15 120 different concepts can be recorded.

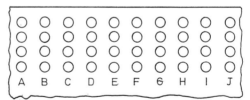

Figure 8.13

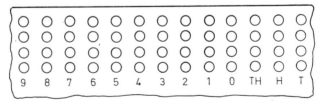

Figure 8.14

In the case of *Figure 8.14*, any number from 1 to 9999 can be coded with four punches in the field. Thousands are punched into the inner row, hundreds into the third row, tens into the second row and units into the outer row. The right-hand three columns are to control the duplication of numbers.

Four-letter codes may be used in a similar way, provided that codes are allowed for duplication. The code for I.E.E.E. would require four punches, I., E., and first and second duplication punches for the remaining E.'s.

Evolving a classification system

Users of edge-notched cards either will have determined the classification system to be used before recording starts or will develop a classification as time progresses. It is usually undesirable to attempt a rigid classification at the outset, since the final selection of terms to be used may not be known at this stage. In order to overcome possible future difficulties, it is better to use edge-notched cards as ordinary file cards in the first instance and then to notch them later when the structure of the file is understood.

As the file develops, the subordination of terms may be required. This may be achieved by, for example, superimposed coding. This method permits two concepts to be added together as subordinate terms to a third; thus, in the case where random numbers are being assigned to subjects:

ALUMINIUM could be coded	2, 14, 39	Numbers from
LEAD could be coded	7, 19, 54	table of
METALS could be coded	15, 27, 31	random numbers

In this case, when the term 'Metals' is added to the dictionary, it is coded on to all of the 'Aluminium' and 'Lead' cards at the same time. By use of this technique, quite complex 'trees' of terms can be constructed if desired and no prior classification of the information is called for.

TECHNICAL WRITING

Any serious student of the art of technical writing will soon discover that there exists a wealth of published literature on the subject. Several new books are published each year and at any one time there may be as many as 100 books in print devoted to aspects of the subject. Books are available which offer expert advice on such topics as style, vocabulary, illustration, abbreviations and editing. However, because the recommendations to be applied to technical literature writing in one subject are patently unsuitable for another, these books conflict somewhat in the advice offered.

Since this brief account can do no more than make fairly general recommendations, a select list of relevant books, etc., is given in Appendix 3 to this chapter. Of these, 'Notes for Authors', published by the Geological Society of London, and *General Notes on the Preparation of Scientific Papers*, published by the Royal Society, are particularly recommended.

Readability and general style

It is rather unfortunate that many style manuals are composed largely of long lists of words and phrases to be avoided. The choice of style and vocabulary is a matter which should be left to the author, since style is largely a manifestation of personality. Provided that consistently long words, sentences and paragraphs are avoided, together with jargon, clichés, ungrammatical English, pomposity, repetition, bad punctuation and so on, the text should be naturally pleasant to read.

Readability can be measured easily. The Gunning Fog Index, Flesch Check System and Dale–Chall Formula are all measures of readability. The Gunning Fog Index, for example, is quoted in *The Technique of Clear Writing*, by Robert Gunning (McGraw-Hill, 1968). The index is obtained from word and sentence length and is supposed to correspond to the number of years of education needed to understand the passage tested. The method is as follows:

(a) Find the average number of words per sentence in a passage at least five sentences long.
(b) Find the percentage of words with three syllables or more in the same section. (Exclude capitalised words, combinations of words or two-syllable words, made longer by adding 'd' or 'ed', such as 'included'.)
(c) Add the two figures and multiply the result by 0.4.

The Flesch Check System is quoted in *How to Read, Speak and Think More Effectively*, by R. F. Flesch (Harper, 1960), and the Dale–Chall Formula is explained in *A Formula for Predicting Readability* (Ohio State University, Bureau of Educational Research, 1948).

Readability is affected by many factors. Most authorities recommend that technical literature should be written in the third person and with a judicious mixture of active and passive voice. Passages entirely in the active voice give the impression of conceit, while those entirely in the passive voice may tend to be pompous.

Many hazards will handicap the unwary author. Detailed and sometimes humorous accounts of these can be found in most books on the subject of technical literature writing. Authors are warned against, for example, the omission of subjects from the secondary clause, elegant variation (calling the same thing by different names), over-capitalisation and the dangling modifier. The last seems, at first glance, to be modifying the wrong word.

Layout

In the absence of alternative detailed instructions, texts should be typewritten throughout in double spacing on one side of international A4-size paper. It is usual, however, for learned societies and commercial organisations to give clear instructions to intending authors. It is always good policy to look for up to date instructions before the article is written, since these may be precise and yet varied. Universities give firm requirements for theses and commercial firms produce 'house style' manuals for reports. Such instructions may stipulate rules for the use of, for example, footnotes, trade names, references, illustrations and tabular matter or units.

The layout of a paper, report or thesis should, where possible, conform to national or local standards. Where no standard is specified, the following arrangement is recommended:

Title.
Name(s) and place(s) of work of author(s).
Summary/Abstract.
Contents list (where appropriate).
Introduction.
Aims and methods.
Results.
Conclusions.
Discussion.
References.
Appendices.

British Standards have recently been prepared for the presentation of both research and development reports (BS 4811) and theses (BS 4821). There are also large numbers of other standards of interest to technical authors. It is well worth examining the latest edition of the *British Standards Yearbook*, since some standards may very well be useful in the particular context of the paper you may be preparing at the time. Some of the more useful British Standards are listed in Appendix 3 to this chapter.

Title. The title should be specific and brief and constructed so as to include keywords suitable for future information retrieval. It should always be borne in mind that uninformative titles result in considerable information loss. An article with the title 'Fossil Footprints in Arizona' is uninformative. The same article with the title 'Late Triassic Footprints of *Ceolophysis* in the Cinle Formation of Northern Arizona' has enough information to enable the reader to decide whether he wishes to read the article or not. Following the title it is important to include the author's name, initials and qualifications, together with the name and address of the institution where the work was carried out.

Summary/Abstract. Abstracts have long been required as part of a university thesis and it is now common for journal editors to demand one. The purpose of an abstract is (a) to help people to decide whether a paper is worth reading and (b) to expedite the work of abstracting services. Abstracts should disclose the most important information about the main article. They should be about 200 words long and factual. Particular attention should be paid to names, data and formulae. An abstract should be written in the third person and intelligible in itself without reference to the main article. It should be assumed that the reader has a basic knowledge of the subject. All newly observed facts, treatments, apparatus, theories, techniques, minerals, etc., should be stressed, together with words such as 'brief', 'exhaustive', 'theoretical', or 'experimental', to describe the treatment of the subject. It is important to explain the objectives of the investigations and the whole abstract should be suitable for quoting or indexing by information services. Specific references and citations should not be included.

Introduction. It is vitally important that the introduction should be kept short and that it should whet the appetite of the reader. Some scientists of the older school tend to approve of the scholarly historical introduction, with its lengthy account of the glorious advances of the past, from which the latest paper is evidently intended to gather some additional lustre. Extensive historical accounts are, however, well provided for today by the increasing number of review serials, and are not therefore to be recommended as parts of original research papers.

The perfect introduction should fall into three brief parts: the general field of interest, the findings of previous workers (which will be challenged or amplified) and the specific question to which the paper is to be devoted, together with an emphasis on new methods.

Aims and methods. It has already been stated that, as far as the abstract is concerned, it should be assumed that the reader has a basic knowledge of the subject. In the aims and methods section it should be assumed that the reader is a trained investigator with considerable experience. If this were not the case, the article would become intolerably long and perhaps resemble a training manual.

When the author has a picture of the amount of detail to be offered, he should then consider breaking the text into easily digestible sections. If a succession of techniques is being described, then each technique could form a sub-heading. Where a technique may be repeated by readers, then sufficient detail must be given to enable this to be done. If an established method or technique is being used, the principle upon which the method is based should be stated. The method may be unfamiliar or a modification may have been used, in which case it is necessary to give a reference so that the original method can be checked. Alternatively, the method should be described in full.

Results, conclusions and discussion. The results section is usually quite straightforward. Conclusions should consider the results in relation to other work on the subject and in the discussion section the meaning of the results should be assessed.

Results should not be explained in lengthy prose, especially if tables and figures speak for themselves. Results should point out the salient features, e.g. '*x*' is greater than '*y*'. Sometimes the results section will also contain conclusions; however, journal editors are inclined to differ on this. Sometimes it may be stipulated that the conclusions appear at the front of the paper, immediately following the abstract. Where the results section omits conclusions, it is likely that the discussion section following may have to be considerably lengthened, since it will not be intelligible if the results are not repeated. Ideally the results section will be understood by itself and should indicate the trend, if any, of the author's thoughts.

The discussion section is one of the most important parts of the paper and often includes extended discussion of the observations in comparison with other people's work. Where no extended discussion is possible, because of the results, then the results and discussion section may be combined. It is quite possible that a controversial issue may arise in this section of the paper; if it does, it should be discussed lucidly and fairly. Authors should also bear in mind that unexpected results should not be hidden in the discussion, especially where there is a possibility of further investigation.

References and footnotes. Two methods of presenting references are in general use:

(1) Names and dates are given in the body of the text and the references listed alphabetically at the end of the paper.
(2) A system in which numbers are inserted in the text, thus: [3] or (3) —and references to these are given at the foot of the page or at the end of the paper.

Authors should make sure that all references given in the text are in the list at the end of the paper (or the footnotes) and vice versa —also that names and dates given in both places agree.

Frequently rules regarding references are laid down by publishers. These should be followed carefully. It is unreasonable to expect an editor to put references into the conventional form. If no system is laid down by the publishers, authors are advised to use the Harvard system, which is as follows.

References are collected at the end of the paper in alphabetical order, each giving:

(a) Name(s), followed by initial(s), of author(s).
(b) Year of publication in parentheses. (If several papers by the same authors in one year are cited, then a, b, c, etc., are placed after the year of publication.)
(c) Title of the paper.
(d) Title of the journal, abbreviated in accordance with the *World List of Scientific Periodicals* (see Chapter 3) or, for new journals, in accordance with *World List* principles, and underlined to indicate italics.
(e) Volume number in arabic numerals underlined with a wavy line to indicate bold type face. No prefix 'vol.' is required.
(f) The numbers of the first and last pages in arabic numerals without prefix 'p.'. Illustrations not included in pagination, and important, may be added as, e g., Fig. 4–65.

Thus references will appear in print in the following form:

King, C. A. M. (1960). 'The Churnet Valley'. *East Midland Geogr.,* **14**, 33–39.

When reference is made to a section of a book, the following method is recommended:

Hencl, V. (1965). In *Mineral Processing* (ed. A. Roberts), 483–499. Pergamon Press.

Footnotes should be avoided in the main text, since they are distracting and are rather costly to print. It may, however, be conventional to give references as footnotes, and when this is so, reference is made by the superscript symbols * † ‡ § and ¶, in that order.

Use of numbers

Since the bulk of scientific and technical information is quantitative, numbers are frequently used, at which time certain conventions should be observed.

Generally speaking, one-digit numbers should be written out, as also should numbers in a statement of approximation, round numbers and ordinal numbers from first to ninth. Numbers used to begin sentences should also be written out. When numbers for quantity and dimensions appear in the same expression, it is better to write out the quantity and use numerals for the dimension, e.g. thirteen 4-dozen boxes. If a large number is confusing to read, then numerals and words are better combined to give clearer expression.

One-digit numbers used for calculation, reference purposes and in tables and charts should not be written out; neither should quantities immediately followed by a unit of measurement, sums of money, fractions and dates. It is also usual to use numerals to number a page or express a percentage or temperature and for time intervals, ratios and proportions. Compound numbers should be hyphenated for the range of 21 to 99 and where a number is used as an adjective, e.g. 10-oz containers.

It is always important to state units, and if non-metric units are used, metric equivalents should also be given at least once in each paper. For values less than unit, zero should be inserted before the decimal point and large numerical values involving many zeros should be expressed as powers of ten, e.g. 3.92×10^7. Percentages may require definition—for example, in describing solutions, where percentage by weight and percentage by volume (e.g. w/w or v/v) must be distinguished.

Illustrations, tables, photographs, etc.

Illustrations are costly to reproduce and should be kept to the essential minimum. Those which have been published elsewhere must carry evidence of permission of any copyright holder. Editors and publishers usually have firm requirements for illustrations of all types. Tables, figures and photographs should be kept separate from the typescript, but their approximate position in the text should be indicated. A high standard of draughtsmanship is required for drawings, and if an author is not capable of high-quality drawings, it is better that he get a good draughtsman to do the work for him. Half-tone figures are particularly expensive and, unless absolutely necessary, should be avoided.

Proof reading

When proofs have been made, they are checked by authors and editors and corrected if necessary. At this stage new material may not be added, except to correct any errors of fact or logic that have escaped earlier notice. Standard marks for proof correcting are given in British Standard 1219 (1958)—*Recommendations for Proof Correction and Copy Preparation*. Correction should be made as legibly as possible in ink, not pencil.

APPENDIX 1 : A SELECT BIBLIOGRAPHY ON THE USE OF PUNCHED CARDS
FOR PERSONAL RECORD KEEPING

Casey, R. S. and Bailey, C. F. 'Punch Card Techniques and Applications.' *Journal of Chemical Education*, **23** (10), 495–499 (1946)
Cox, G. J., Bailey, C. F. and Casey, R. S. 'Punch Cards for a Chemical Bibliography'. *Chemical and Engineering News*, **23** (18), 1623–1626 (1945)
Cox, G. J., Casey, R. S. and Bailey, C. F. 'Recent Developments in Keysort Cards'. *Journal of Chemical Education*, **24** (2), 65–70 (1947)
Foskett, A. C. *A Guide to Personal Indexes Using Edge-Notched and Peek-a-Boo Cards*, 2nd edn. Clive Bingley (1970)
Hardy, F. E. M. 'The Capacity of Edge-Punched Cards for Information Storage'. *Office Magazine*, **5** (54), 452–453 (1958)
Hinde, P. T. 'Using Edge-Notched Cards for Personal Interest Literature Files'. *Journal of Chemical Education*, **42** (10), 566–569 (1965)
Macdougall, A. I. 'Marginal Punched Cards'. *Lancet*, **269** (6899), 1089 (1955)
Maclean Smith, J. 'Marginal Punched Cards and their Usefulness in Medical Practice'. *Lancet*, **269** (6896), 919–921 (1955)
Price, E. M. S. 'Card Indexing'. *Metal Industry*, **64** (22), 354-356 (1944)
Soper, A. K. 'Some Observations on the Use of Punched Cards for a Personal Information File'. *Aslib Proceedings*, **7** (4), 251-258 (1955)

APPENDIX 2 : SOME SUPPLIERS OF EDGE-NOTCHED CARDS

G. Anson & Company Limited, 255-259 High Road, Ilford, Essex
Copeland-Chatterson Company Limited, Stroud, Gloucestershire
Manifoldia Limited, Manisort Division, Bromford Lane, West Bromwich, Staffordshire

Royal McBee (U.K.) Limited, Lilton House, Goswell Road, London, EC1
E-Z Sort Systems Limited, 45 Second Street, San Francisco, California
94105, USA

APPENDIX 3: SOME REFERENCE WORKS ON TECHNICAL WRITING

British Standards Institution. *Recommendations for Proof Correction and
Copy Preparation*, BS 1219 (1958)
British Standards Institution. *Table of Symbols for Printers' and Authors'
Proof Corrections*, BS 1219C (1958)
British Standards Institution. *Bibliographical References*, BS 1629 (1950)
British Standards Institution. *Presentation of Numerical Values*, BS 1957
(1953)
British Standards Institution. *Letter Symbols, Signs and Abbreviations*,
BS 1991 (1967)
Cooper, B. M. *Writing Technical Reports*. Penguin (1964)
COSATI. *Guidelines to Format Standards for Scientific and Technical Re-
ports Prepared by or for the Federal Government*. CFSTI (1968)
Crouch, W. G. and Zeller, R. L. *Guide to Technical Writing*, 3rd edn. Ronald
Press (1964)
Geological Society of London. 'Notes for Authors'. *Proceedings of the Geo-
logical Society*, No. 1627 (1965).
Gilman, W. *The Language of Science: A Guide to Effective Writing*. Har-
court Brace and World (1961)
Gowers, E. *Fowler's Modern English Usage*, 2nd edn. Oxford University
Press (1965)
Graves, H. F. and Hoffman, L. S. S. *Report Writing*, 4th edn. Prentice-
Hall (1965)
Hays, R. *Principles of Technical Writing*. Addison-Wesley (1965)
Houp, K. W. and Pearsall, T. E. *Reporting Technical Information*. Glencoe
Press (1968)
Jones, W. P. *Writing Scientific Papers and Reports*, 5th edn. Wm. C. Brown
(1965)
King, L. S. and Roland, C. G. *Scientific Writing*. American Medical Asso-
ciation (1968)
Kirkpatrick, T. W. and Breese, M. H. *Better English for Technical Authors*.
Leonard Hill (1961)
Menzel, D. H., Jones, H. M. and Boyd, L. G. *Writing a Technical Paper*.
McGraw-Hill (1961)
Miller, W. J. and Saidla, L. A. *Engineers as Writers*. Van Nostrand (1953)
Mitchell, J. H. *Handbook of Technical Communication*. Wadsworth (1962)
Mitchell, J. H. *A First Course in Technical Writing*. Chapman and Hall
(1967)
Mitchell, J. H. *Writing for Professional and Technical Journals*. Wiley
(1968)
Morris, J. E. *Principles of Scientific and Technical Writing*. McGraw-Hill
(1966)
Naft, S. and De Sola, R. *International Conversion Tables*. Cassell (1965)
Nelson, J. R. *Writing the Technical Report*. McGraw-Hill (1952)
O'Hayre, J. *Gobbledygook Has Gotta Go*. USGPO (1966)
Peterson, M. S. *Scientific Thinking and Scientific Writing*. Reinhold (1961)
Rathbone, R. R. *Communicating Technical Information*. Addison-Wesley
(1966)

Rathbone, R. R. and Stone, J. B. *A Writer's Guide for Engineers and Scientists.* Prentice-Hall (1962)

Robertson, W. S. and Siddle, W. D. *Technical Writing and Presentation.* Pergamon (1966)

Royal Society. *General Notes on the Preparation of Scientific Papers.* Royal Society (1965)

Shearring, H. A. and Christian, B. C. *Reports and How to Write Them.* Allen and Unwin (1965)

Sherman, T. A. *Modern Technical Writing,* 2nd edn. Prentice-Hall (1966)

Society of Technical Writers and Publishers. *An Annotated Bibliography on Technical Writing, Editing, Graphics, and Publishing.* The Society (1965)

Strunk, W. *The Elements of Style.* Macmillan (1962)

Trelease, S. F. *How to Write Scientific and Technical Papers.* MIT Press (1969)

Turner, R. P. *Technical Report Writing.* Holt, Rinehart and Winston (1965)

Turner, R. P. *Technical Writer's and Editor's Stylebook.* Foulsham, Slough (1965)

UNESCO. *Guide for the Preparation of Scientific Papers for Publication.* UNESCO (1968)

Waldo, W. H. *Better Report Writing.* Reinhold (1965)

Ward, R. R. *Practical Technical Writing.* Knopf (1968)

Weisman, H. M. *Technical Report Writing.* Chas. E. Merrill (1966)

Woodford, F. P. *Scientific Writing for Graduate Students.* Rockefeller University Press (1968)

9

Stratigraphy (historical geology) including regional geology

W. A. S. Sarjeant and A. P. Harvey

Stratigraphy has both a narrow and a broad definition, both of which are current. In its narrower definition it is the study of strata coupled with the interpretation of the evidence derived from such study. For most European geologists, however, the scope of stratigraphy is very much broader—no less than the reconstruction of environmental conditions during all the successive periods of the earth's history. This definition makes 'stratigraphy' synonymous with the American term 'historical geology'. Geography has been defined as the 'study of the totality of conditions prevailing on the surface of the earth'. On a similar basis, one can formulate a comparable definition for stratigraphy—*the study of the totality of conditions prevailing upon the earth's surface during successive periods of the history of the earth.*

Stratigraphy thus draws upon all other branches of geology; indeed, it can be said to embrace all other aspects of the subject. (For this reason, many stratigraphic texts are entitled 'The Geology of . . .' a particular area). Geochronology, the division of stratigraphy specifically concerned with the establishment of absolute dates for geological events, nowadays relies heavily upon geophysics. The establishment of relative dates by a process of correlation between one sequence of rocks and another depends primarily on fossils (and increasingly upon microfossils)—stratigraphic palaeontology. Fossils can also be employed in the interpretation of past environmental conditions, along with evidence afforded by the sediments themselves; this branch of geology—palaeontology—is dealt with in Chapter 10. Structural geology, petrology, sedimentology and oceanography, likewise dealt with in other chapters, are also of special relevance.

Because of this broad coverage, it is understandably difficult to decide when and by whom the first stratigraphic studies were made. However, the earliest major figure in the history of stratigraphy was Nils Steensen, better known as Nicholas Steno (1638–86), who was not only the first to work out a geological succession but also the first to attempt a geological history for a region (Tuscany). The earliest geological map seems to have been that published by Christopher Packe in 1743. Another pioneer in geological map-making was Jean-Etienne Guettard (1715–86), whose map of northern France (1746) has the geological boundaries clearly drawn and who, in a later map (1751), attempted a correlation based on lithology between the rocks of France and England.

One of the two major keys to stratigraphy was provided by James Hutton (1726–97), who recognised that the processes of weathering, erosion and sedimentation in modern rivers and seas could produce such rocks as sandstones and shales and that volcanic processes could give rise to basalts, etc. He concluded that there was no need to invoke major convulsions of the earth to explain how rocks originated: that 'the present is the key to the past'. His observations and deductions were developed by Charles Lyell (1797–1875) into the *Principles of Geology* laid down in three volumes published between 1830 and 1833 (see Chapter 17). The second major key was the realisation that strata could be correlated on the basis of the fossils they contained. This was first recognised by J. L. Girard-Soulevie (1780), but his work attracted scant attention and it was left to William Smith (1769–1839) to establish this to the satisfaction of the scientific world. His maps and sections initiated intensive study of stratigraphy. Smith's principles were first employed on a universal scale by Alcide d'Orbigny (1802–57). Although this work was over-ambitious at the time when it was done, d'Orbigny's system of stratigraphic sub-divisions is now employed internationally. Modern stratigraphy stems wholly from the work done before 1850. Although techniques have improved and interpretations become more refined, the approaches initiated by these pioneers are exactly those that are employed today.

In the following sections an attempt has been made to cite the major sources of stratigraphic information. General works are considered first, followed by works concerned with world stratigraphy. Regional geology is dealt with under geographic sub-divisions, and for each country, ocean or island group the principal books, bibliographies and journals are listed. Attention has been generally confined to works in the English language or in German, Spanish and French, the three foreign languages most commonly known to

English-speaking scientists. The dates of commencement given for periodicals and series relate to the first use of that particular title, very minor changes and changes of country names being ignored. It is therefore possible that a particular journal will have a much longer run than is at first suggested. Mention of reprints, except in exceptional circumstances, has been avoided, as so many books and journals have now received this treatment. Finally the user's attention is drawn to the general introductory chapters of the book, which contain much valuable information even to those only interested in limited fields.

GENERAL WORKS

Works dealing with the principles of stratigraphy as such are remarkably few. Only two can be recommended: *Stratigraphic Principles and Practice*, by J. Marvin Weller (Harper, 1960), and *Stratigraphy: An Introduction to Principles*, by D. T. Donovan (Murby, 1966).

Geochronology

The specific aspect of geochronology is very well covered by the literature. The development of this subject is dealt with in two works: *The Discovery of Time*, by S. Toulmin and J. Goodfield (Harper, 1965), and *Growth of a Prehistoric Time Scale Based on Organic Evolution*, by W. B. N. Berry (Freeman, 1968).

The techniques of geochronology are described in detail in *Potassium-Argon Dating. Principles, Techniques and Applications to Geochronology*, by G. B. Dalrymple and M. A. Lanphere (Freeman, 1969); *Applied Geochronology*, by E. Hamilton (Academic Press, 1965); *Radiometric Dating for Geologists*, edited by E. I. Hamilton and R. M. Farquar (Interscience, 1968); and *Radiocarbon Variations and Absolute Chronology*, edited by I. U. Olsson (Almqvist and Wiksell, 1970), Nobel Symposium No. 12.

A review of the results of attempts at absolute age dating is given in *The Phanerozoic Time Scale: A Symposium*, edited by W. B. Harland *et al.* (Geological Society of London, 1964). This was in fact issued as Vol. **120S** of the *Quarterly Journal of the Geological Society of London*. However, it should be stressed that the dates allotted in this work have already been modified in varying degrees by subsequent studies as yet only recorded in research journals. More generalised treatments of geochronology are given in several works, of which the most important are: *Dating the Past: An Introduction to Geochronology*, by F. E. Zeuner (Methuen, 1952); *Time in Stratigraphy*, by A. B. Shaw (McGraw-Hill, 1964); and *Geologic Time*, by D. L. Eicher (Prentice-Hall, 1968).

Origin of the earth's atmosphere and oceans

The origin of the earth's atmosphere and oceans is still a subject of discussion. Papers on this topic will be found in a variety of journals, notably *Nature* and *Scientific American*. The only recent book devoted to the subject is *The Origin and Evolution of Atmospheres and Oceans*, edited by P. J. Brancazio and A. G. W. Cameron (Wiley, 1964).

Palaeoclimatology

The following works relate to the more general aspects of the changes of climate through geological time (palaeoclimatology): *Climate Through the Ages*, by C. E. P. Brooks (Benn, 1949); *Descriptive Palaeoclimatology*, edited by A. E. M. Nairn (Interscience, 1961); *Climates of the Past*, by M. Schwarzbach (Van Nostrand, 1963); *Problems in Palaeoclimatology*, edited by A. E. M. Nairn (Wiley, 1964); and *Causes of Climatic Change*, edited by J. M. Mitchell (American Meteorological Society, 1968). *Proceedings of the VIIth INQUA Congress*, **5**.

The only journal in the field is *Palaeogeography, Palaeoclimatology, Palaeoecology* (1965–   ).

Stratigraphical palaeontology

Particular emphasis to the relation of palaeontology and stratigraphy is given by: *Histoire Géologique de la Biosphère. La Vie et les Sédiments dans les Géographies Successives*, by H. and G. Termier (Masson, 1952); *Stratigraphical Palaeontology*, by E. Neaverson (Clarendon Press, 1955); *Stratigraphy and Life History*, by M. Kay and E. H. Colbert (Wiley, 1965); and 'Essays in Paleontology and Stratigraphy', edited by C. Teichert and E. L. Yochelson (*Special Publication, University of Kansas, Department of Geology*, **2**, 1967).

The following textbooks on stratigraphy also allocate considerable space to palaeontology. These all originate in the USA and therefore the texts have a strong bias to North America. The illustrations, however, are usually of a very high standard.

An Introduction to Historical Geology with Special Reference to North America, by W. J. Miller (6th edn., Van Nostrand, 1952)
Introduction to Historical Geology, by R. C. Moore (2nd edn., McGraw-Hill, 1958)
Principles of Stratigraphy, by A. W. Grabau (Dover Books, 1960, reprint of 1924 edition)
Historical Geology, by A. O. Woodford (Freeman, 1965)
Historical Geology, by C. O. Dunbar and D. M. Waage (3rd edn., Wiley, 1969)

Sedimentology

This particular facet of geology has a section devoted to it elsewhere in this book (see Chapter 11) but it is relevant to mention two works here. They are *Stratigraphy and Sedimentation*, by W. C. Krumbein and L. L. Sloss (Freeman, 1951), and *Ancient Sedimentary Environments. A Brief Survey*, by R. C. Selley (Chapman and Hall, 1970).

Stratigraphic techniques and mapping

The application of computer techniques to stratigraphy is discussed in *Computer Applications in Stratigraphical Analysis*, by J. W. Harbaugh and D. Merriam (Wiley, 1968).

Geological maps are dealt with in Chapter 7 but it is relevant to mention in this section that for the general techniques of geological mapping reference can still be made to three classic works: *Structural and Field Geology*, by J. Geikie (6th edn., Oliver and Boyd, 1953); *Methods in Geological Surveying*, by E. Greenly and H. Williams (Murby, 1930); and *Elements of Field Geology*, by G. W. Himus and G. S. Sweeting (3rd edn., University Tutorial Press, 1972).

Submarine geology

Submarine geomorphology and stratigraphy are attracting increasing attention and new techniques and data are continually forthcoming. Although now rather dated, two texts are worth mentioning: *Marine Geology*, by P. H. Kuenen (Wiley, 1950), and

The Earth Beneath the Sea, by F. P. Shepard (Johns Hopkins Press, 1959). For more up to date information, reference should be made to the journal *Marine Geology* (1964–   ) and the various oceanographic works (see Chapter 15).

Stratigraphic terminology and nomenclature

In recent years there has been a significant increase in the number of papers on stratigraphic classification and terminology and in the production of codes of practice. The former is the subject of three review articles: 'Remarques sur la Nomenclature Stratigraphique', by P. F. Burollet (*Sciences de la Terre*, **5**, 117–136, 1959); 'Concepts of Stratigraphical Classification and Terminology', by L. Størmer (*Earth-Science Reviews*, **1**, 5–28, 1966); and 'Stratigraphic Classification: A Critical Review', by W. J. Verwoerd (*Transactions and Proceedings of the Geological Society of South Africa*, **67**, 263–282, 1964).

A most useful paper treating with terminology is: 'Stratigraphic Terminology and Nomenclature, A Guide for Editors and Authors', by F. W. B. van Eysinga (*Earth-Science Reviews*, **6**, 267–288, 1970).

Various attempts have been made to introduce an international system of nomenclature. Among these is 'Towards an International Code of Stratigraphic Nomenclature', by P. C. Sylvester-Bradley, in *Essays in Paleontology and Stratigraphy. Raymond C. Moore Commemorative Volume* (University of Kansas Press, 1967).

Various national attempts at codes of practice include: *Code of Stratigraphic Nomenclature*, American Commission on Stratigraphic Nomenclature (American Association of Petroleum Geologists, 1970); 'A Critical Review of Stratigraphic Terminology as applied in South Africa', by J. F. Truswell (*Transactions of the Geological Society of South Africa*, **70**, 81–116, 1967); and *Recommendations on Stratigraphic Classification Prepared on Behalf of the British National Committee for Geology by the Society*, Geological Society of London (Royal Society, 1968). Reference could also be made to 'Recommendations on Stratigraphical Usage' (*Proceedings of the Geological Society of London*, No. 1656, 139–166, 1969).

The first journal devoted entirely to an exchange of information on topics of general stratigraphical interest, *Newsletter on Stratigraphy*, commenced publication in 1970 under the editorship of Gerd Lüttig. Another source of continuing information is the *Geological Newsletter* of IUGS.

Anyone who wishes to obtain a picture of the geology of a

country other than his own is greatly helped by use of the *Lexique Stratigraphique International*, a long series of inexpensive volumes initiated by the 1956 International Geological Congress (Copenhagen) and published by the Centre National de la Recherche Scientifique in Paris. These volumes not only provide an excellent dictionary of stratigraphic terms for their respective areas but also contain very useful bibliographies. The following have so far been published.

EUROPE
Volume I: 1a Greenland; 1bcd Iceland, Faeroes, Spitzbergen; 2ab Norway, Finland; 2c Sweden; 2d Denmark; 3a England, Wales, Scotland (a series of volumes dealing with different parts of the stratigraphic column not all of which have yet been published); 3b Ireland; 4a France, Netherlands, Belgium, Luxemburg (a series of volumes); 5a–h Germany; 6a Poland; 6b Czechoslovakia; 7abc Switzerland; 8 Austria; 9 Hungary; 10a Spain; 10b Portugal; 11 Italy; 12a Yugoslavia; 13a Romania; 13b Bulgaria.

USSR
Volume II: In four volumes.

ASIA
Volume III: 1 China (in two volumes); 2a Korea; 2b Manchuria; 3ab Japan–Ryukyu Islands; 4 Taiwan; 5 Philippines; 6a Indochina; 6bcd Malaysia, Thailand, Burma; 7abc Indonesia, British Borneo, Malaya; 8abc India, Pakistan, Nepal, Bhutan, Burma, Ceylon; 9a Afghanistan; 9c Turkey; 9d Cyprus; 10a Iraq; 10bI Saudi Arabia; 10bII Aden and Dhufar; 10cI Lebanon, Syria, Jordan; 10cII Israel.

AFRICA
Volume IV: 1a Morocco; 1bc Algeria, Tunisia; 2 Sahara, French and Portuguese West Africa; 3 British West Africa, Sierra Leone, Gambia [Ghana], Nigeria, Cameroun; 4a Libya; 4b Egypt, Sudan; 5 French, British and Italian Somaliland, Ethiopia, Eritrea; 6 French Equatorial Africa, French Cameroun, Spanish Guinea, Sao Tomé; 7a Belgian Congo; 7b Angola; 8a Kenya, later edition than 8ab; 8ab Kenya, Uganda; 8c Tanganyika; 9 Nyasaland, Northern and Southern Rhodesia; 10a Mozambique; 10b South Africa; 11 Madagascar; 12 General introduction to the stratigraphy of Africa; 13 Geochronology of Africa.

LATIN AMERICA
Volume V: 2a Central America; 2b Antilles; 2c Cuba; 3 Venezuela; 4a Colombia; 5a Ecuador; 5b Peru; 7 Chile; 9a Uruguay; 9bc Paraguay, Falkland Islands; 10b British, French and Dutch Guyana.

OCEANIA
Volume VI: 2 Oceania *sensu stricto*; 3a New Guinea; 4 New Zealand; 5a Australia: Queensland; 5b Australia: New South Wales; 5c Australia: Victoria; 5d Australia: Tasmania; 5e Australia: South Australia; 5f Australia: Western Australia; 5g Australia: Northern Territory.

NORTH AMERICA

Volume VII: 1 United States of America. Consists of three volumes which are reprints of 'Lexicon of Geologic Names of the United States for 1936–1960', by G. C. Keroher *et al. (Bulletin of the United States Geological Survey*, No. 1200, 1966). This in turn has been updated by a series entitled 'Changes in Stratigraphic Nomenclature', by the US Geological Survey and issued in the *Bulletin of the United States Geological Survey*. The changes for 1961–67 are cumulated in *Bulletin* No. 1350 (1970).

Also forming part of the *Lexique Stratigraphique International* is a series of works devoted to the major stratigraphic names, their significance and constituent zones entitled *Termes Stratigraphiques Majeurs* and together constituting Vol. VIII. So far published are: *Infracambrian*, by R. Furon; *Aptien*, by J. Sornay; and *Pennsylvanian*, by H. R. Wanless.

Naturally enough the *Lexique* is not the only source of information on stratigraphic terms, other lists having been published as books and as papers in journals. These include 'Annotated Lexicon of Quaternary Stratigraphical Nomenclature in East Africa', by W. W. Bishop, in *Background to Evolution in Africa*, edited by W. W. Bishop and J. D. Clark (University of Chicago Press, 1967); and *Indice Bibliografico de Estratigrafía Argentina*, by A. V. Borrello (Comisión de Investigación Cientifíca, La Plata, 1965).

WORLD STRATIGRAPHY AND THE GEOLOGIC SYSTEMS

The first great attempt at an account of the stratigraphy of the earth as a whole was made by Edvard Suess in the closing years of the nineteenth century. His work was originally published in German, but appeared in English as *The Face of the Earth* (Clarendon Press, 1904–24).

The subsequent vast expansion of knowledge has made the task of synthesising information on world stratigraphy increasingly difficult. Only two authors have attempted this single-handed (though Kummel's work is too much orientated towards undergraduate teaching to altogether qualify). The works are: *Stratigraphic Geology*, by M. Gignoux (Freeman, 1955), and *History of the Earth*, by B. Kummel (2nd edn., Freeman, 1970). Kummel presents some very useful international correlation charts but his work has a strong North American slant. Gignoux's book, a somewhat shaky translation from the original French, is more international in outlook.

Attempts at the almost equally difficult task of synthesising in greater detail the stratigraphy of the different geological systems were begun by F. A. Quenstedt in his classic *Der Jura* (1858). At present two series of volumes—one entitled *The Geologic Systems* and edited by K. Rankama (Interscience), the other *Handbuch der Stratigraphischen Geologie*, under the editorship of F. Lotze and published by Enke in Stuttgart—are attempting a comprehensive coverage of the whole geological column, though neither is yet complete. In addition, particular systems have been the subject of colloquia at national or international level. Books and international colloquia together with the few journals are listed below.

Quaternary

BOOKS: *The Quaternary Era, with Special Reference to its Glaciation.* I. *Geomorphology.* II. *Palaeontology, Cultures, Stratigraphy*, by J. K. Charlesworth (2 vols., Arnold, 1957)
Ice Ages. Their Nature and Effects, by I. Cornwall (Baker, 1970)
Glacial and Quaternary Geology, by R. F. Flint (Wiley, 1971)
The Quaternary of the United States, by D. G. Frey and H. E. Wright (Princeton University Press, 1965)
Histoire des Congrés, by M. I. Neustadt. Issued as a supplement to the journal *Bulletin de l'Association Française pour l'Etude Quaternaire*, 1969. This is a complete bibliography of all the works published by and for the various INQUA congresses and is therefore an important source for information about publications dealing with Quaternary matters
The Geologic Systems. The Quaternary, edited by K. Rankama. So far published are Vol. 1. *Denmark, Norway, Sweden, Finland* and Vol. 2. *British Isles, France, Germany and the Netherlands* (Interscience, 1965–67)
Pleistocene Geology and Biology with Special Reference to the British Isles, by R. G. West (Longmans, 1968)
Das Eiszeitalter. Grundlinien einer Geologie des Quartärs, by P. Woldstedt (3 vols., Enke, 1961–65)
Handbuch der Stratigraphischen Geologie. 2. Quartär, by P. Woldstedt (Enke, 1969)
Quaternary Geology and Climate, edited by H. E. Wright (National Academy of Sciences, 1969). *Proceedings of the VII INQUA Congress*, **16**
The Pleistocene Period; Its Climate, Chronology and Faunal Successions, by F. E. Zeuner (Hutchinson, 1959)
JOURNALS: *Quartär* (1938–)
Quaternaria (1954–)
Quaternary Research (1970–)
The Quaternary Research (1957–)

Tertiary

BOOKS: *Handbuch der Stratigraphischen Geologie. 3/II. Tertiär*, by A. Papp and E. Thenius (Enke, 1959)
Fundamentals of Mid-Tertiary Stratigraphical Correlation, by F. Eames *et al.* (Cambridge University Press, 1962)

'Colloque sur l'Eocène (Paris, May 1968)' (*Mémoires du Bureau de Recherches Géologiques et Minières*, **58–59, 69,** 1968–69)
'Colloque sur le Paléogène (Bordeaux, September 1962)' (*Mémoires du Bureau de Recherches Géologiques et Minières*, **28,** 1964)

Cretaceous

BOOKS: *El Sistema Cretácico; Un Symposium Sobre el Cretácico en el Hemisferio Occidental y su Correlación Mundial*, edited by L. B. Kellum (International Geological Congress, 20th Session, Mexico, 1956, 1959–). Three volumes have been issued so far: I. *Etapys y Correlación—Europa*; II. *Africa, Asia, Lejano Oriente*; III. *Groenlandia, America, Antillas*

Jurassic

BOOKS: *Jurassic Geology of the World*, by W. J. Arkell (Oliver and Boyd, 1956)
Handbuch der Stratigraphischen Geologie. 4. Der Jura, by H. Hölder (Enke, 1964)
Colloque du Jurassique (Luxemburg, 1962), edited by P. L. Maubeuge (Ministère des Arts et des Sciences, Luxemburg, 1964)

Permian and Triassic

BOOK: *The Permo-Triassic Formations: A World Review*, by R. L. Sherlock (Hutchinson, 1948)

Carboniferous

BOOKS: *Compte Rendu. Congrès pour l'Avancement des Etudes de Stratigraphie (et de Géologie) Carbonifère.* This is a series of conference proceedings dating from 1927, the last being the sixth congress held in Sheffield, England, in 1967, proceedings being published in several volumes from 1969 onwards
'Colloque sur la Stratigraphie du Carbonifère, Liege 13–20 Avril 1969', edited by M. Steel and R. H. Wagner (*Les Congrés et Colloques de l'Université de Liege*, **55,** 1970)
JOURNAL: *Bibliography of Carboniferous Geology* (1959–)

Silurian–Devonian

BOOKS AND REVIEWS: *International Symposium on the Devonian System* (Alberta Society for Petroleum Geology, 1967)
2. Internationale Arbeitstagung über die Silur/Devon Grenze und die Stratigraphie von Silur und Devon. Bonn-Bruxelles, 1960. Symposiums-band, edited by H. K. Erben (Schweizerbart'sche, 1962)

'Colloque sur le Devonien Inférieur et ses Limites', by R. Laffitte (*Mémoires du Bureau de Recherches Géologiques et Minières*, **33**, 1967)
'The Series of the Redefined Silurian System', by A. Martinsson (*Lethaia*, **2**, 153–161, 1969)

Cambrian

BOOKS: *Cambrian Geology and Paleontology*, by C. D. Walcott (2 vols., Smithsonian Institution, 1910). Almost entirely concerned with North America
El Sistema Cambrico; Su Paleogeografía y el Problema de su Base: Symposium (2 vols., International Geological Congress, 20th Session, Mexico, 1956)
Cambrian of the New World, edited by C. H. Holland (Wiley/Interscience, 1971). This is the first of a series entitled *Lower Palaeozoic Rocks of the World*

Pre-Cambrian

BOOKS: *Handbuch der Stratigraphischen Geologie. 12. Präkambrium. I. Nordliche Halbkugel*, edited by F. Lotze and K. Schmidt; II. *Südliche Halbkugel*, edited by P. Bankwitz et al. (Enke, 1966–68)
The Geologic Systems. The Precambrian, edited by K. Rankama (Interscience, 1963–70). So far published are: I. *Denmark, Norway, Sweden, Finland*; II. *Spitsbergen, British Isles, Greenland, Canada*; III. *India, Ceylon, Seychelles, Madagascar, Congo, Rwanda, Burundi*; IV. *United States of America and Mexico*

REGIONAL GEOLOGY

INTERNATIONAL

There is only one series, *Beiträge zur Regionalen Geologie der Erde* (1961–), which attempts to be a world regional geology. Many national journals, however, have numbers of papers on areas other than that in which they are published—for example, *Quarterly Journal of the Geological Society of London, Geological Magazine* and *Bulletin of the American Association of Petroleum Geologists*.

The various International Geological Congresses have produced series of volumes which usually include papers dealing with the areas or region in which the congress was held. The first congress was held in 1878 and a list of the more recent meetings is given below. Each congress has been responsible for issuing a number of volumes of reports as well as guidebooks to the various excursions undertaken.

XV Congress, South Africa, 1929
XVI Congress, USA, 1935
XVII Congress, USSR, 1937
XVIII Congress, Great Britain, 1948
XIX Congress, Algeria, 1952
XX Congress, Mexico, 1956
XXI Congress, Denmark, 1960
XXII Congress, India, 1964
XXIII Congress, Czechoslovakia, 1968
XXIV Congress, Canada, 1972

Other international conferences also make their impact on the literature: for example, *Gondwana Stratigraphy, IUGS Symposium in Buenos Aires, 1967* (UNESCO, 1969) and the similar *Reviews Prepared for the First Symposium on Gondwana Stratigraphy* (IUGS, ?1970). There is even a newsletter which aims at being useful bibliographically, *Gondwana Newsletter* (1969–).

The geology of the former colonial empires of the western European countries was treated in a number of books and the national journals of those countries. For the British Empire, the most comprehensive work is *The Geology of the British Empire*, by F. R. C. Reed (2nd edn., Arnold, 1949). Important journals covering the British Empire are *Bulletin of the Imperial Institute* (1903–1948), which was continued as *Colonial Geology and Mineral Resources* and finally as *Overseas Geology and Mineral Resources* (1958–). As well as containing relevant papers, excellent bibliographies will be found in some issues.

The stratigraphy of the former French colonies is treated in *Bibliographie Géologique de la France d'Outre-Mer*, compiled by F. Blondel and G. Dumain (2 vols., Bureau d'Etudes Géologiques et Minières Coloniales, Paris, 1941–52).

The geology of the Portuguese colonies is to be found in *Bibliógrafia Geologica do Ultramar Português*, compiled by F. Gonçalves and J. Caseiro (Junta de Investigações do Ultramar, Lisbon, 1959) and in the journal *Memórias Junta de Investigações do Ultramar, Serie Geologica* (1945–).

The literature on regional stratigraphy is remarkably diffuse in the European countries, the USA and the USSR but much sparser in other parts of the world. In most countries maintaining an official geological survey the publications of that body are the prime sources of information. In the section that follows, books, current journals and bibliographies are listed; however, in the case of the UK the treatment is more detailed in order to show the diffuse nature of the literature of stratigraphy and regional geology.

A useful bibliography of bibliographies will be found in the section entitled 'Compilation of Bibliographies of Use to Paleontologists and Stratigraphers', by B. Kummel in *Handbook of Paleontological Techniques*, edited by B. Kummel and D. Raup (Freeman, 1965).

EUROPE

BOOKS: *Geologie von Europa*, by S. von Bubnoff (3 vols., Borntraeger, 1926–35)
Geologic Evolution of Europe, by R. Brinkmann (Enke, 1960)
The Geology of Western Europe, by M. G. Rutten (Elsevier, 1969). This includes a section entitled 'Sources of Geological Information in Western European Countries'
Geologie der Alpen. Stratigraphie, Paläogeographie, Tektonik, by M. P. Gwinner (Schweizerbart'sche, 1971)
Various major areas of Europe are treated within the following works:

Handbuch der Regionalen Geologie. IV, pt. 3. *Fennoskandia (Norwegen, Schweden, Finnland)*, by A. G. Hogböm (Winter, 1913)
Géologie de la Méditerranée Occidentale. A series of publications resulting from the 14th International Geological Congress of 1926 and published from 1929
Geologie von Mitteleuropa, by P. Dorn (3rd edn., Schweizerbart'sche, 1966)
'Geological Perspectives in the North Sea Area', by T. Sörgenfrei (*Meddelelser fra Dansk Geologisk Forening*, **19**, 160–169, 1969)

ALBANIA

BIBLIOGRAPHIES: *Bibliografia Geologica e Geografico-Fisica della Regione Albanese*, by M. Magnani (2nd edn., Reale Officio Geologico d'Italia, 1941). A supplement is provided by 'Bibliografi e Botimeve Shqipe për Gjeolgjinë Minierat dhe Nafter 1944–1969', by V. Meko and A. Papa (*Përmbledhje é Studimesh*, **2** (15), 129–156, 1970)
JOURNALS: *Anuar Fakulteti i Gjeologjisë dhe i Minierave* (1962–   )
Buletini Universitetit Shtetëror të Tiranës (1962–   )
Përmbledhje Studimesh, Instituti i Studimeve dhe Projektimeve Gjeologo-Minerale (1965–   )

AUSTRIA

BOOKS: *Handbuch der Regionalen Geologie.* II, pt. 5a. *Die Österreichischen und Deutschen Alpen bis zum Alpino-Dinarischen Grenze (Ostalpen)*, by F. Heritsch (Winter, 1915)
Geologie von Österreiche, edited by F. X. Schaffer (2nd edn., Deuticke, 1951)
BIBLIOGRAPHY: 'Geologische Literatur in Österreich 1945–1950', by C. Exner and J. Windbrechlinger (*Verhandlungen der Geologischen Bundesanstalt*, Sonderbad B, 1951). This is regularly updated in *Verhandlungen der Geologischen Bundesanstalt* (1945–   )
JOURNALS: *Abhandlungen der Geologischen Bundesanstalt* (1852–   )
Annalen des Naturhistorischen Museums in Wien (1866–   )
Jahrbuch der Geologischen Bundesanstalt (1850–   )
Mitteilungen der Geologischen Gesellschaft in Wien (1908–   )
Verhandlungen der Geologischen Bundesanstalt (1945–   )

Note. Earlier issues, pre-1918, of the above journals also contain papers dealing with areas of Italy, Yugoslavia, Poland and Czechoslovakia which were then part of the Austro-Hungarian Empire.

BELGIUM

BOOKS: *Prodrome d'une Description Géologique de la Belgique*, edited by P. Fourmarier (Société Géologique de Belgique, 1954)
BIBLIOGRAPHY: *Abstracts of Belgian Geology and Physical Geography, 1967–* , edited by L. Walschot (Geological Institute of the University, Ghent, 1969–)
JOURNALS: *Annales de la Société Géologique de Belgique* (1874–)
Bulletin de la Société Belge de Géologie de Paléontologie et d'Hydrologie (1887–)
Mémoires de l'Institut Géologique de l'Université de Louvain (1913–)
Mémoires pour servir à l'Explication des Cartes Géologiques et Minières de la Belgique (1955–)
Publications. Centre National de Géologie Houillère (1959–)

BULGARIA

BIBLIOGRAPHY: *Abstracts of Bulgarian Scientific Literature, Geology and Geography, 1957–* (1959–)
'Prêglad na Literaturata po Geologiyata Paleontologiyata, Mineralogiyata, Petrografiyata . . . na Bŭlgariya 1828–1928', by N. Nikolov and W. G. Radev (*Sbornik na Bŭlgarskata Akademiya na Naukite*, **23**, 1928)
JOURNALS: *Izvestiya na Geologicheskiya Institut* (*Strashimir Dimitrov*) (1951–). Currently issued in several series
Spisanie na Bŭlgarskogo Geologichesko Druzhestvo (1927–)

CYPRUS

REVIEW: 'A Synopsis of the Stratigraphy and Geological History of Cyprus', by F. R. S. Henson *et al.* (*Quarterly Journal of the Geological Society of London*, **105**, 1–41, 1949)
BIBLIOGRAPHY: 'A Revised Bibliography of Cyprus Geology', by Th. M. Pantazis (*Bulletin of the Geological Survey of Cyprus*, **2**, 57–81, 1969)
JOURNALS: *Bulletin of the Geological Survey of Cyprus* (1963–)
Memoirs of the Geological Survey of Cyprus (1959–)
Reports of the Geological Survey of Cyprus (1955–)

CZECHOSLOVAKIA

BOOKS: *Regional Geology of Czechoslovakia*, edited by J. Svoboda *et al.* (3 vols., Geological Survey of Czechoslovakia, Prague, 1966–68). I. *The Bohemian Massif.* II. *The West Carpathians.* III. *Maps*
BIBLIOGRAPHY: *Mineralogicko Geologická Bibliografie CSSR 1930–* (Ustrendi Ustav Geologicky, 1934–). An attempt is now being made to cover the pre-1930 literature, volumes with the same title having been issued for 1897–1918 (1969) and 1919–27 (1970)
JOURNALS: *Acta Geologica et Geographica Universitatis Comenianae. Geologica* (1958–)
Acta Universitatis Carolinae. Geologica (1954–)
Časopis pro Mineralogii a Geologii (1956–)
Geologické Práce Spravy (1954–)
Geologický Sborník (1950–)
Knihovna Ústredního Ústavu Geologického (1949–)
Rozpravy Ústredního Ústavu Geologického (1950–)

Sborník Géologických Věd. Rada G. Geologie (1963–)
Sborník Géologických Věd. Rada ZK. Západné (1964–)
Věstník Ústredního Ústavu Geologického (1951–)
*Zprávy o Géologických Výskumoch. Geologický Ustav Dionýza Stúra.
Bratislava* (1963–)
Zprávy o Géologických Výzkumoch Vroce (1955–)

Many works concerned with the stratigraphy of Czechoslovakia are included among the publications of the XXIII International Geological Congress held in Prague in 1968.

DENMARK
BOOKS: *Handbuch der Regionalen Geologie.* I, pt. 2. *Dänemark,* by N. V. Ussing (Winter, 1910)
Summary of the Geology of Denmark, by J. Anderson *et al.* (Danmarks Geologiske Undersøgelse, Copenhagen, 1928)
Both these works are now rather out of date.
BIBLIOGRAPHY: 'Litteraturfortegnelse Omfattende Skrifter af Geologisk eller Lignende Natur og som ved Emne Forfatter eller Udgivelsessted er Knyttede til Danmark og Grønland samt Island'. (List of Papers of Geology and Similar Matters which by Subject, Author and Place of Publication are Related to Denmark, Greenland and Iceland). In *Bulletin of the Geological Society of Denmark* (1904–)
JOURNALS: *Meddelelser fra Dansk Geologisk Forening* (1894–)
Danmarks Geologiske Undersøgelse, Series I to V (1890–)

In addition the publications of the XXI International Geological Congress (held in Copenhagen) include many works relating to Danish stratigraphy.

FINLAND
BOOKS: Apart from *Suomen Geologia,* by K. Rankama (Kirjayhtymä, 1964), which is in Finnish, no work deals with the stratigraphy of the whole country.
BIBLIOGRAPHIES: *Guide to the Publications of the Geological Survey of Finland 1879–1960,* by M. Okko and M. Hannikainen (Geologinen Tutkimuslaitos, 1960)
'Geologische Bibliographie Finnlands 1555–1933', by A. Laitakari (*Bulletin de la Commission Géologique de Finlande,* **108**, **231**, 1934, 1968)
JOURNALS: *Bulletin de la Commission Géologique de la Finlande* (1895–)
Bulletin of the Geological Society of Finland (1968–)
Suomen Geologinen Kartta (1952–)

FRANCE
BOOKS: The most useful comprehensive treatments of French stratigraphy are:

Géologie de la France, by L. de Launay (Librairie Armand Colin, 1921)
Géologie de la France, by R. Abrard (Payot, 1948)
Géologie de la France, by J. Goguel (Presses Univ. Français, 1950)

There are two series of commercially published books covering the geology of various regions of France:

Géologie Régionale de la France (Hermann, 1942–)
Guides Géologiques Régionaux (Masson, 1968–)

The former has volumes devoted to the Paris basin, Provence, La Touraine, Aquitain, Le Bas-Vivarais, Normandy, Alps of Provence, Le Massif des Maures de Toulin à Saint Raphael, and L'Esterel et la Massif de Tanneron; the latter, at present, the Paris basin, Alpes-Savoie et Dauphine, Provence, Bourgogne Morvan and Massif Central.

Other works on the important Paris basin region are:

Région de Paris. Excursions Géologiques et Voyages Pédagogiques, by A. F. de Lapparent (Hermann, 1964)
Geological Guidebook. International Field Institute. Paris Basin (National Academy of Sciences, 1965)

Official publications include *Bulletin du Service de la Carte Géologique de la France* (1889–), explanations provided with the map sheets, and the massive *Mémoires pour Servir à l'Explication de la Carte Géologique Détaillée de la France* (1874–).

A series of colloquia on particular parts of the geological column have been held and are planned for the future. The resultant volumes, some of which have been listed earlier, are in most cases international in scope albeit with France prominent. The following, however, are almost exclusively devoted to France:

'Colloque sur le Trias de la France et des Régions Limitrophes' (*Mémoires du Bureau de Recherches Géologique et Minières*, **15**, 1963)
'Colloque sur le Lias Français', edited by J. Roger (*Mémoires du Bureau de Recherches Géologiques et Minières*, **4**, 1961)
'Colloque sur le Crétacé Inférieur' (*Mémoires du Bureau de Recherches Géologiques et Minières*, **34**, 1965)
Colloque sur le Crétacé Supérieur Français. Compte Rendu du Congrés des Sociétés Savantes, Dijon, 1959 (Gauthier-Villars, 1959)

BIBLIOGRAPHIES: Three bibliographies may prove of use to those interested in the geology of France:

'Bibliographie Géologique et Minière des Départements du Puy-de-Dôme du Cantal de la Haute-Loire et de l'Allier' (*Annales de la Faculté des Sciences Université de Clermont-Ferrand*, **1**, 1959)
'Bibliographie Géologique du Centre-Ouest de la France' (*Travaux de l'Institut de Géologie et d'Anthropologie Préhistorique*, **3**, 1962)
'Bibliographie Géologique, Minéralogique et Paléontologique du Nord-Ouest de la France (Bretagne, Basse-Normandie, Maine, Anjou, Vendée)'. The latest part is in *Bulletin de la Société des Sciences Naturelles de l'Ouest de la France*, **66** (1968). A full list of previous bibliographies is also given
JOURNALS: *Annales Scientifiques de l'Université de Besançon. Series 2. Géologie* (1950–)
Annales de la Société Géologique du Nord (1870–)
Bulletin du Bureau de Recherches Géologiques et Minières (1961–)
Bulletin du Centre de Recherches de Pau (1967–)
Bulletin d'Information des Géologues du Bassin de Paris (1964–)
Bulletin de l'Institut de Géologie du Bassin d'Aquitaine (1966–)
Bulletin du Service de la Carte Géologique d'Alsace et de Lorraine (1920–)
Bulletin du Service de la Carte Géologique de la France (1889–)
Bulletin de la Société Géologique de France (1830–)
Bulletin de la Société Géologique et Minéralogique de Bretagne (1920–)
Cahiers Géologiques (1950–)

Compte Rendu Sommaire des Séances de la Société Géologique de France (1890–)
Documents du Laboratoire de Géologie de la Faculté des Sciences de Lyon (1962–)
Géobios (1968–)
Géologie Alpine (1967–)
Mémoires pour Servir à l'Explication de la Carte Géologique Détaillée de la France (1874–)
Mémoires de la Société Géologique de France (1833–)
Mémoires de la Société Géologique et Minéralogique de Bretagne (1924–)
Nouvelles Archives du Muséum d'Histoire Naturelle de Lyon (1946–)
Travaux de l'Institut de Géologie et d'Anthropologie Préhistorique de la Faculté des Sciences de Poitiers (1959–)

Many papers of importance will also be found in general natural history and science journals and there are in fact many other geological publications in France besides those listed above.

GERMANY
BOOKS: The following works summarise the geology of Germany:

Geologie von Deutschland und den Angrenzenden Gebieten, by R. Lepsius (2 vols., Engelmann, 1887–1910)
Grundzüge der Geologie von Deutschland, by W. von Seidlitz (Fischer, 1933)
Geologie von Deutschland und Einigen Randgebieten, by G. Knetsch (Enke, 1963)
Grundriss der Geologie der Deutschen Demokratischen Republik (Akademie Verlag, 1968–)

Geological surveys in Germany were originally instituted by the separate states prior to unification and responsibility for geological mapping remained with these state surveys until brought under a central authority, the Reichs-stelle für Bodenforschung, in 1938. Since the Prussian survey had been the largest, its serial publications were chosen for continuation after this consolidation. Since the collapse of the Nazi state in 1945 the Western and Eastern divisions of Germany have followed dissimilar geological paths. In the East (DDR) there is a unified state survey, the Geologische Landes-sanstalt, while in the West (DFR) the state surveys have been reactivated. In consequence of its complex political history, Germany has produced an extremely large number of geological publications, many of them short-lived.

BIBLIOGRAPHIES: Bibliographic coverage of the country is really very poor, continuing bibliographies being hidden in various journals. A few examples are given below:

'Die Geologische Literatur über Sachsen', in *Jahrbuch des Staatlichen Museums für Mineralogie und Geologie zu Dresden*
'Geologisches Schrifttum über Nordost-Bayern', in *Geologische Blätter für Nordost-Bayern*
'Hessisches Geologisches Schrifttum', in *Notizblatt des Hessischen Landes-amtes für Bodenforschung zu Wiesbaden*

JOURNALS: *Abhandlungen der Deutschen Akademie der Wissenschaften zu Berlin, Klasse für Chemie, Geologie und Biologie* (1955–)

Abhandlungen des Geologischen Landesamtes in Baden Württemburg (1953–)
Abhandlungen der Hessischen Landesamtes für Bodenforschung (1950–)
Abhandlungen des Staatlichen Museums für Mineralogie und Geologie zu Dresden (1966–)
Abhandlungen des Zentralen Geologischen Institut (1965–)
Bericht der Geologischen Gesellschaft in der Deutschen Demokratischen Republik. Series A & Sonderheft (1955– , 1963–)
Clausthaler Geologische Abhandlungen (1964–)
Erlanger Geologische Abhandlungen (1952–)
Freiberger Forschungshefte (1952–)
Fortschritte in der Geologie von Rheinland und Westfalen (1958–)
Geologica Bavarica (1952–)
Geologica et Palaeontologica (1967–)
Geologie (1962–)
Geologische Blätter für Nordost-Bayern und Angrenzende Gebiete (1951–)
Geologische Rundschau (1951–)
Geologisches Jahrbuch (1943–)
Göttinger Arbeiten zur Geologie und Paläontologie (1969–)
Hallesches Jahrbuch für Mitteldeutsche Erdgeschichte (1949–)
Jahrbuch für Geologie (1967–)
Jahresbericht und Mitteilungen des Oberrheinischen Geologischen Vereins (1911–)
Jahresbericht des Niedersächsischen Geologischen Vereins (1908–)
Jahreshefte des Geologischen Landesamt in Baden-Württemburg (1955–)
Meyniana (1952–)
Mitteilungen der Bayerischen Staatssammlung für Paläontologie und Historische Geologie (1961–)
Mitteilungen aus dem Geologischen Staatsinstitut in Hamburg (1935–40, 1949–)
Mitteilungen der Zentralen Geologischen Institut (1965–)
Neues Jahrbuch für Geologie und Paläontologie. Abhandlungen and *Monatshefte* (1950–)
Notizblatt des Hessischen Landesamtes für Bodenforschung zu Wiesbaden (1950–)
Oberrheinische Geologische Abhandlungen (1940–)
Zeitschrift der Deutschen Geologischen Gesellschaft (1849–)
Zitteliana. Abhandlungen der Bayerischen Staatssammlung für Paläontologie und Historische Geologie (1969–)

The following state surveys produce maps but not a regular publication: Geologisches Landesamt Hamburg, Geologisches Landesamt Rheinland-Pfaltz, Geologisches Landesamt des Saarlandes and Geologisches Landesamt Schleswig-Holstein.

GIBRALTAR
Gibraltar is adequately treated in *The Geology of the British Empire*, by F. R. C. Reed (2nd edn., Arnold, 1949)

GREECE
BOOKS: *Guide to the Geology and Culture of Greece*, edited by P. Norton (Petroleum Exploration Society of Libya, 1965)
Etude Géologique de l'Epire (Grèce Nord-Occidentale) (Technip, 1966)

BIBLIOGRAPHY: *Geological and Physicogeographical Bibliography of Greece*, compiled by D. Haralambous (Institute for Geology and Subsurface Research, 1961)
JOURNALS: *Geological and Geophysical Research* (1951–)
Geology of Greece (Eidikai Meletai epi tes Geologias tos Hellados) (1951–)

HUNGARY

BOOKS: *Regionalen Geologie der Erde*, 8. *Geologie von Ungarn*, by L. Trunko (Borntraeger, 1969)
BIBLIOGRAPHY: 'Repertoire Bibliographie des Publications du Domaine des Sciences Géologiques en Hongrie'. This has been issued, under various titles, since 1900 in *Földtani Közlöny*
JOURNALS: *Acta Geologica, Academiae Scientiarum Hungaricae* (1952–)
Annales Universitatis Scientiarum Budapestinensis de Rolando Eötvös Nominatae. Sectio Geologica (1957–)
Évi Jelentés à Magyar Allami Földtani Intézet (1925–)
Ezhegodnik Vengerskogo Geologicheskogo Instituta (1960–)
Földtani Közlöny (1871–)
Geologica Hungarica, Seria Geologica (1914–)

IRELAND

BOOKS: *The Geology of Ireland: An Introduction*, by J. K. Charlesworth (Oliver and Boyd, 1953)
Historical Geology of Ireland, by J. K. Charlesworth (Oliver and Boyd, 1963)
Regional Geology of Northern Ireland, by H. E. Wilson (Geological Survey of Northern Ireland, 1972)

The Geological Survey of Ireland, Dublin, issues *Sheet Memoirs* to accompany its one inch to one mile geological maps; it also issued a series of *Emergency Period Pamphlets* during and immediately after World War II. Other publications are listed below.
BIBLIOGRAPHY: 'The Geology of Northwest and Central Donegal', by W. S. Pitcher and M. O. Spencer (*Proceedings of the Geological Society of London*, No. 1645, 332–337, 1968)
JOURNALS: Apart from the *Bulletin of the Geological Survey of Ireland* mentioned above, there are no journals specifically concerned with Irish geology. The ones listed below, however, will be found to contain significant geological papers:

Bulletin Geological Survey of Ireland (1970–)
Information Circulars Geological Survey of Ireland (1969–)
Irish Geography (1947–)
Irish Naturalists' Journal (1925–)
Proceedings of the Royal Irish Academy, B (1902–)
Scientific Proceedings of the Royal Dublin Society A (1877–)
Special Paper Geological Survey of Ireland (1971–)

ITALY

BOOKS: *Italian Mountain Geology*, by C. S. Du Riche Preller (3 parts in 2 vols., Wheldon and Wesley, 1923–24). Parts I-II. *Northern Italy and Tuscany*. Part III. *Central and Southern Italy*

Italy 1964. International Field Institute Guidebook (American Geological Institute, 1964)
Le Dolomiti: Geologia dei Monti tra Isarco e Piave, by P. Leonardi (2 vols., Consiglio Nazionale delle Richerche, Giunta Provinciale di Trento, 1967). This contains English summaries
'Geology and History of Sicily' (*Annual Field Conference Petroleum Society of Libya*, **9**, 1967)
BIBLIOGRAPHIES: 'Bibliografia Geologica Italiana', in *Giornale di Geologia*, **6** (2), 1931. This covers the literature up to 1930

A most useful series which can be used quite readily by non-Italian speakers is:

Bibliografia Geologia d'Italia (Consiglio Nazionale delle Richerche, 1956–). I. *Lazio*, II. *Lombardia*, III. *Campania*, IV. *Calabria*, V. *Paglia*, VI. *Friuli*, VII. *Umbria*, VIII. *Sicilia*, IX. *Marche*
JOURNALS: *Atti dell' Istituto di Geologia della Università di Genova* (1963–)
Atti dell' Istituto Geologico della Università di Pavia (1943–)
Bollettino, Servicio Geologico d'Italia (1945–)
Bollettino della Società Geologica Italiana (1882–)
Formazioni Geologiche. Studi Illustrativi della Carta Geologico d'Italia 1968–)
Geologica Romana (1962–)
Giornale di Geologia (1926–)
Memorie Descrittive della Carta Geologica d'Italia (1886–)
Memorie Geopaleontologiche della Università di Ferrara (1964–)
Memorie dell' Istituto Geologico e Mineralogico della Università di Padova (1912–)
Memorie della Società Geologica Italiana (1933–)
Revista Italiana di Palaeontologia et di Stratigrafia (1895–); *Memoria* (1934–)

LUXEMBURG
JOURNALS: *Bulletin du Service de la Carte Géologique de Luxembourg* (1937–)
Publications du Service de la Carte Géologique de Luxembourg (1937–)

MALTA
BOOKS: *Geology of the Maltese Islands*, by H. P. T. Hyde (Author, 1955)
'Geology and Structure of the Maltese Islands', by M. R. House *et al.*, in *Malta: Background for Development* (Newcastle, 1961)

NETHERLANDS
BOOKS: *Geologie van Nederland*, by F. J. Faber (4 vols., J. Noorduijn en Zoon, 1948–60)
Geological History of the Netherlands: Explanation to the General Map of the Netherlands: on the Scale of 1:20 000, by A. J. Pannekoek (Nederlandsch Geologisch Mijnbouwkundig Genootschap, 1956)

In addition reference should be made to *Geologie en Mijnbouw*, **13** (6) (1951), which is a special issue devoted to Dutch stratigraphy.
BIBLIOGRAPHY: An example of the bibliographic coverage of Dutch geology

is afforded by 'Geological Bibliography of the Netherlands 1949–1964: Geological Bibliography of the Netherlands 1968' (*Mededelingen Rijks Geologische Dienst. N.S.*, **19**, 1968). The first of these contains details of previous compilations and there have been supplements in this journal. At present the supplements have separate pagination and an index will be provided in the future.

JOURNALS: *Geologie en Mijnbouw* (1939–)
Grondboor en Hamer (1960–)
Jaarverslag van de Geologische Stichting (1963–)
Leidse Geologische Mededelingen (1925–)
Mededelingen van de Geologische Stichting (1941–)
Mededelingen van de Werkgroep voor Tertiaire en Kwartaire Geologie (1964–)
Verhandelingen der Koninklijke Nederlandsche Akademie van Wetenschappen, Aftdeling Natuurkunde (1936–)
Verhandelingen van het Koninklijke Nederlands Geologisch-Mijnbouwkundig Genootschap (1955–)

NORWAY

BOOKS: 'The Geology of Norway', by O. Holtedahl (*Norges Geologiske Undersøgelse*, **208**, 1960)
BIBLIOGRAPHY: 'Litterature Géologique Concernant Norvège avec Svalbarde et la Térritoire des Recherches de la Mer Glaciale et la Mer Norvégienne et Groenlandienne', compiled by H. Rosendahl (*Norsk Geologisk Tidsskrift*, **13**, 129–301, 1934)
JOURNALS: *Avhandlinger ugitt av det Norske Videnskapsakademi i Oslo. Matematisk-Naturvidenskapelig Klasse* (1925–)
Arbok for Universitetet i Bergen, Matematisk-Naturvitenskapelig Serie (1961–)
Norges Geologiske Undersøgelse (1891–)
Norsk Geologisk Tidsskrift (1905–)

POLAND

BOOKS: *Geology of Poland*, edited by S. Sokołowski (Geological Institute, (1970–). Volume I, *Stratigraphy*, part I, *Pre-Cambrian and Palaeozoic*. Further volumes are planned. Volume II will cover palaeontology
BIBLIOGRAPHY: *Retrospektywna Bibliografia Geologiczna Polski*, by R. Fleszarowa (Wydanwnictwa Geologiczne, 1957–). The literature from 1750 is covered and the bibliography can be used (albeit with some difficulty) by non-Polish readers
Bibliografia Geologiczna Polski (1914–). Issued irregularly
JOURNALS: *Acta Geologica Polonica* (1950–)
Annales Universitatis Mariae Curie-Skłodowska. B. Geographia, Geologia, Mineralogia et Petrographia (1946–)
Biuletyn Instytut Geologiczny (1938–)
Bulletin de l'Académie Polonaise des Sciences. Série des Sciences Géologiques et Géographiques (1960–)
Geologia Sudetica (1964–)
Kwartalnik Geologiczny (1957–)
Prace Geologiczne Polska Akademia Nauk Oddział w Krakowie (1961–)
Prace. Instytut Geologiczny. Warszawa (1953–)
Przeglad Geologiczny (1953–)
Rocznik Polskiego Towarzystwa Geologicznego (1923–)
Studia Geologica Polonica (1958–)

Studia Societatis Scientiarum Torunensis. Sectio C. Geographia et Geologia (1953–)

PORTUGAL

No comprehensive treatment of Portuguese geology is currently available.
BIBLIOGRAPHIES: A very sound bibliographic coverage, quite readily comprehensible to non-Portuguese readers, is: *Geologia de Portugal. Ensaio Bibliográfico*, by L. de Menezes Acciaouli (2 vols., Direcção-General de Minas e Serviços Geológicos, 1957). Also of value is: *Catálogo das Publicações dos Serviços Geológicos de Portugal 1865–1968*, by M. de F. Beato (Serviços Geológicos de Portugal, 1969)
JOURNALS: *Boletim do Museu e Laboratório Mineralógico e Geológico da Faculdade de Ciências Universidade di Lisboa* (1931–)
Boletim da Sociedade Geológica e Portugal (1941–)
Comunicações do Commissão dos Trabalhos do Serviço Geológico de Portugal (1883–)
Memórias e Notícias. Museo e Laboratório Mineralógio e Geológico de Universidade de Coimbra e do Centre de Estudos Geológicos (1921–)
Memórias. Serviços Geológicos de Portugal (1959–)
Revista da Faculdade de Ciências, Universidade de Lisboa (1950–)

ROMANIA

BOOK: The only recent account of Romanian geology is *Geologia Republicii Populare Romîne*, by N. Oncescu (Editura Tehnicá, 1960)
BIBLIOGRAPHY: *Bibliografia Geologicá a României*, originally published in 1926 with supplements issued in 1929, 1939, 1962 and 1969 (Institut Geol. Romaniei)
JOURNALS: The present situation with regard to serial publications is very complex. The current position is not made any easier by the fact that Romanian serials have in the past usually had a number of title changes.

Analele Universitătii Bucureşti. Seria Ştiinţele Naturii, Geologie-Geografie (1964–)
Anuarul Comitetului de Stat al Geologiei (1967–)
Dări de Seamă ale Şedinţelor (1911–)
Memoriile Comitetului Geologic (1956–)
Revue Roumaine de Géologie Géophysique et Géographie. Série de Géologie (1964–)
Studii şi Cercetări de Geologie, Geofizica, Geografie. Seria Geologie (1959–)
Studii Tehnice şi Economice Institutului Geologic. Seria H. Geologia Cuaternarului (1965–) and *Seria J. Stratigrafie* (1963–)

SPAIN

BOOK: The only volume which reviews Spanish geology as a whole is now largely out of date. This is: *Handbuch der Regionalen Geologie*. III, pt. 3. *La Péninsule Ibérique. A: Espagne*, by R. Douvillé (Winter, 1911). Other volumes have been published in Spanish in recent years on selected areas.
BIBLIOGRAPHIES: 'Bibliografía Geológica Española' is issued in the journal *Acta Geologica Hispanica*, for example, 1964–65, in **2** (2), 25–52 (1967) 'Bibliografía Geológico Minera de la Provincia de Cordoba' (*Memorias del Instituto Geológico y Minero de España*, **74**, 1970)

JOURNALS: *Acta Geológica Hispanica* (1966–)
Boletín Geológico y Minero (1967–)
Breviora Geológica Astúrica (1957–)
Cursillos y Conferencias. Instituto 'Lucas Mallada' de Investigaciones Geológicas (1960–)
Estudios Geológicos. Instituto 'Lucas Mallada' de Investigaciones Geológicas (1945–)
Explicación Mapa Geológico de España (1941–)
Memorias del Instituto Geológico y Minero de España (1911–)
Publicaciones Extranjeras sobre Geología de España (1944–)
Studia Geológica (1971–)
Trabajos de Geología (1967–)

SWEDEN

BOOKS: There is no book on Swedish geology available in the four major languages under consideration. The following work, in Swedish, is currently standard: *Sveriges Geologi*, by N. H. Magnusson *et al.* (3rd edn., Nordstedt, 1963)

BIBLIOGRAPHIES: There have been various bibliographies of Swedish geology, an example being: 'Swedish Geological Literature 1958–1963', edited by W. Larsson (*Sveriges Geologiska Undersökning*, **C630**, 1968). It is possible to trace the earlier compilations from this.

JOURNALS: *Arkiv för Mineralogi och Geologi* (1949–)
Bulletin of the Geological Institution of the University of Upsala (1892–)
Geologiska Föreningens i Stockholm Förhandlingar (1872–)
Stockholm Contributions in Geology (1957–)
Sveriges Geologiska Undersökning. Afhandlinger och Uppsatser (1868–)

SWITZERLAND

BOOKS: Despite intensive British concern with Swiss geology throughout the history of its investigation, there has been no English book on the stratigraphy of Switzerland this century. The most comprehensive accounts are:

Geologie der Schweiz, by A. Heim (Tauchnitz, 1919–22)
Geologie der Schweizer Alpen, by J. Cadisch (2nd edn., Wepf, 1953)
Geologischer Führer der Schweiz (9 parts, Wepf, 1967)
Proceedings of the Geologists' Association, **79**, 1–127 (1968) contains two useful papers by E. R. Oxburgh on the geology of the Eastern Alps

The publications of the Schweizerische Geologische Kommission, which takes the place in Switzerland of a geological survey, are listed in *Jubileebook: Schweizerische Geologische Kommission 1860–1960* (Basel, 1960).

JOURNALS: *Beiträge zur Geologischen Karte der Schweiz* (1862–)
Bulletin des Laboratoires de Géologie, Minéralogie, Géophysique et du Musée Géologique de l'Université de Lausanne (1901–)
Bulletin der Vereinigung Schweizerischer Petroleum-Geologen und-Ingenieure (1926–)
Eclogae Geologicae Helvetiae (1888–)

TURKEY

REVIEW: A general account is provided by 'Introduction à la Géologie et à l'Hydrologie de la Turquie', by R. Furon, in *Mémoires Muséum National d'Histoire Naturelle, Ser. C*, **3** (1953)
'Geology and History of Turkey', edited by A. S. Campbell (*Annual Field*

Conference Petroleum Society of Libya, **13**, 1971

BIBLIOGRAPHIES: Two bibliographies of Turkish geology, which can be used (albeit with some difficulty) by non-Turkish readers, are contained in *Türkiye Jeoloji Kurumu Bülteni*, **1** (2), 101–135 (1947) and **3** (1), 165–168 (1951). Official compilations are *Maden Tetkik ve Arama Enstitüsü Dergisi Makaleler Biblyografyasi 1936–1958* (1959) and *1959–1962* (1963)

JOURNALS: *Bulletin of the Mineral Research and Exploration Institute* (1936–)
Maden Tetkik ve Arama Enstitüsü Yayinlarindan (1937–)
Türkiye Jeoloji Kurumu Bülteni (1947–)

UNITED KINGDOM

The literature on the geology of the UK is vast and stretches far back in time. For information about the earliest works nothing can provide better information than the following series of papers, all of which are by J. Challinor:

'The Early Progress of British Geology. I. From Leland to Woodward 1538–1728' (*Annals of Science*, **9**, 124–153, 1953); 'II. From Strachey to Mitchell 1719–1788' (*ibid.*, **10**, 1–19, 1954); 'III. From Hutton to Playfair 1788–1802' (*ibid.*, **10**, 107–148, 1954)
'The Progress of British Geology During the Early Part of the Nineteenth Century' (*ibid.*, **26**, 177–234, 1970)
These have, in part, now been issued as a book, *The History of British Geology. A Bibliographical Study*, by J. Challinor (David and Charles, 1971).

A further list of early works is to be found in *Geology and Mining Books Published before 1840*, by M. Hadcroft (Wigan and District Mining and Technical College, 1970).

The following are the most useful general texts on British stratigraphy:

Handbook of the Geology of Great Britain, edited by J. W. Evans and C. J. Stubblefield (Murby, 1929)
The Physiographic Evolution of Britain, by L. J. Wills (Arnold, 1929)
A Palaeogeographical Atlas of the British Isles and Adjacent Parts of Europe, by L. J. Wills (Blackie, 1951)
Stanford's Geological Atlas of Great Britain, by T. Eastwood (5th edn., Stanford, 1963)
Outlines of Historical Geology, by A. K. Wells and J. F. Kirkaldy (6th edn., Murby, 1966)
The Stratigraphy of the British Isles, by D. Rayner (Cambridge University Press, 1967)
The Geological History of the British Isles, by G. M. Bennison and A. E. Wright (Arnold, 1969)

In addition there are two important works treating with particular systems:

The Jurassic System in Great Britain, by W. J. Arkell (Clarendon Press; reprinted 1969)
Pleistocene Geology and Biology with Special Reference to the British Isles, by R. G. West (Longmans, 1968)

The Institute of Geological Sciences and its predecessor, the Geological Survey of Great Britain, are the most fertile sources of publications on British geology. The various parts of the popular *British Regional Geology* are listed below. For more specific information reference should be made to the *Memoir* series. Sheet memoirs accompany most published maps, and a series of district memoirs dealing with somewhat larger areas have also been published. The district memoirs dealing with water supply frequently contain borehole data of considerable importance, and the series *Summaries of Progress* now continued as the *Annual Reports* contain items of interest. Shorter papers by officers of the Survey and Institute are contained in *Bulletin of the Geological Survey of Great Britain* (1939–) and *Reports of the Institute of Geological Sciences* (1969–). A useful index to the former is that by D. M. Gregory, *Index to Papers Published in the Bulletin of the Geological Survey of Great Britain, Nos. 1–30, 1939–1969* (Institute of Geological Sciences, Leeds, 1969).

For up to date information about the publications of the Institute reference should be made to *Government Sectional List*, No. 45, issued by HMSO. The Institute itself issued in 1937 a *List of Memoirs, etc.*, and this is still of considerable value in locating definite information about the earlier publications. The *Classified Geological Photographs* (1963) catalogue is also of considerable value to those concerned with teaching.

The publishing policy of the Geological Society of London has recently been the subject of an overhaul, with the result that the long-familiar *Quarterly Journal of the Geological Society of London* has given way to the *Journal of the Geological Society* (1971–). The *Proceedings*, which have had a more complicated history, have also ceased publication, although *Memoirs* and the *Special Publications* continue. These are prime sources of information on British stratigraphy, as are the *Proceedings of the Geologists' Association* (1859–). The latter body is also producing (1958–) a long series of *Guides* to areas of special geological interest to celebrate its centenary. Stratigraphical papers are also featured in the *Transactions of the Royal Society of Edinburgh* and the *Proceedings* and *Philosophical Transactions* of the Royal Society. Many other papers appear in general science and natural history journals. The many journals produced by university geological societies or departments are not mentioned below but some contain valuable accounts of local geology.

For its more recent meetings the British Association for the Advancement of Science has sponsored the production of regional surveys, each of which contains an account of the geology of the area in question. (Sales of these volumes were handled by local organisations in most cases.) The volumes are:

1949 *A Scientific Survey of North-Eastern England*, edited by P. C. C. Isaac and R. E. A. Allan

1950 *Birmingham and its Regional Setting. A Scientific Survey*

1951 *Scientific Survey of Southeast Scotland*, edited by A. G. Ogilvie *et al.*

1952 *Belfast in its Regional Setting*, edited by E. E. Evans *et al.*

1953 *A Scientific Survey of Merseyside*, edited by W. Smith

1954 *The Oxford Region*, edited by A. F. Martin and R. W. Steel

1955 *Bristol and its Adjoining Counties*, edited by C. M. MacInnes and W. F. Whittard

1956 *Sheffield and its Region*, edited by D. L. Linton

1957 *A View of Ireland*, edited by J. Meenan and D. A. Webb

1958   *The Glasgow Region: A General Survey*, edited by R. Miller and J. Tivy
1959   *York: A Survey*, edited by G. F. Willmot *et al.*
1960   *Cardiff Region: A Survey*, edited by J. G. C. Anderson *et al.*
1961   *Norwich and its Region*
1962   *Manchester and its Region*, edited by C. F. Carter
1963   *The Northeast of Scotland*, edited by A. C. O'Dell and J. Mackintosh
1964   *A Survey of Southampton and its Region*, edited by F. J. Monkhouse
1965   *The Cambridge Region*, edited by J. A. Steers
1966   *Nottingham and its Region*, edited by K. C. Edwards
1967   *Leeds and its Region*, edited by M. W. Beresford and G. R. J. Jones
1968   *Dundee and District*, edited by S. J. Jones
1969   *Exeter and its Region*, edited by F. Barlow
1970   *Durham County and City with Teesside*, edited by J. C. Dewdney
1971   *Swansea and its Region*, edited by W. G. V. Balchin
1972   *Leicester and its Region*, edited by N. Pye

It should also be remembered that various other organisations are responsible for publishing general books which contain a great deal of local geology—for example, local natural history societies, the Forestry Commission, National Parks and the Nature Conservancy.

A recent bibliography of considerable value in locating general works on the local geology of Britain is 'Directory of British Geology. I. A Provisional Annotated Bibliography and Index of Geological Exursion Guides and Reports for Areas in Britain, C. England,' by D. A. Bassett (*Welsh Geological Quarterly*, **3** (3/4), 3–91, 1969)

Southern England
The following books relating to the geology of southern England are sufficiently recent to have retained their value:

Handbook of Cornish Geology, by E. H. Davison (Royal Geological Society of Cornwall, 1926)
Outlines of Sussex Geology and Other Essays, by E. A. Martin (Archer, 1932)
Geology of London and Southeast England, by G. M. Davies (Murby, 1939)
Bristol and Gloucester District. British Regional Geology (2nd edn., HMSO, 1948)
The Weald, by S. W. Wooldridge and R. Goldring (Collins, 1953)
The Dorset Coast: A Geological Guide, by G. M. Davies (2nd edn., Black, 1956)
Hampshire Basin and Adjoining Areas. British Regional Geology (3rd edn., HMSO, 1960)
London and the Thames Valley. British Regional Geology (3rd edn., HMSO, 1960)
An Introduction to the Geology of Cornwall, by R. M. Barton (Barton, 1964)
Present Views of Some Aspects of the Geology of Cornwall, edited by K. F. G. Hosking and G. J. Shrimpton (Royal Geological Society of Cornwall, 1964)
Dartmoor Essays, edited by I. G. Simmons (The Devonshire Association, 1964)
The Wealden District. British Regional Geology (3rd edn., HMSO, 1965)
Ancient Purbeck, by J. B. Calkin (Friary Press, 1968)

South-West England. British Regional Geology (3rd edn., HMSO, 1969)
Geological Highlights of the West Country, by W. A. Macfadyen (Butterworths, 1970)
Geology Explained in South Devon and Dartmoor, by J. W. Perkins (David and Charles, 1971)

In addition there is a long series of works (many of them published by Barton) dealing with the extractive industries, especially mining, in south-west England, and a smaller number with such industries in the south-east. Examples are *Purbeck Shop*, by E. Benfield (Cambridge University Press, 1940), on Purbeck stone; and *Wealden Iron*, by E. Straker (reprint of 1931 edition by David and Charles, 1969).

The *Proceedings of the Geologists' Association* especially contains papers on southern England. Other journals from this region in which geological papers figure are:

Transactions of the Royal Geological Society of Cornwall (1818–)
Proceedings of the Dorset Natural History and Archaeological Society (1928–)
Proceedings of the Ussher Society (1962–)
Report and Transactions of the Devonshire Association for the Advancement of Science (1862–)
Transactions and Proceedings, Torquay Natural History Society (1922–)

The English Midlands
The most comprehensive and useful volumes covering this area are:

The Palaeogeography of the Midlands, by L. J. Wills (2nd edn., Liverpool University Press, 1950)
The Geology of the East Midlands, edited by P. C. Sylvester-Bradley and T. D. Ford (Leicester University Press, 1968)

Of more local interest are:

Geology of the Ancient Rocks of Charnwood Forest, Leicestershire, by W. W. Watts (Leicestershire Literary and Philosophical Society, 1947)
The Geology of Oxford, by W. J. Arkell (Clarendon Press, 1947)
East Yorkshire and Lincolnshire. British Regional Geology (2nd edn., HMSO, 1948)
The Welsh Borderland. British Regional Geology (2nd edn., HMSO, 1948)
The Geology of Lincolnshire, by H. H. Swinnerton and P. E. Kent (Lincolnshire Naturalists' Union, 1949)
The Geology of Norfolk, edited by G. P. Larwood and B. M. Funnell (Norfolk and Norwich Naturalists' Society, 1961)
East Anglia and Adjoining Areas. British Regional Geology (4th edn., HMSO, 1961)
'A Contribution to the Geological History of Suffolk I, II, III', by H. E. P. Spencer (*Transactions of the Suffolk Naturalists' Society*, **13**, 197-209, 290–313, 366–389, 1966–67)
Geology Explained in the Severn Vale and Cotswolds, by W. Dreghorn (David and Charles, 1967)
'Geology of North Herefordshire', by W. J. Norton (*Transactions of the Woolhope Naturalists' Field Club*, **39**, 23–30, 1967)
Wren's Nest National Nature Reserve (The Nature Conservancy, 1967)

Geology Explained in the Forest of Dean and Wye Valley, by W. Dreg-horn (David and Charles, 1968)
Central England. British Regional Geology (3rd edn., HMSO, 1969)
'Lincolnshire Geology in its Regional Setting', by P. E. Kent (*Transactions of the Lincolnshire Naturalists' Union*, **17**, 135–139, 1970)

This has always been an area characterised by the intensive activity of local societies both geological and natural history. In consequence a variety of journals, the majority of them now extinct, contain material relating to the geology of this region. The most important are:

The Mercian Geologist (1964–   )
North Staffordshire Journal of Field Studies (1961–   )
Proceedings of the Birmingham Natural History Society (1894–   )
Proceedings of the Cotteswold Naturalists' Field Club (1847–   )
Transactions and Annual Report of the North Staffordshire Field Club (1915–60)
Transactions of the Leicester Literary and Philosophical Society (1835–79, 1889–   )
Transactions of the Lincolnshire Naturalists' Union (1893–94, 1905–   )
Transactions of the Norfolk and Norwich Naturalists' Society (1869–   )
Transactions of the Woolhope Naturalists' Field Club (1852–   )

It should be noted that *Mercian Geologist*, **2** (2) consists of a 'Bibliography of the Geology of the Peak District of Derbyshire (to 1965)', by T. D. Ford and M. H. Mason, an invaluable source work for that region; a supplement was issued in the same journal, **4** (2) (1972). A somewhat more limited bibliography is: *Bibliography of Local References: Birmingham and the Surrounding Region: Geology and Geomorphology* (Geographical Association, Birmingham Branch, 1968).

Although the extractive industries have long been prominent in the Midlands, books dealing with them are few; Nellie Kirkham's *Derbyshire Lead Mining through the Centuries* (Barton, 1968) and Donovan Purcell's *Cambridge Stone* (Faber, 1967) merit mention, however. The publications of the Peak District Mines Historical Society and of the Shropshire Mining Club also contain much of geological interest.

Northern England
Books dealing with this area include:

Geology of Yorkshire: An Illustration of the Evolution of Northern England, by P. F. Kendall and H. E. Wroot (2 vols., privately published, 1924)
The Glacial Geology of Holderness and the Vale of York, by S. Melmore (privately published, 1935)
Geology and Scenery of the Countryside Around Leeds and Bradford, by H. C. Versey (Murby, 1948)
East Yorkshire and Lincolnshire. British Regional Geology (2nd edn., HMSO, 1948)
The Pennines and Adjacent Areas. British Regional Geology (3rd edn., HMSO, 1954)
Geology of the Appleby District, by H. C. Versey (4th edn., Whilehead, Appleby, 1960)
The Limestone Series of West Cumberland, by E. H. Shackleton (Cumberland Geological Society, 1962)

Northern England. British Regional Geology (3rd edn., HMSO, 1963)
The Geology of Moor House. A National Nature Reserve in North-East Westmorland, by G. A. L. Johnson and K. C. Dunham (HMSO, 1963)
Geological Excursions in the Sheffield Region and the Peak District National Park, edited by R. Neaves and C. Downie (Northend, Sheffield, 1967)
Lakeland Geology, by E. H. Shackleton (Dalesman Publishing Company, 1969)

Three bibliographies worthy of mention are:

'Silurian Successions in NW England—A Bibliography', by P. G. Llewellyn (*Journal of the Society for the Bibliography of Natural History*, **5**, 41–56, 1968)
A Bibliography of Lake District Geology and Geomorphology, by R. A. Smith (Cumberland Geological Society, 1965)
'North and West Lancashire and the Isle of Man. A Bibliography of the Geology and Physical Geography', by J. Thorpe (*University of Lancaster Library Occasional Papers*, **5**, 1972)

Geological activity in this area has long been concentrated in a few strong societies. Their journals (four of them successively produced by two societies) are:

Geological Journal (1964–   )
Journal of Earth Sciences (1970–   ) [formerly *Transactions of the Leeds Geological Association*]
Journal of the Manchester Geological Association (1925–49)
Liverpool and Manchester Geological Journal (1951–63)
Proceedings of the Cumberland Geological Society (1962–   )
Proceedings of the Liverpool Geological Society (1874–1950)
Proceedings, North-East Lancashire Group of the Geologists' Association (1964–   )
Proceedings of the Yorkshire Geological Society (1879–   )
The Sorby Record (1958–   )
Transactions of the Natural History Society of Northumberland, Durham and Newcastle (1904–   )

Once again, extractive industries have a long history that has remained largely undocumented, although *A History of Lead Mining in the Pennines*, by A. Raistrick and B. Jennings (Longmans, 1965), *Mines and Mining in the English Lake District*, by J. Postlethwaite (privately published, 1913), and *Mining in the Lake Counties*, by W. T. Shaw (Dalesman Publishing Company, 1970) merit attention.

Scotland
The most comprehensive work dealing with Scottish geology is *The Geology of Scotland*, edited by G. Y. Craig (Oliver and Boyd, 1965). Works dealing with more limited areas are:

The Rocks of West Lothian, by H. C. Cadell (Oliver and Boyd, 1925)
The West Highlands and the Hebrides: A Geologist's Guide for Amateurs, by A. Harker (Cambridge University Press, 1941)
Midland Valley of Scotland. British Regional Geology (2nd edn., HMSO, 1948)

South of Scotland. British Regional Geology (2nd edn., HMSO, 1948)
Geological Excursion to the Glasgow District, by D. A. Bassett (Geological Society of Glasgow, 1958)
Geological Excursion Guide to the Assynt District of Sutherland, by M. MacGregor and J. Phemister (2nd edn., Oliver and Boyd, 1958)
Northern Highlands. British Regional Geology (3rd edn., HMSO, 1960)
Edinburgh Geology: An Excursion Guide (Oliver and Boyd, 1960)
Tayside Geology, by F. Walker (Dundee Museum, 1961)
Tertiary Volcanic Districts. British Regional Geology (3rd edn., HMSO, 1961)
The Geology and Scenery of Strathearn, by F. Walker (Dundee Museum, 1963)
Excursion Guide to the Geology of Arran, by M. MacGregor (Glasgow Geological Society, 1965)
Arthur's Seat: A History of Edinburgh's Volcano, by G. P. Black (Oliver and Boyd, 1966)
Grampian Highlands. British Regional Geology (3rd edn., HMSO, 1966)
Fife and Angus Geology: An Excursion Guide, by A. R. MacGregor (Blackwood, 1968)
Tertiary Volcanic Rocks of Ardnamurchan: An Excursion Guide, by W. A. Deer (Geological Societies of Edinburgh and Glasgow, 1969)
Carboniferous Volcanic Rocks of the Midland Valley of Scotland: An Excursion Guide, edited by B. G. J. Upton (Geological Societies of Edinburgh and Glasgow, 1969)

Geological papers on Scotland are frequently featured in the *Transactions of the Royal Society of Edinburgh.* Other Scottish journals concerned with geology are:

Proceedings of the Geological Society of Glasgow (1966–)
Scottish Journal of Geology (1965–)
Scottish Journal of Science (1965–)
Transactions of the Edinburgh Geological Society (1866–1963)
Transactions of the Geological Society of Glasgow (1860–1963)
Transactions, Journal and Proceedings of the Dumfriesshire and Galloway Natural History and Antiquarian Society (1862–)
Transactions and Proceedings of the Perthshire Society of Natural Science (1886–)

There are also a number of books and papers devoted to the extractive industries, especially coal and metal mining.
Two useful bibliographies should also be mentioned. They are:

'Directory of British Geology I. A Provisional Annotated Bibliography and Index of Geological Excursion Guides and Reports for Areas in Britain. A. Scotland', by D. A. Bassett (*Welsh Geological Quarterly,* **2** (3), 1967)
'Theses on Scottish Geology 1960–1968', by W. D. Ian Rolfe (*Scottish Journal of Geology,* **6**, 401–407, 1970); '1968–1969' (*ibid.,* **7**, 367–368, 1971)

Wales
Admirable bibliographic works dealing with Welsh geology have been produced by Dr. D. A. Bassett. They are:

*Bibliography and Index of **Geology and Allied Sciences** for Wales and the*

Welsh Borders 1536–1896, by D. A. Bassett (National Museum of Wales, 1963)
Bibliography and Index of Geology and Allied Sciences for Wales and the Welsh Borders 1897–1958, by D. A. Bassett (National Museum of Wales, 1961). This was added to in *Liverpool and Manchester Geological Journal*, **3** (1), *Geological Journal*, **4** (1) and **5** (1), and then in the *Welsh Geological Quarterly*, **1** (4), etc.
A Sourcebook of Geological, Geomorphological and Soil Maps for Wales and the Welsh Borders, by D. A. Bassett (National Museum of Wales, 1967)
'Directory of British Geology. I. A Provisional Annotated Bibliography and Index of Geological Excursion Guides and Reports for Areas in Britain. B. Wales and the Welsh Borders', by D. A. Bassett (*Welsh Geological Quarterly*, **3**, 1968)

In other respects, however, Wales is much less well served with geological literature than England and Scotland, and no comprehensive account of its geology is available. Works of more limited compass are:

The Middle Silurian Rocks of North Wales, by P. G. H. Boswell (Arnold, 1949)
North Wales. British Regional Geology (3rd edn., HMSO, 1961)
'A Review of Geological Research in Cardiganshire', by J. Challinor (*Welsh Geological Quarterly*, **4**, 3-37, 1969)
The Pre-Cambrian and Lower Palaeozoic Rocks of Wales, edited by A. Wood (University of Wales Press, 1969)
South Wales. British Regional Geology (3rd edn., HMSO, 1970)
In a series of publications of the National Museum of Wales Dr. F. J. North has reviewed the history of stratigraphy in Wales and has dealt with the major extractive industries (coal, slate, non-ferrous metals): his works well merit consultation. His papers also appear in another important journal for this area, *Transactions of the Cardiff Naturalists' Society* (1901–). The only journal entirely devoted to geology in Wales, although it is often of more value than just to Wales, is the *Welsh Geological Quarterly* (1965–).

Channel Islands
Only three items merit attention, as so many of the works are now out of date. They are:

'Outline and Guide to the Geology of Guernsey', by R. A. Roach (*Report and Transactions of the Société Guernésiaise*, **17**, 751–775, 1966)
'A Bibliography of the Geology of the Bailiwick of Guernsey', by J. N. Van Leuven (*Report and Transactions of the Société Guernésiaise*, **18**, 427-434, 1970)
A Bibliography on the Geology of Alderney, by A. D. Squire (Chelsea College, Department of Geology, 1972)

YUGOSLAVIA
BOOKS: Only one account of the geology of Yugoslavia is available in the languages considered: *Excursion to Yugoslavia* (Petroleum Exploration Society of Libya, 1962)
BIBLIOGRAPHY: *Geološka Bibliogràfija Yugoslavije od XIV Veka do 1944 godine. Bibliographie Géologique de la Yougoslavie du XIVe Siècle à*

1944, by S. Milojevic (Institut Bibliographique de la Yougoslavie, 1954)
JOURNALS: *Geološki Anali Balkanskog Poluostrva* (1889–)
Geološki Glasnik, Cetinjè (1956–)
Geološki Glasnik, Sarajevo (1955–)
Geološki Vjesnik, Zagreb (1947–)
Glasnik Prirodnjačkog Muzeja u Beogrady A. Mineralogija, Geologija, Paleontologija (1948–)
Letopis Slovenske Akademije Znanosti in Umetnosti (1952–)
Posebna Izdanja. Geološkiš Glasnika Geološki Zavod v Sarajevu (1953–)
Posebna Izdanja. Geološki Institut. Srpska Akademija Nauka (1951–)
Posebna Izdanja. Geološki Zavod Skopje (1965–)
Prirodoslovna Istraživanja (Acta Geologica) (1956–)
Prvi Kolokvij o Geologiji Dinaridov (1968–)
Trudovi na Geološki Zavod na Narodna Republika Makedonija (1947–)
Vesnik Zavoda za Geološka i Geofizička Istraživanja (1953–)
Zapisnici Srpskog Geološkog Društva (1897–1907, 1948–)

USSR
The twin problems of language and a vast volume of geological publications confront the geologist interested in the stratigraphy of the USSR. An additional hazard is the prevalent tendency to produce stratigraphical works in proceedings of colloquia or other non-serial volumes. Although some journals are available in partial or cover-to-cover translations (see Chapter 6), this is of little help to the stratigrapher, since the stratigraphic papers are but rarely translated.
BOOKS AND REVIEWS: The only recent general account is the rather brief *The Geology of the USSR. A Short Outline*, by D. V. Nalivkin (Pergamon, 1960). A translation by N. Rast, edited by Professor Westoll, of *The Geology of the USSR*, by D. V. Nalivkin, was announced for issue in 1970 but has so far not appeared. Earlier works, too outdated now to be of any great value, include:

'Geologie von Sibirien', by W. A. Obrutschew (*Fortschritte der Geologie und Paläontologie*, **15**, 1926)
Handbuch der Regionalen Geologie. V, pt. 3. *Armenien*, by F. Oswald (Winter, 1912)
Handbuch der Regionalen Geologie. V, pt. 5. *Kaukusus*, by A. F. von Stahl (Winter, 1923)
Atlas de la Géologie Transcaucasusienne, by P. Bonnet (Paris, 1933–37)

The principal treatments in Russian are contained in two series of volumes: *Geologiya SSSR* and *Stratigrafiya SSSR* (1963–). Useful summaries of the geology together with extensive bibliographies are given in a series entitled *Geologicheskaya Izuchennost' SSSR*. There are of course numerous recent monographs in Russian on the various states of the USSR.
BIBLIOGRAPHIES: A most useful source of reference to the various bibliographies is *Guide to Russian Reference Books. V. Science, Technology and Medicine*, by K. Maichel (Hoover Institution on War, Revolution and Peace, Stanford University, 1967). Additional material includes:

Geologicheskaya Literatura SSSR, which covers the literature from 1934

Eesti Geoloogia Bibliograafia 1840–1959, by D. Kaljo (Eesti Riiklik Kirjastus, 1960)

JOURNALS: The following is a select list of the more important journals:

Byulleten Moskovskogo Obshchestva Ispÿtatelei Prirodÿ. Otdel Geologicheskiĭ (1922–)
Doklady Akademii Nauk Armyanskoĭ SSR (1944–)
Doklady Akademii Nauk Azerbaĭdzhanskoĭ SSR (1945–)
Doklady Akademii Nauk Belorusskoĭ SSR (1957–)
Doklady Akademii Nauk SSSR. Seriya Geologiya (1965–)
Doklady Akademii Nauk SSSR. Sibirskoe Otdelenie (1962–)
Doklady Akademii Nauk Tadzhikskoĭ SSR (1951–)
Doklady Akademii Nauk Uzbekskoĭ SSR (1948–)
Eesti NSV Teaduste Akadeemia Geoloogia Instituudi Uurimused (1956–)
Geologiya i Geofizika (1960–)
Geologicheskii Zhurnal (1934–)
Geoloogiline Kogumik Eesti NSV Teaduste Akadeemia Loodusuurijate Selts (1961–)
Geoloogilised Märkmed Loodusuurijate Selts Eesti NSV Teaduste Akadeemia (1960–)
Izvestiya Akademii Nauk Azerbaĭdzhanskoĭ SSR. Seriya Nauko Zemle (1966–)
Izvestiya Akademii Nauk SSSR. Seriya Geologicheskaya (1936–)
Izvestiya Akademii Nauk Tadzhikskoĭ SSR. Otdelenie Geologo-khimicheskikh i Tekhnicheskikh Nauk (1959–63)
Materialy po Geologii Tyan'Shanya (1961–)
Ocherki Regional'noi Geologii SSSR (1957–)
Paleontologiya i Stratigrafiya Pribaltiki i Belorussii (1966–)
Trudÿ Geologicheskogo Instituta. Akademiya Nauk Gruzinskoĭ SSR. Geologicheskaya Seriya (1942–)
Trudÿ Geologicheskogo Instituta. Akademiya Nauk SSSR (1932–38, 1956–)
Trudÿ Instituta Geologicheskikh Nauk Akademiya Nauk SSSR (1939–56)
Trudÿ Instituta Geologii. Akademiya Nauk Tadzhikskoĭ SSR (1956–)
Trudÿ Instituta Geologii. Akademiya Nauk SSSR Ural'skii Filial (1962–)
Trudÿ Instituta Geologii i Geofiziki. Sibirskoe Otdelenie. Akademiya Nauk SSSR (1960–)
Trudÿ Instituta Geologii Gosudarstvennyi Geologicheskii Komitet SSSR (1965–)
Trudÿ Instytutu Geologichnykh Nauk. Seriya Stratigrafii i Paleontologii. Akademiya Nauk Ukrains'koĭ RSR (1950–)
Trudÿ Litovskiĭ Nauchno-Issledovatel'skiĭ Geolograzvedochnyĭ Institut (1965–)
Trudÿ Vsesoyuznogo Nauchno-Issledovatel'skogo Geologicheskogo Instituta (1950–)
Trudÿ Vsesoyuznogo Neftyanogo Nauchno-Issledovatel'skogo Geologo-Razvedochnogo Instituta (VNIGRI) (1939–)
Uchenye Zapiski Nauchno-Issledovatel'skii Institut Geologii Arktiki, Regionalnaya Geologiya (1963–)
Uzbekskii Geologicheskii Zhurnal (1958–)
Voprosy Geologii Kuzbassa (1956–)

Reference should also be made to the extensive and useful series of guidebooks and papers prepared for the XVII International Geological Congress, Moscow, 1937.

ASIA

BOOKS: Only one comprehensive work on Asian stratigraphy has been attempted and this was never completed. This is *Geologie von Asien*, by K. Leuchs (2 vols., Borntraeger, 1935–37). I. *Uberblick uber Asien. Nordasien.* II. *Zentralasien.*

Other regional works are:

Handbuch der Regionalen Geologie. V, pt. 2. *Kleinasien*, by A. Phillippsen (Winter, 1918)

A Sketch of the Geography and Geology of the Himalaya Mountains and Tibet, by S. G. Bureard and A. M. Meron (rev. edn., Government of India Press, 1934)

Geological Map of Asia and the Far East. Explanatory Brochure (UNESCO, 1961)

Geology and Palaeontology of South East Asia (University of Tokyo Press, 1964–). This is a series edited by T. Kobayashi and R. Toriyama, and contains both original and reprinted papers

Regional Geology of the Himalayas, by A. Gansser (Wiley, 1965)

'Addresses and Detailed Papers Contributed to the 1st Himalayan Geology Seminar (1966)' (*Publication of the Centre of Advanced Study in Geology*, 3, 1967). Other congress papers have been issued in later numbers of this journal

Geology and Mineral Resources of the Far East, edited by T. Ogura (University of Tokyo Press, 1967–). Covers Manchuria, Korea and China

Works of interest will also be found under Indonesia.

BIBLIOGRAPHIES: *Bibliography of Levant Geology Including Cyprus, Hatay, Israel, Jordania, Lebanon, Sinai and Syria*, by M. A. Avnimelech (Israel Program for Scientific Translations 1965, 1969)

Selected Geologic Bibliography. Regional Descriptions and Maps. 4. Middle East (CGMW, 1964)

'Bibliography of Himalayan Geology', by M. N. Saxena (*Publication of the Advanced Centre of Palaeontology and Himalayan Geology*, 2, 1966)

ADEN

Two recent works are:

'Geology of the Arabian Peninsula. Eastern Aden Protectorate and Part of Dhufar', by Z. R. Beydoun (*Professional Paper, United States Geological Survey*, No. 560H, 1966)

'Geology of the Arabian Peninsula, Aden Protectorate', by J. E. G. W. Greenwood and D. Bleackley (*Professional Paper, United States Geological Survey*, No. 560C, 1967)

AFGHANISTAN

BOOKS: No comprehensive book dealing specifically with Afghan stratigraphy has been located. The following references, however, provide some information:

'Géologie du Plateau Iranien (Perse-Afghanistan-Béloutchistan)', by R.

Furon (*Mémoires Muséum National d'Histoire Naturelle*, **7**, 1941)
'Zur Geologie von Südost-Afghanistan', by O. Ganss (*Beihefte zum Geologischen Jahrbuch*, **84**, 1970)

JOURNAL: *Bulletin of the Afghan Geological and Mineral Survey* (1964–)

BAHRAIN
'Geology of the Arabian Peninsula. Bahrain', by R. P. Willis (*Professional Paper, United States Geological Survey*, No. 560E, 1967)

BHUTAN
'A Preliminary Note on the Geology of Bhutan Himalaya', by S. P. Nautiyal *et al.* (*Proceedings of the 22nd International Geological Congress*, **11**, 1–14, 1964)

BURMA
BOOKS: *Geology of Burma*, by H. L. Chhibber (Macmillan, 1934)
A Manual of the Geology of India and Burma, revised by E. H. Pascoe (3 vols., Government of India, 1950–64)
Geology of India and Burma, by M. S. Krishnan (3rd edn., Higginbothams, 1956)
Geology of India, Pakistan and Burma, by R. C. Mehdiratta (2nd edn., Atma Ram, 1962)
JOURNAL: *Union of Burma Journal of Science and Technology* (1968–)

CEYLON
BOOKS: 'The Geology and Mineral Resources of Ceylon', by J. D. Fernando (*Bulletin of the Imperial Institute*, **7**, 1948)
An Introduction to the Geology of Ceylon, by P. G. Cooray (National Museums of Ceylon, 1967)
BIBLIOGRAPHY: *Bibliography of Ceylon Geology*, by D. N. Wadia (Department of Mineralogy, Colombo, 1943)
JOURNALS: *Memoirs of the Geological Survey of Ceylon* (1959–)
Spolia Zeylanica (1903–)
Proceedings of the Ceylon Association for the Advancement of Science (1957–)

CHINA
The literature on Chinese geology is, for linguistic and political reasons, largely inaccessible to Western readers. Most books and bibliographies available, with the exception of the odd translation, are now out of date. Some coverage of the more recent literature is provided by the principal geological bibliographic serials but much remains to be learned and no cover-to-cover translations of Chinese serials are at present available.
BOOKS: *Stratigraphy of China*, by A. W. Grabau (2 vols., Geological Survey of China, 1923–24). I. *Paleozoic and Older*. II. *Mesozoic*
A Regional Stratigraphic Table of China (Joint Publications Research Service, 1963). Translation JPRS 18539 of Chinese original dated 1958
The Geology of China, by Ta Chang (Joint Publications Research Service, 1963). Translation JPRS 19209 of a Chinese original dated 1959
BIBLIOGRAPHIES: *Bibliography of Chinese Geology: Bibliography of Geology and Allied Sciences of Tibet and Regions to the West of Chinshachiang*, compiled by T. C. Tseng (Geological Survey of China, 1946)
Bibliography of Geology and Geography of Sinkiang, compiled by V. S.

Chi (Hsiao-Fung Li, Nanking, 1947)
Bibliography of Chinese Geology up to 1934, compiled by T. I. Young (National Academy, Peking, 1935)
Bibliography of Chinese Geology for the Years 1936–1940, compiled by Y. Chi (Geological Survey of China, 1947)
JOURNALS: Those in Western languages include:

Acta Geologica Sinica (1922–66)
Bulletin of the Geological Society of China (1922–51)
Geological Bulletin (1930–48)
Memoirs of the Geological Survey of China (1919–47)
Memoirs of the National Research Institute of Geology. Shanghai (1930–47)
Monographs of the National Research Institute of Geology. Shanghai (1930–47)
Scientia Sinica (1954–66)

HONG KONG
BOOKS: *The Geology of the British Empire*, by F. R. C. Reed (2nd edn., Arnold, 1949)
The Geology of Hong Kong, by S. G. Davis (Government Printer, Hong Kong, 1952)
Economic Geology of Hong Kong, edited by S. G. Davis (Hong Kong University Press, 1964)

INDIA
BOOKS: *A Manual of the Geology of India and Burma*, revised by E. H. Pascoe (3 vols., Government of India, 1950–64)
Geology of India and Burma, by M. S. Krishnan (3rd edn., Higginbothams, 1956)
Geology of India, by D. N. Wadia (3rd edn., Macmillan, 1953)
Elements of Indian Stratigraphy, by S. K. Borooah (Darrsons, 1962)
Geology of India, Pakistan and Burma, by R. C. Mehdiratta (2nd edn., Atma Ram, 1962)
Geology of India, by A. K. Dey (National Book Trust, 1968)

Covering a smaller area are:

A Handbook to the Geology of Mysore State, by B. R. Fao (Bangalore Printing and Publishing Co., 1962)
A Geología de Goa (Ministerio do Ultramar, Lisbon, 1960)

BIBLIOGRAPHY: *A Bibliography of Indian Geology and Physical Geography*, by T. H. D. La Touche (5 vols., Geological Survey of India, 1917–26)
JOURNALS: *Bulletin of the Geological Survey of India* (1950–   )
Journal of the Geological Society of India (1959–   )
Journal of the Indian Geoscience Association (1964–   )
Memoirs of the Geological Survey of India (1856–   )
Quarterly Journal of the Geological Mining and Metallurgical Society of India (1926–   )
Records of the Geological Survey of India (1868–   )
Transactions of the Mining, Geological and Metallurgical Institute of India (1906–   )

INDONESIA
BOOKS: 'The Geology of the Netherlands Indies', by H. Stauffer, in

P. Honig and F. Verdoorn (editors) 'Science and Scientists in the Nether-
lands Indies' (*Natuurwetenschappelijk Tijdschrift voor Nederlandsch-Indië*,
102, 1945)
Geological Explorations in the Islands of Celebes, edited by H. W. Brouwer
(North-Holland, 1947)
Structural History of the East Indies, by J. H. G. Umbgrove (Cambridge
University Press, 1949)
The Geology of Indonesia, by R. W. van Bemmelen (2nd edn., 2 vols.,
Nijhoff, 1970). Extensive coverage of a much wider area than is suggested
by the title—for example, British part of Borneo, Malay Peninsula, Anda-
man and Nicobar Islands
BIBLIOGRAPHY: *Geologisch-Mijnbouwkundige Bibliografie van Neder-*
landsch-Indie (1924–51)
'Geologisch-Mijnbouwkundig Genootschap voor Nederland en Koloniën'.
Reprint from *Verhandelingen van het Geologisch-Mijnbouwkundig*
Genootschap voor Nederland en Koloniën, Geologische Serie, **1**, 31–248
(1912), with additions. The last part, i.e. Deel IV, Afleveringen 5–10, is
entitled *Geologisch-Mijnbouwkundige Bibliografie van Indonesië*
JOURNALS: *Bulletin of the Geological Survey of Indonesia* (1964–)
Contributions from the Department of Geology, Institute of Technology,
Bandung (1959–)
Indonesian Journal for Natural Science (1951–57)
Laporan Tahunan. Djawatan Geologi Republik Indonesia (1960–)

IRAN
BOOKS: *Handbuch der Regionalen Geologie*. V, pt. 6. *Persien*, by A. F.
von Stahl (Winter, 1911)
'Géologie du Plateau Iranien (Perse-Afghanistan-Béloutchistan)', by R. Furon
(*Mémoires Muséum National d'Histoire Naturelle*, **7**, 1941)
Geological Map of Iran 1:2 500 000 with Explanatory Notes (National
Iranian Oil Company, 1959)
'Stratigraphic Lexicon of Iran. 1. Central, North and East Iran', by J.
Stöcklin (*Report of the Geological Survey of Iran*, **18**, 1971)

In addition the following merit consultation: 'Contribution to the Strati-
graphy and Tectonics of the Iranian Ranges', by H. Böckh *et al.*, in *The*
Structure of Asia, by J. G. Gregory (Methuen, 1929, pp. 58–176); and
'Geology of Eastern Iran', by F. G. Clapp (*Bulletin of the Geological*
Society of America, No. 51, 1–101, 1950)
BIBLIOGRAPHY: 'A Preliminary Bibliography of the Natural History of
Iran', by R. L. Burgess *et al.* (*Science Bulletin, Pahlavi University, College*
of Arts and Sciences, **1**, 1966)
'Bibliography of the Geology of Iran', by N. C. Rosen (*Special Publication*
of the Geological Survey of Iran, **2**, 1969)

A useful list of recent publications is to be found in

'Das UN-Projekt "Geological Survey Institute Iran" Organisation und
Arbeitsergebnisse 1962–1968', by A. Ruttner and O. Thiele (*Verhandlungen*
der Geologischen Bundesanstalt (2), 143–158, 1969)
JOURNALS: *Report of the Geological Survey of Iran* (1964–)
Bulletin of the Iranian Petroleum Institute (1964–)
Publication of the National Iranian Oil Company Geological Laboratories
(1964–)

IRAQ

BOOKS: With one notable exception, the only general accounts are considerably out of date.

'Geology of the Arabian Peninsula, Southwestern Iraq', by K. M. Al Naqib (*Professional Paper, United States Geological Survey*, No. 560G, 1967)
Handbuch der Regionalen Geologie. V, pt. 4. *Syrien, Arabien und Mesopotamien*, by M. L. P. Blanckenhorn (Winter, 1914)

BIBLIOGRAPHY: *A Guide to the Literature of Iraq's Natural Resources* (UNESCO, 1969). Covers the period 1833–1968
JOURNAL: *Journal of the Geological Society of Iraq* (1968–)

ISRAEL

BOOK AND REVIEW: The works cited below are both somewhat out of date:

Structure and Evolution of Palestine with Comparative Notes on some Neighbouring Countries, by L. Picard (2nd edn., 1958). The original edition was issued as *Bulletin of the Geological Department of the Hebrew University*, **4** (1943)
'The Geology and Mineral Resources of Palestine', by S. H. Shaw (*Bulletin of the Imperial Institute*, **46**, 87–103, 1948)
JOURNALS: *Bulletin of the Geological Survey of Israel* (1950–)
Israel Journal of Earth Sciences (1963–)
Summary of Activities of the Geological Survey of Israel (1949–)

JAPAN

BOOKS: *Geology and Mineral Resources of Japan* (2nd edn., Geological Survey of Japan, 1960)
Geology of Japan, by F. Takai *et al.* (University of California Press, 1963)
Geological Development of the Japanese Islands, by M. Minato *et al.* (University of Tokyo Press, 1965)
BIBLIOGRAPHIES: *Alphabetical List of Papers and Reports Published in Japan on Stratigraphy, Palaeontology and Marine Ecology for 1947–1951*, compiled by S. Hanzawa and K. M. Hatai (US Geological Survey, Military Geology Branch, Tokyo, 1951)
Bibliography of the Geology of Japan 1873–1955, by H. Fujimoto (Chijin Shokan, 1956)
JOURNALS: *Bulletin of the Geological Survey of Japan* (1886–1937, 1950–)
Chikyu Kagaku (Earth Science) (1949–)
Contributions from the Institute of Geology and Palaeontology, Tohoku University (1924–)
Japanese Journal of Geology and Geography (1922–)
Journal of Earth Sciences, Nagoya University (1969–)
Journal of the Faculty of Science, Hokkaido University, Series 4. Geology and Mineralogy (1930–)
Journal of the Faculty of Science, University of Tokyo, Section 2. Geology Mineralogy, Geography, Geophysics (1926–)
Journal of the Geological Society of Japan (1935–)
Journal of Geosciences, Osaka City University (1962–)
Journal of Science of the Hiroshima University, Series C. Geology and Mineralogy (1951–)
Memoirs of the Faculty of Science, Kyushu University, Series D. Geology (1940–)

Memoirs of the Faculty of Science, University of Kyoto, Series Geology and Mineralogy (1967–)
Report of the Geological Survey of Japan (1922–)
Science Reports of the Tohoku University, Series 2. Geology (1912–)
Science Reports, Yokohama National University, Section II. Biological and Geological Sciences (1952–)

JORDAN
BOOKS: *Handbook of the Geology of Jordan*, by D. J. Burdon (Government of the Hashemite Kingdom of Jordan, 1959)
Beiträge zur Regionalen Geologie der Erde. 7. Geologie von Jordanien, by F. Bender (Borntraeger, 1968)

Reference could also be made to F. Bender *et al.*, 'Beiträge zur Geologie Jordaniens' (*Beihefte Geologischen Jahrbuch*, **81**, 1969)

KOREA
The geology of Korea is covered in the series of volumes entitled *Geology and Mineral Resources of the Far East*, edited by T. Ogura (University of Tokyo Press, 1967–). Other relevant references are:

'A Sketch of Korean Geology', by T. Kobayashi (*American Journal of Science*, **26** (5), 585–606, 1933)
Geology of South Korea, by T. Kobayashi (University of Tokyo Press, 1953)

JOURNAL: *Bulletin of the Geological Survey of Korea* (1957–)

KUWAIT
'Geology of the Arabian Peninsula. Kuwait', by D. I. Milton (*Professional Paper, United States Geological Survey*, No. 560F, 1967)

LEBANON
Beiträge zur Regionalen Geologie der Erde. 6. Geologie von Syrien und der Libanon, by R. Wolfart (Borntraeger, 1967)

MALAYSIA
BOOKS: *The Geology of Malaya*, by J. B. Scrivenor (Macmillan, 1931)
'Geology of the Colony of North Borneo', by M. Reinhard and E. Wenk (*Bulletin of the Geological Survey Department of the British Territories in Borneo*, **1**, 1951)
'The Geology of Sarawak, Brunei and the Western Part of North Borneo', by P. Leicht and F. W. Roe (*Bulletin of the Geological Survey Department of the British Territories in Borneo*, **3**, 1960)
BIBLIOGRAPHY: 'Bibliography and Index of Geology of West Malaysia', by D. J. Gobbett (*Bulletin of the Geological Society of Malaysia*, **2**, 1969)
'Bibliography of West Malaysia and Singapore Geology, Supplement 1968', by D. J. Gobbett (*Newsletter of the Geological Society of Malaysia*, **20**, 7–11, 1969)
JOURNALS: *Annual Report of the Geological Survey, Borneo Region, Malaysia* (1949–)
Bulletin of the Geological Society of Malaysia (1968–)
Bulletin of the Geological Survey, Borneo Region, Malaysia (1951–)
District Memoirs of the Geological Survey, Federation of Malaya (1937–)

Memoirs of the Geological Survey, Borneo Region, Malaysia (1954–)
Professional Papers of the Geological Survey Department, Federation of Malaya (1962–)
Report of the Geological Survey, Borneo Region, Malaysia (1963–)

MANCHURIA
The geology of Manchuria is discussed in *Geology and Mineral Resources of the Far East*, edited by T. Ogura (University of Tokyo Press, 1967–)

MONGOLIA
BOOKS: Only one work in the western languages is available on Mongolian geology: *Geology of Mongolia*, by C. P. Berkey and F. K. Morris (American Museum of Natural History, 1927). Two Russian works may be of use: *Materialy po Geologii Mongolskoĭ Narodnoi Republiki*, by N. A. Marinov (Nedra, 1966), and *Stratigrafiya i Tektonika Mongolskoĭ Narodnoi Republiki* (Nedra, 1970).

NEPAL
There is no major work but the following may be of value:

Recherches Géologiques dans l'Himalaya du Népal, Region du Makalu, by P. Bordet (CNRS, 1961)
'On the Geology of Central West Nepal—A Preliminary Note', by T. W. A. Bodenhausen *et al.* (*Proceedings of the 22nd International Geological Congress*, **11**, 101–122, 1964)
'Report on the Geological Survey of Nepal, 1', by T. Hagen (*Denkschriften der Schweizerischen Naturforschenden Gesellschaft*, **86**, 1969)
'Geological Investigations in West Nepal and their Significance for the Geology of the Himalayas', by W. Frank and G. R. Fuchs (*Geologische Rundschau*, **59** (2), 552–580, 1970)

OMAN, QATAR AND PERSIAN GULF SHEIKHDOMS
The stratigraphy of this area is briefly reviewed in *The Geology of the British Empire*, by F. H. C. Reed (2nd edn., Arnold, 1949). Also worthy of mention are:

'The Geology and Tectonics of Oman and Parts of Southeastern Arabia', by M. G. Lees (*Quarterly Journal of the Geological Society of London*, **84**, 585–670, 1929)
The Geology and Mineral Resources of Dhufar Province, Muscat and Oman, by C. S. Fox (Calcutta, 1947)
'The Geology of Oman', by D. M. Morton (*Proceedings of the 5th World Petroleum Congress*, Section 1, Paper 14, 1959)

PAKISTAN
BOOKS: Reference should be made to India for the more general works. One entirely devoted to Pakistan is: *Reconnaissance Geology of Part of West Pakistan* (Hunting Survey Corporation Limited, Toronto, 1961)
'Stratigraphic Boundary Problems: Permian and Triassic of West Pakistan', edited by B. Kummel and S. Teichert (*Special Publication, Department of Geology, University of Kansas*, **4**, 1970)
BIBLIOGRAPHIES: 'Preliminary Bibliography and Index of the Geology of Pakistan' (*Records of the Geological Survey of Pakistan*, **12**, 1965)
Geological Bibliography of the Former Province of Sind (Jamshoro, 1968)

JOURNALS: *Geological Bulletin of the Department of Geology, University of Peshawar* (1966–)
Geological Bulletin of the Punjab University (1961–)
Geonews (1968–)
Memoirs of the Geological Survey of Pakistan (1956–)
Records of the Geological Survey of Pakistan (1948–)

PHILIPPINES
BOOKS: *Handbuch der Regionalen Geologie.* VI, pt. 5. *The Philippine Islands*, by W. D. Smith (Winter, 1910)
Geology and Mineral Resources of the Philippine Islands, by W. D. Smith (Bureau of Science of the Philippines Department of Agriculture and Natural Resources, 1951)
'Geology and Oil Possibilities of the Philippines', by G. W. Corby *et al.* (*Technical Bulletin, Philippines Department of Agriculture and Natural Resources*, **21**, 1951)
BIBLIOGRAPHIES: *Bibliography of Philippine Geology, Mining and Mineral Resources*, compiled by J. Teves (Philippine Bureau of Mines, 1953)
'Bibliography of Philippine Palaeontology and Stratigraphy 1861–1957', compiled by B. A. Daleon (*The Philippine Geologist*, **12**, 16–32, 1957)
JOURNALS: *The Philippine Geologist* (1947–)
Report of the Bureau of Mines, Republic of the Philippines (1952–)

SAUDI ARABIA
BOOKS AND REVIEWS: Some of the more important works on this area are:

Handbuch der Regionalen Geologie. V, pt. 4. *Syrien Arabien und Mesopotamien*, by M. L. P. Blanckenhorn (Winter, 1915)
Structure Géologique de l'Arabie, by P. Lamare (Beranger, 1936)
'Esquisse Géologique de l'Arabie Séoudite' (*Bulletin de la Société Géologique de France*, **7** (6), 653–697, 1957)
'The Arabian Shield', by G. R. Brown and R. O. Jackson (*Report of the 21st International Geological Congress*, **9**, 69–77, 1960)
'Geology of the Arabian Peninsula: Sedimentary Geology of Saudi Arabia', by R. W. Powers *et al.* (*Professional Paper, United States Geological Survey*, No. 560D, 1966)

SINGAPORE
'Note on the Geology of the Republic of Singapore', by M. Mainguy (*Technical Bulletin, Economic Commission for Asia and the Far East*, **2**, 83–85, 1969)

SYRIA
BOOK: *Beiträge zur Regionalen Geologie der Erde.* 6. *Geologie von Syrien und der Libanon*, by R. Wolfart (Borntraeger, 1967)

TAIWAN (FORMOSA)
BOOK: *Physiography and Geology of Taiwan*, by V. C. Juan (China Culture Publishing Foundation, 1954)
BIBLIOGRAPHY: *Bibliography of Geology of Taiwan*, compiled by T. P. Yen *et al.* (Geological Survey of Formosa, 1947)
JOURNALS: *Acta Geologica Taiwanica* (1947–)
Bulletin of the Geological Survey of Taiwan (1947–)
Petroleum Geology of Taiwan (1962–)

Proceedings of the Geological Society of China (1957–)

THAILAND

BOOK: 'Geologic Reconnissance of the Mineral Deposits of Thailand', by G. F. Brown *et al.* (*Bulletin of the United States Geological Survey*, No. 984, 1951). This contains a stratigraphic summary and a useful bibliography. Reference should also be made to two papers by T. Kobayashi: 'Geology of Thailand' and 'Palaeontology of Thailand 1916–62' (*Geology and Palaeontology of South-East Asia*, **1**, 1–15, 17–29, 1964).

VIETNAM

JOURNAL: *Archives Géologiques du Viet-Nam* (1957–)

YEMEN

BOOK: 'Geology of the Arabian Peninsula: Yemen', by F. Geukens (*Professional Paper, United States Geological Survey*, No. 560B, 1966)

AFRICA

BOOKS: Africa as a whole is surprisingly well covered by general stratigraphic texts. In contrast, the coverage of the component countries is extremely varied and some of Africa's states do not maintain any sort of geological survey. Reports of the Association of African Geological Surveys will be found in the proceedings volumes of the International Geological Congress. The following are the general texts:

Geologie Afrikas, by E. Krenkel (3 vols., Borntraeger, 1925–34)
Regionalen Geologie der Erde. 1. sect. 5. *Afrika*, by G. Hennig (Akademische Verlag, 1938)
Geologie und Bodenschätze Afrikas, by E. Krenkel (Akademische Verlag, 1957)
Geology of Africa, by R. Furon (Oliver and Boyd, 1963)
The Mineral Resources of Africa, by N. de Kun (Elsevier, 1965). This work includes a summary of the stratigraphy for each of the countries covered

In addition, the following works have a broad geographic scope:

The Stratigraphic History of Africa, South of the Sahara, by S. H. Haughton (Oliver and Boyd, 1963)
Report on the Geology and Geophysics of the East African Rift System, edited by A. M. Hunter *et al.* (East African Committee for Co-operation in Geophysics, Nairobi, Kenya, 1965)
The Geochronology of Equatorial Africa, by L. Cahen and N. J. Snelling (North-Holland, 1966)
'Sahara-Correlazioni Geologico-Lithostratigrafiche fra Sahara Centrale ed Occidentale', by L. K. Ratschiller (*Memorie del Museo Tridentino di Scienze Naturali*, **16**, 1967)

The geology of the many former British African possessions is summarised in *The Geology of the British Empire*, by F. R. C. Reed (2nd edn., Arnold, 1949), and more succinctly in 'The Geology of the British African Colonies', by F. Dixey and E. S. Willbourn (*Report of the XVIII International Geological Congress, London, 1948*, **14**, 1951).
There have been a number of conferences devoted to African geology,

one of the latest being issued as 'Résumés des Communications Présentées au 5e Colloque de Géologie Africaine, Clermont-Ferrand, 9–11 Avril, 1969', in *Annales de la Faculté des Sciences de l'Université de Clermont*, **41** (1969).
BIBLIOGRAPHIES: The continent as a whole has received excellent attention from many bibliographers. The most important general publications are: *Geological Bibliography of Africa* (UN Economic Commission for Africa, Standing Committee on Industry and Natural Resources, 1963. Document E/CN.14/INR/48) and *Selected Geologic Bibliography. Regional Descriptions and Maps*. I. *Africa* (CGMW 1964). Covering more restricted areas are: *Geological Bibliography of Africa South of the Sahara* (Commission for Technical Co-operation in Africa South of the Sahara). I. *Bibliography of the Karroo System*, by S. H. Haughton (1956). II. *Bibliography of the Jurassic and Cretaceous Systems of Africa South of the Sahara* (1959). This series has been continued by 'Abstracts of African Mining, Geology and Metallurgy (African Geology South of the Sahara)' in *Journal of the Nigerian Mining and Metallurgical Society*, now called *Journal of Mining and Geology*.

Since Africa was so recently composed mostly of European colonies, the publications of the colonial geological surveys are especially relevant. It should be noted in particular that the former Afrique Equatoriale Française now comprises four nations—the Central African Republic, the Republic of the Congo (Brazzaville), Chad and Gabon. Also, the former Afrique Occidentale Française covered the present Mali, Chad, Niger, Mauretania, Senegal and Upper Volta. Relevant bibliographies covering these areas are:

Bibliographie Géologique de l'Afrique Centrale (édition provisoire). Pt. I: *To 1935*; Pt. II: *1935–1944*; Pt. III: *1945–1949* (Association des Services Géologiques Africains, 1937–51)
'Bibliographie Géologique de l'Afrique Equatoriale Française du Cameroun et des Régions Limitrophes', compiled by M. E. Denaeyer (*Annales Academie des Sciences Coloniales*, **6**, 1933).
Bibliographie et Carte Géologique de l'Afrique Occidentale Française, compiled by L. Masvier (Direction-Fédérale des Mines et de la Géologie, Dakar, 1952)
'Bibliographie Géologique de la France d'Outre-Mer', by F. Blondel (*Publications du Bureau d'Etudes Géologiques et Minières Coloniales*, **11**, 1941)
'Bibliographie Géologiques de la France d'Outre Mer (Général Algerie, Tunisie, Maroc) tome I' (*Publications du Bureau d'Etudes Géologiques et Minières Coloniales*, **20**, 1952)
Bibliografia Geologica Italiana dell'Africa Italiana Sino al 1948 Incluso, by A. Desio (Ufficio Studi Ministero dell'Africa Italiana, Rome, 1950)
Bibliografia Geologia do Ultramar Português, by F. Gonçalves and J. Caseiro (Junta da Investigações do Ultramar, 1959)
JOURNALS: Journals covering the former Afrique Equatoriale Française and Afrique Occidentale Française are:
Bulletin de la Direction Fédérale des Mines Gouvernement Général de l'Afrique Occidentale Française (1943–55)
Notice Explicative Carte Géologique de Reconnaissance (Gouvernement Général de l'Afrique Equatoriale Française (1935–48)

The desert areas of northern Africa are well served by *Publications du Centre de Recherches sur les Zones Arides. Série Géologie* (1964–)

ALGERIA

BOOKS AND REVIEWS: The best summary accounts of Algerian stratigraphy are to be found in the excursion guides and papers of the XIX International Geological Congress, held in Algeria in 1952. Other works are:

La Géologie Algérienne et Nord-Africaine depuis 1830, by J. Savornin (Ancienne Maison Bastide-Jourdan, Algiers, 1931)
'Les Traits Essentials de la Géologie Algérienne', by M. Kieken, in *Livre à la Mémoire du Professeur Paul Fallot*, tome I, 545–614 (Société Géologique de France, 1962)

BIBLIOGRAPHY: 'Bibliographie de l'Algérie du Sud (Sahara) et des Régions Limitrophes', by O. Merabet (*Bulletin Service de la Carte Géologique de l'Algérie*, N.S., 37, 1968)
JOURNALS: *Bulletin du Service de la Carte Géologique de l'Algérie* (1953–). Contains, since 1954, periodical bibliographic summaries of Algerian geology
Publications du Service de la Carte Géologique de l'Algérie (1953–)

ANGOLA

A brief outline is contained in 'Carte Géologique de l'Angola, Notice Explicative', by F. Mouta (*Report of the XVIII International Geological Congress, London, 1948*, 14, 1951)
BIBLIOGRAPHIES: 'Bibliografia Geológica de Angola', by A. T. S. Ferreira da Silva (*Memoría Direccão Provincial dos Serviços de Geologia e Minas*, 10, 1971)
Bibliografia Geologia do Ultramar Português, by F. Gonçalves and J. Caseiro (Junta de Investigações do Ultramar, 1959)
JOURNAL: *Boletim dos Serviços de Geologia e Minas Angola* (1960–)

BOTSWANA

BOOK: A general account is 'An Outline of the Geology of the Bechuanaland Protectorate', by A. Poldervaart and D. Green in *Comptes Rendus XIX Congrès Géologique International*, 20 (1954)
BIBLIOGRAPHY: *Annotated Bibliography and Index of the Geology of Botswana to 1966* (Geological Survey of Botswana, 1968)
JOURNALS: *Annual Report of the Geological Survey of Botswana* (1953–)
Records of the Geological Survey of Botswana (1956–)

BURUNDI

BIBLIOGRAPHY: *Bibliographie Géologique du Congo et du Ruanda-Urundi* (1955–68). Also known for part of the time under a slightly different title, this bibliography covers the period 1818–1966.

CAMEROUN

BOOK: *Notice Explicative de la Carte Géologique du Cameroun au 1: 1 000 000*, by J. Gazel *et al.* (Direction des Mines et de la Géologie, 1956)
BIBLIOGRAPHY: 'Bibliographie Géologique du Cameroun', compiled by A. Denaeyer and F. Blondel (*Bulletin Société Camerounaises*, 6, 1944)
JOURNALS: *Bulletin de la Direction des Mines et de la Géologie Territoire du Cameroun* (1953–)
Bulletin du Service Géologique Territoire du Cameroun (1954–)
Rapport Annuel du Service Géologique Territoire du Cameroun (1954–)

CENTRAL AFRICAN REPUBLIC
BOOK: *Notice Explicative de la Carte Géologique de l'Afrique Equatoriale Française. 1:2 000 000*, by G. Gerard (Direction des Mines et de la Géologie de AEF, 1958)
JOURNALS: *Notice Explicative Carte Géologique de Reconnaissance République Centrafricaine* (1960–)

CHAD
BOOKS: *Excursion to Chad* (Petroleum Exploration Society of Libya. 3rd Annual Field Conference, 1961)
South Central Libya and Northern Chad. A Guide Book to Geology and Prehistory, edited by J. J. Williams (Petroleum Exploration Society of Libya. 8th Annual Field Conference, 1966)

CONGO (BRAZZAVILLE)
BOOK: 'Notice Explicative de la Carte Géologique de la République du Congo Brazzaville au 1:500 000', by P. Dadet (*Mémoires du Bureau de Recherches Géologiques et Minières*, **70**, 1969)
JOURNALS: *Bulletin de l'Institut de Recherches Scientifiques au Congo* (1950–63)
Cahiers de l'Institut de Recherches Scientifiques au Congo (1964–)

CONGO (KINSHASA)
BOOKS: *Géologie du Congo Belge*, by L. Cahen (Vaillant-Carmanne, Liege, 1954)
Géologie et Géographie du Katanga, by M. Robert (Hayez, Brussells, 1956)
BIBLIOGRAPHY: *Bibliographie Géologique du Congo et du Ruanda-Urundi 1955–68*. Also known for some time under a slightly different title, this bibliography covers the literature from 1818 to 1966.
JOURNALS: *Annales Musée Royale de l'Afrique Centrale. Sciences Géologiques* (1961–)
Publications de l'Université Officielle du Congo à Lubumbashi (1966–)

DAHOMEY
BOOKS AND REVIEWS: 'Le Précambrien du Dahomey', by R. Pougnet (*Bulletin Direction des Mines et de la Géologie* (Dakar), **22**, 1959)
'Vue d'Ensemble sur le Bassin Sédimentaire Côtier du Dahomey-Togo', by M. Slansky (*Bulletin Société Géologique Française*, **8** (5), 555–580, 1958)
Notice Explicative de la Carte Géologique du Dahomey au 1:1 000 000, by P. Aicard (Bureau de Recherches Géologiques et Minières, Paris, 1960)

EGYPT (UNITED ARAB REPUBLIC)
BOOKS: *Geology of Egypt*, by W. F. Hume (2 vols., in 4 parts, Geological Survey, Cairo, 1925–37)
'Review of Egyptian Geology', by R. Said and E. M. El-Shazly (*Egyptian Reviews of Science*, **1**, 1957)
The Geology of Egypt, by R. Said (Elsevier, 1962)
Guidebook to the Geology and Archaeology of Egypt, edited by F. A. Reilly (Petroleum Exploration Society of Libya, 5th Annual Field Conference, 1964)
BIBLIOGRAPHIES: *A Bibliography of Geology and Related Sciences Concerning Egypt up to the End of 1939*, compiled by E. H. Koldani (Government Printer, Cairo, 1941). Supplemented to 1957 by 'Review of Egyptian Geology' (see above)

JOURNALS: *Journal of Geology of the UAR* (1960–)
Papers of the Geological Survey and Mineral Research Department, UAR (1958–)

ETHIOPIA
BOOKS: *Saggio di una Carta Geologica dell'Eritrea, della Somalia e dell'Etiopia alla Scala di 1:2 000 000 con Note Illustrative*, by G. Stefanini (Instituto Geografico Militare, Florence, 1933)
Geologia dell'Africa Orientale, by G. Dainelli (4 vols., Reale Accademia d'Italia, 1943)
The Geology of Ethiopia, by P. A. Mohr (University College of Addis Ababa, 1963)
JOURNALS: *Bulletin of the Geophysical Observatory, Haile Selassie I University* (1959–)
Contributions of the Geophysical Observatory, Haile Selassie I University (1963–)

FRENCH TERRITORY OF THE AFARS AND ISSAS
BOOKS AND REVIEWS: 'Etude de Géologie et de Géographie Physique sur la Côte Française des Somalis', by M. Dreyfuss (*Revue de Géographie Physique et Géologie Dynamique*, **4**, 1931)
'Itineraires Géologiques en Somalie Française', by E. Aubert de la Rue (*Revue de Géographie Physique et Géologie Dynamique*, **12**, 1939)
Notice Explicative de la Carte Géologique de la Côte Françaises des Somalis 1:400 000, by H. Besairie (Imprimerie Nationale, Paris, 1946)

GABON
BOOK: 'Carte Géologique de la République Gabonaise au 1:1 000 000' (*Mémoires du Bureau de Recherches Géologiques et Minières*, **72**, 1970)

GAMBIA
BOOK: 'Report on a Rapid Geological Survey of Gambia', by W. G. Cooper (*Bulletin of the Geological Survey of the Gold Coast*, **3**, 1925)

GHANA
BOOKS: 'Geology of the Gold Coast and Western Togoland with a Revised Geological Map', by N. R. Junner (*Bulletin of the Geological Survey of the Gold Coast*, **11**, 1940)
The Geological Evolution of the Gold Coast, by D. A. Bates (Commission for Technical Co-operation in Africa, London, 1957)
BIBLIOGRAPHY: 'A Bibliography of Gold Coast Geology, Mining and Archaeology to March 1937', compiled by W. T. James (*Bulletin of the Geological Survey of the Gold Coast*, **9**, 1937)
JOURNALS: *Bulletin of the Geological Survey of Ghana* (1925–)
Report of the Geological Survey of Ghana (1913–)

GUINEA
BOOK: 'Géologie de la Guinée Française', by R. Furon (*Publications du Bureau d'Etudes Géologiques et Minières Coloniales*, **19**, 1943)

IVORY COAST
BOOK: 'Reconnaissance Pétrolière (Géologiques et Géophysique) du Bassin Sédimentaire de la Côte d'Ivoire', by P. Maugis (*Bulletin de la Direction des Mines et Géologie Afrique Occidentale Française*, **19**, 1955)

JOURNAL: *Rapport Annuel de la Direction des Mines et de la Géologie* (1966–   )

KENYA
BOOKS: *An Outline of the Geology of Kenya*, by S. M. Cole (Pitman, 1950) 'The Geology and Mineral Resources of Kenya', edited by W. Pulfrey (*Bulletin of the Geological Survey of Kenya*, **2**, 1960)
BIBLIOGRAPHY: 'Bibliography of the Geology of Kenya, 1859–1968', by N. P. Dosaj and J. Walsh (*Bulletin of the Geological Survey of Kenya*, **10**, 1970)
JOURNALS: *Annual Report of the Geological Survey of Kenya* (1949–   )
Bulletin of the Geological Survey of Kenya (1954–   )
Memoirs of the Geological Survey of Kenya (1953–   )

LESOTHO
BOOK: *Report on the Geology of Basutoland*, by G. M. Stockley (Government Printer, Maseru, 1947)

LIBERIA
BOOK: *Handbuch der Regionalen Geologie*. VI, pt. 6A. *English Colonies on the West Coast of Africa and Liberia*, by J. Parkinson (Winter, 1913)
JOURNALS: *Bulletin of the Geological Society of Liberia* (1966–   )
Bulletin of the Geological Survey of Liberia (1967–   )

LIBYA
BOOKS: 'Geology and Mineral Resources of Libya. A Reconnaissance', by G. H. Goudarzi (*Professional Paper, United States Geological Survey*, No. 660, 1970)

The excellent series of guidebooks produced by the Petroleum Exploration Society of Libya unexpectedly contains only one partial account of Libya itself, namely, *South-Central Libya and Northern Chad. A Guidebook to the Geology and Prehistory*, edited by J. J. Williams (Petroleum Exploration Society of Libya. 8th Annual Field Conference, 1966). A conference entitled 'Symposium on the Geology of Libya' was held in 1969 and the papers were published by the University of Libya in 1971.
BIBLIOGRAPHY: *A Bibliography of Libya*, by R. W. Hill (University of Durham, 1959)

MALAGASY (MADAGASCAR)
BOOKS AND REVIEWS: 'Monographie Géologique de Madagascar', by H. Besaire (*Lexique Stratigraphique International*, **4** (11), supplément, 1960) 'Geology and Geomorphology of Madagascar in Comparison with Eastern Africa', by F. Dixey (*Quarterly Journal of the Geological Society of London*, **116**, 255–268, 1960)
BIBLIOGRAPHIES: Various attempts seem to have been made at a bibliography of the geology of Madagascar—for example, in *Publications du Service Géologique*, **26, 115, 135, 145**.
JOURNALS: *Annales Géologiques de Madagascar* (1958–   )
Documentation du Bureau Géologique, Comité National Malgache de Géologie (1964–   )
Documentation du Bureau Géologique, Service Géologique, Republique Malgache (1956–   )
Mémoires de l'Académie Malgache (1926–   )

MALAWI

BOOK: 'The Geology and Mineral Resources of Nyasaland', by W. G. G. Cooper (*Bulletin of the Geological Survey of Nyasaland*, **6**, 1952)
JOURNALS: *Annual Report of the Geological Survey Department, Malawi* (1963–)
Bulletin of the Geological Survey Department, Malawi (1965–)
Memoirs of the Geological Survey Department, Malawi (1963–)
Records of the Geological Survey Department, Malawi (1961–)

MALI

BOOK: 'Contribution à l'Etude Géologique de Soudan Oriental', by H. Radier (*Bulletin Direction Fédéral des Mines et de Géologie* (Dakar), **26**, 1959)
BIBLIOGRAPHY: 'Bibliographie de l'Algérie du Sud (Sahara et des Régions Limitrophes)', by O. Merabet (*Bulletin du Service de la Carte Géologique de l'Algerie*, N.S., **37**, 1968)

MAURITANIA

BOOK AND REVIEW: *Notice Explicative de la Carte Géologique de l'A.O.F. Feuille 7. Mauritanie* (Direction Fédéral des Mines et de Géologie, Dakar, 1957)
'Les Mauritanides et leur Avant-Pays' (*Bulletin de la Société Géologique de France*, **11** (7), 133–272, 1969)
BIBLIOGRAPHY: 'Bibliographie de l'Algérie du Sud (Sahara et des Régions Limitrophes)', by O. Merabet (*Bulletin du Service de la Carte Géologique de l'Algérie*, N.S., **37**, 1968)

MOROCCO

BOOK: 'Géologie du Maroc. Histoire Géologique du Domaine de l'Anti-Atlas', by G. Choubert (*Notes et Mémoires du Service des Mines et de la Géologie du Maroc*, **100**, 1952)
BIBLIOGRAPHY: 'Bibliographie Analytique des Sciences de la Terre: Maroc et Régions Limitrophes (Depuis le Début des Recherches Géologiques à 1964)' (*Notes et Mémoires de Service des Mines et de la Carte Géologique du Maroc*, **182**, 1965); '1965–1969' (ibid., **212**, 1970)
JOURNALS: *Mémoires de la Sociéte des Sciences Naturelles et Physiques de Maroc N.S. Géologie* (1955–)
Mines et Géologie: Bulletin Trimestriel de la Direction des Mines et de la Géologie (1958–)
Notes et Mémoires du Service des Mines et de la Carte Géologique du Maroc (1927–)

MOZAMBIQUE

BOOKS: 'Noticia Explicativa do Esboço Geológico de Moçambique 1:2 000 000, by A. J. De Freitas (*Boletim. Serviços de Indústria, Minas e Geologia*, **23**, 1957)
A Geologia e o Desenvolvimento Econômico e Social de Moçambique, by A. J. De Freitas (Imp. Nac. Moçambique, 1959)
BIBLIOGRAPHY: *Bibliografia Geológica do Ultramar Portugues*, by F. Gonçalves and J. Caseiro (Junta de Investigações do Ultramar, 1959)
JOURNALS: *Boletim. Serviços de Indústria, Minas e Geologia* (1937–)
Memórias do Instituto de Investigação Científicas de Moçambique (1959–)
Revista de Ciencias Geológicas Universidade de Lourenco Marques, Ser. A. (1968–)

NIGER
BOOK: 'Essai de Description des Formations Géologiques de la Republique du Niger', by J. Greigert and R. Pougnet (*Mémoires du Bureau de Recherches Géologiques et Minières*, **48**, 1967)
BIBLIOGRAPHY: 'Bibliographie de l'Algérie du Sud (Sahara et des Régions Limitrophes)', by O. Merabet (*Bulletin du Service de la Carte Géologique de l'Algérie*, N.S., **37**, 1968)

NIGERIA
BOOKS AND REVIEWS: *The Geology and Geography of Northern Nigeria*, by J. D. Falconer (Macmillan, 1911)
'A Review of Nigerian Stratigraphy', by C. M. Tattam (*Annual Report of the Geological Survey of Nigeria*, 1943)
Aspects of the Geology of Nigeria, by R. A. Reyment (Ibadan University Press, 1965)
JOURNALS: *Annual Report of the Geological Survey of Nigeria* (1930–)
Bulletin of the Geological Survey of Nigeria (1921–)
Journal of Mining and Geology (1968–)
Records of the Geological Survey of Nigeria (1956–)

PORTUGUESE GUINEA
'Geologia da Guiné Portuguesa', by J. E. Teixeira, in *Curso de Geologia do Ultramar*, Vol. 1, pp. 53–104 (Junta de Investigações do Ultramar, 1968)
BIBLIOGRAPHY: *Bibliografia Geológica do Ultramar Português*, by F. Gonçalves and J. Caseiro (Junta de Investigações do Ultramar, 1959)
JOURNAL: *Boletim Cultural da Guiné Portuguesa* (1946–)

RHODESIA
BOOK: 'An Outline of the Geology of Southern Rhodesia', edited by W. H. Swift (*Bulletin of the Geological Survey of Southern Rhodesia*, **50**, 1961)
BIBLIOGRAPHY: 'Rhodesian Geology: A Bibliography and Brief Index to 1968', by C. C. Smith and H. E. van der Heyde (*Occasional Paper National Museum of Rhodesia*, **4**, 323–575, 1971)
JOURNALS: *Annual Report of the Geological Survey of Rhodesia* (1911–)
Bulletin of the Geological Survey of Rhodesia (1913–)

RWANDA
BIBLIOGRAPHIES: *Bibliographie Géologique du Congo et du Ruanda-Urundi* (1955–68). Also known for some time under a slightly different title, this bibliography covers the period 1818–1966. Forming a continuation of this is a compilation by A. Bertossa, 'Liste des Travaux Ayant Trait à la Géologie du Rwanda et des Régions Limitrophes Publies en 1965 et 1966 et Compléments pour la Période 1960–1964' (*Bulletin du Service Géologique de la Republique Rwandaise*, **3**, 1966)
JOURNAL: *Bulletin du Service Géologique de la Republique Rwandaise* (1964–)

SAO THOMÉ AND PRINCIPE
REVIEWS: 'L'Ile de Sao Thomé', by A. Chevalier (*La Géographie*, **13**, 1906)
'Notas Sobre e Geologia das Ilhas de Sao Thomé e do Principe', by C. Teixera (*Estudios Coloniais*, **1**, 1948–49). See also *Beiträge zur Regionalen Geologie der Erde*. 10. *Geology of the South Atlantic Islands*, by R. C. Mitchell-Thomé (Borntraeger, 1970)

SENEGAL
JOURNALS: *Annales de la Faculté des Sciences Université de Dakar* (1959–)
Annuaire de l'Association Sénégalaise pour l'Etude du Quaternaire de l'Ouest Africain (1967–)
Bulletin de Liaison. Association Sénégalaise pour l'Etude du Quaternaire de l'Ouest Africain (1967–)

SIERRA LEONE
BOOK: 'The Geology and Mineral Resources of Sierra Leone', by J. D. Pollet (*Colonial Geology and Mineral Resources*, **2**, 3–28, 1951)
JOURNALS: *Bulletin of the Geological Survey of Sierra Leone* (1958–)
Report of the Geological Survey of Sierra Leone (1918–)

SOMALIA
BOOKS: *Handbuch der Regionalen Geologie.* VII, VIIIa, pt. 26. *Abes-somalien* (*Abessinien und Somalien*), by E. Krenkel (Winter, 1926)
Geologia dell'Africa Orientale, by G. Dainelli (Reale Accademia d'Italia, 1943)
The Geology of British Somaliland, by W. A. Macfadyen (Government Press, 1933)
A General Survey of the Somaliland Protectorate 1944–1950 (*CDW Scheme D.484*), by J. A. Hunt (Crown Agents, London, 1951)
JOURNAL: *Report of the Geological Survey of the Somali Republic* (1956–)

SOUTH AFRICA
BOOKS: *The Geology of South Africa*, by A. L. Du Toit (3rd edn., Oliver and Boyd, 1954)
Geology of Southern Africa, by E. D. Mountain (Books of Africa, 1968)
Geological History of Southern Africa, by S. H. Haughton (Geological Society of South Africa, 1969). This is intended to include knowledge gained since the third edition of A. L. Du Toit's book in 1954
BIBLIOGRAPHIES: 'Bibliography of South African Geology', compiled by A. J. Hall (*Memoirs of the Geological Survey of South Africa*, **18**, **22**, **25**, **30**, **37**, 1922–39)
Bibliography and Subject Index of South African Geology, 1957– (Geological Survey of South Africa, 1959–)
JOURNALS: *Annals of the Geological Survey of South Africa* (1962–)
Annals of the South African Museum (1898–)
Annals of the Transvaal Museum (1908–)
Bulletin of the Geological Survey of South Africa (1934–)
Bulletin of the Transvaal Museum (1955–)
Memoirs of the Geological Survey of South Africa (1912–)
Report of the Chamber of Mines' Pre-Cambrian Research Unit, University of Cape Town (1963–)
South African Journal of Science (1909–)
Special Publication of the Geological Society of South Africa (1969–)
Transactions and Proceedings of the Geological Society of South Africa (1934–)

SOUTHWEST AFRICA
Many accounts of the geology of Southwest Africa appear in publications covering the Republic of South Africa but reference may also be made to:
BOOK AND REVIEW: 'Südwestafrika, Geologie und Bergbau', by P. Range (*Zeitschrift der Deutschen Geologischen Gesellschaft*, **39**, 468–509, 1937)

The Pre-Cambrian Geology of South West Africa and Namaqualand, by H. Martin (University of Cape Town Press, 1965)
BIBLIOGRAPHY: 'A Bibliography of Geological and Allied Subjects, South West Africa', by H. Martin (*Bulletin of the Chamber of Mines' Pre-Cambrian Research Unit, University of Capetown*, 1, 1966)

SPANISH GUINEA
BOOKS AND REVIEWS: 'Someras Notas para Contribuir a la Descripsción Geológica de la Zona N.W. de la Isla de Fernando Poo y de Guinea Continental Española', by E. d'Almonte (*Boletín de la Real Sociedad de Geografía*, 44, 190–347, 1906)
Geología y Geografía Física de la Guinea Continental Española, by J. D. de Lizaur y Roldan (Dirección General de Marruecoe y Colonias, Madrid, 1945)

See also *Beiträge zur Regionalen Geologie der Erde. 10. The Geology of the South Atlantic Islands*, by R. C. Mitchell-Thomé (Borntraeger, 1970)

SPANISH WEST AFRICA
REVIEW: 'Lithostratigraphy of the Northern Spanish Sahara', by L. K. Ratschiller (*Memorie del Museo Tridentino di Scienze Naturali*, 18, 1970)

SUDAN
BOOK: *The Geology of the Sudan Republic*, by A. J. Whiteman (Oxford University Press, 1971)
BIBLIOGRAPHIES: 'Sources of Information on the Geology of the Anglo-Egyptian Sudan', compiled by G. Andrew (*Bulletin of the Geological Survey of the Sudan*, 3, 1945). Supplements, between 1950 and 1957, were published in the *Annual Report of the Geological Survey of the Sudan*
JOURNALS: *Bulletin of the Geological Survey of the Sudan* (1911–)
Memoirs of the Geological Survey of the Sudan (1962–)

SWAZILAND
Most major early references are to be found in books and journals on the geology of the Union of South Africa.
BOOKS: *The Geology of Swaziland*, by D. R. Hunter (Swaziland Geological Survey and Mines Department, 1961). A book entitled *The Geology of Swaziland*, by H. J. R. Way, is given as 'in press' in some of the more recent bibliographies.
JOURNALS: *Annual Report of the Geological Survey and Mines Department, Swaziland* (1958–)
Bulletin of the Geological Survey and Mines Department, Swaziland (1961–)

TANZANIA
BOOKS AND REVIEWS: *Report on the Geology of the Zanzibar Protectorate*, by G. M. Stockley (Zanzibar Government, 1928)
'The Geology of the Zanzibar Protectorate and its Relation to the East African Mainland', by G. M. Stockley (*Geological Magazine*, 79, 233–240, 1942)
'Geology and Mineral Resources of Tanganyika', by C. B. Bisset and K. A. Davies (*Bulletin of the Imperial Institute*, 45, 1947)
'Summary of the Geology of Tanganyika. I. Introduction and Stratigraphy',

by A. M. Quennell *et al.* (*Memoir of the Geological Survey of Tanganyika*, **1**, 1956)
BIBLIOGRAPHY: *Bibliography of the Geology and Mineral Resources of Tanzania to December 1967* (Bureau of Resources Assessment and Land Use Planning, Dar-es-Salaam, 1969)
JOURNALS: *Bulletin of the Geological Survey of Tanzania* (1927–)
Memoirs of the Geological Survey of Tanzania (1956–)
Records of the Geological Survey of Tanzania (1951–)

TOGO
BOOKS: 'Geology of the Gold Coast and Western Togoland', by N. R. Junner (*Bulletin of the Geological Survey of the Gold Coast*, **11**, 1940)
'Esquisse Physique et Géologiques du Togo. Gisements de Chromite du Togo. L'Or du Togo. Les Métaux Autres que l'Or et le Chrome du Togo (le Titane, l'Aluminium, le Plomb)' (*Bulletin de la Direction Fédérale des Mines et de la Géologie*, **11**, 1949)
BIBLIOGRAPHY: 'Bibliographie Géologique et Minière du Togo. La Chronique des Mines Coloniales', compiled by V. Kachinsky (*Annales du Bureau d'Etudes Géologiques et Minières Coloniales*, **2** (21), 1933)

TUNISIA
BOOK: *Guidebook to the Geology and History of Tunisia*, edited by L. Martin (Petroleum Exploration Society of Libya, 9th Annual Field Conference, 1967)
BIBLIOGRAPHY: *Bibliographie Géologique de la Tunisie* (1954–)
JOURNALS: *Annales des Mines et de la Géologie* (1947–)
Mémoires du Service de la Carte Géologique de la Tunisie (1934–)
Notes du Service Géologique. Service des Mines de l'Industrie et de l'Energie, Tunisie (1950–)

UGANDA
BOOKS: 'Geology and Mineral Deposits of Uganda', by K. A. Davies and C. B. Bisset (*Bulletin of the Imperial Institute*, **45**, 161–181, 1947)
BIBLIOGRAPHY: 'Bibliography of Uganda Geological Literature' (*Bulletin of the Geological Survey of Uganda*, **3**, 1939)
JOURNALS: *Annual Report of the Geological Survey of Uganda* (1922–)
Bulletin of the Geological Survey of Uganda (1933–)
Memoirs of the Geological Survey of Uganda (1925–)
Occasional Papers of the Geological Survey of Uganda (1926–)
Record of the Geological Survey of Uganda (1950–)
Report of the Geological Survey of Uganda (1959–)

UPPER VOLTA
BOOKS: 'La Géologie et les Resources Minières de la Haute Volta Méridionale', by Y. Sagatsky (*Bulletin de la Direction des Mines et de la Géologie, AOF*, **13**, 1954)
'Etude Géologique des Migmatites et des Granites Précambriens au Nord-Est de la Côte d'Ivoire et de la Haute-Volta Méridionale', by A. Arnould (*Mémoires du Bureau de Recherches Géologiques et Minières*, **3**, 1961)
JOURNAL: *Rapport Annuel de la Direction de la Géologie et des Mines* (1961–)

ZAMBIA
BOOKS AND REVIEWS: 'A Summary of the Provisional Geological Features

of Northern Rhodesia', by T. D. Guernsey and J. A. Bancroft (*Colonial Geology and Mineral Resources*, 1 (2), 1950)
The Geology of the Northern Rhodesian Copperbelt, edited by F. Mendelsohn (MacDonald, 1961)
'The Geology and Mineral Resources of Northern Rhodesia', by W. H. Reeve (*Bulletin of the Geological Survey of Northern Rhodesia*, 3, 1963)
BIBLIOGRAPHIES: 'A Bibliography of Northern Rhodesia Geology', by G. J. Snowball (*Records of the Geological Survey of Northern Rhodesia, for 1959*, 35–76, 1960). Continued by 'An Annotated Bibliography and Index of Northern Rhodesia Geology 1960–61', by G. J. Snowball (*Records of the Geological Survey of Northern Rhodesia*, 9, 27–67, 1963). Continued by the serial *Annotated Bibliography and Index of the Geology of Zambia*. There also exists, by G. J. Snowball, *Annotated Bibliography and Index of the Geology of Zambia 1931–1959* (Geological Survey of Zambia, 1965)
JOURNALS: *Bulletin of the Geological Survey of Northern Rhodesia* (1959–60). None issued since independence
Memoirs of the Geological Survey of Northern Rhodesia (1963–64). None issued since independence
Records of the Geological Survey of Zambia (1966–)
Report of the Geological Survey of Zambia (1965–)

AUSTRALASIA
A comprehensive treatment of the whole region is not available. However, the conference proceedings and other papers contained in the periodical *Pacific Geology* (1968–) should be consulted.

AUSTRALIA
BOOKS: *Australian Stratigraphy*, by R. W. Fairbridge (2nd edn., University of Western Australia Press, 1953)
The Geology of the Commonwealth of Australia, by T. W. Edgeworth David (3 vols., Arnold, 1959)
The Geological Evolution of Australia and New Zealand, by D. A. Brown *et al.* (Pergamon, 1968)
Ancient Australia, by C. F. Laseron (2nd edn., Angus and Robertson, 1969)

Australian geology is also summarised in *The Geology of the British Empire*, by F. R. C. Reed (2nd edn., Arnold, 1949).

There is no over-all Australian geological survey, although the Commonwealth Government's Bureau of Mineral Resources, Geology and Geophysics is in some measure concerned with the whole subcontinent. Geological mapping and also much research is undertaken by state surveys. The relevant publications of the Bureau and the other national journals are listed below, after which books and journals relating to each state are successively considered:

Bulletin of the Bureau of Mineral Resources, Geology and Geophysics (1932–)
Journal of the Geological Society of Australia (1953–)
Reports of the Bureau of Mineral Resources, Geology and Geophysics (1947–)

Special Publication of the Geological Society of Australia (1967–)
There are, of course, many journals which cover all aspects of the natural sciences and these often contain material of interest to the stratigrapher. One which may perhaps be less well-known than some is the *Australian Journal of Marine and Freshwater Research* (1965–).

New South Wales
BOOKS: 'The Geology of New South Wales', edited by G. H. Packham (*Journal of the Geological Society of Australia*, **16** (1), 1969)
Field Geology of New South Wales, by D. F. Branagan and G. H. Packham (Science Press, 1967)
JOURNALS: *Annual Reports of the Geological Survey of New South Wales* (1962–)
Bulletin of the Geological Survey of New South Wales (1922–)
Journal of the Proceedings of the Royal Society of New South Wales (1875–)
Memoirs of the Geological Survey of New South Wales (1887–)
Records of the Geological Survey of New South Wales (1889–)

Northern Territory
BOOK: *The Geology and Palaeontology of Queensland and New Guinea*, by R. I. Jack and R. Etheridge Jnr. (2 vols., Government of Queensland, 1892)

Queensland
BOOKS: *The Geology and Palaeontology of Queensland and New Guinea* (see above)
'The Geology of Queensland', edited by D. Hill and A. K. Denmead (*Journal of the Geological Society of Australia*, **7**, 1960)
Geological Excursions in South East Queensland, by N. C. Stevens (University of Queensland Press, 1965)
Elements of the Stratigraphy of Queensland, by D. Hill and W. G. H. Maxwell (2nd edn., University of Queensland Press, 1967)
JOURNALS: *Papers, Department of Geology, University of Queensland* (1937–)
Publications of the Geological Survey of Queensland (1902–)
Proceedings of the Royal Society of Queensland (1884–)
Queensland Government Mining Journal (1902–1910; 1922–)

South Australia
BOOKS: 'The Geology of South Australia', edited by M. F. Glaessner and L. W. Parkin (*Journal of the Geological Society of Australia*, **5**, 1958)
Handbook of South Australian Geology, edited by L. W. Parkin (Geological Survey of South Australia, 1969)
BIBLIOGRAPHY: *Bibliography of South Australian Geology*, by E. N. Teesdale-Smith (Geological Survey of South Australia, 1959)
JOURNALS: *Bulletin of the Geological Survey of South Australia* (1912–)
Quarterly Geological Notes, Geological Survey of South Australia (1962–)
Report of the Geological Survey of South Australia (1954–)

Tasmania
BOOKS: 'The Geology of Tasmania', edited by A. Spry and M. R. Bank (*Journal of the Geological Society of Australia*, **9** (2), 1962)
'The Geology and Mineral Resources of Tasmania', by I. B. Jennings *et al.*

(Bulletin of the Geological Survey of Tasmania, **50**, 1967)
JOURNALS: *Bulletin of the Geological Survey of Tasmania* (1907–)
Record of the Geological Survey of Tasmania (1913–)
Papers and Proceedings of the Royal Society of Tasmania (1848–)

Victoria
BOOK: *Regional Guide to Victorian Geology,* edited by J. McAndrew and
M. A. H. Marsden (Parkville, 1968)
JOURNALS: *Mining and Geological Journal* (1937–)
Proceedings of the Royal Society of Victoria (1888–)
Memoirs of the Geological Survey of Victoria (1903–)

Western Australia
BOOK: 'The Stratigraphy of Western Australia', by J. R. H. McWhae *et al.*
(Journal of the Geological Society of Australia, **4**, 1958)
JOURNALS: *Bulletin of the Geological Survey of Western Australia* (1898–)
Journal and Proceedings of the Royal Society of Western Australia (1914–)
Report of the Geological Survey of Western Australia (1969–)

New Guinea
BOOKS AND REVIEWS: *The Geology and Palaeontology of Queensland and
New Guinea* (see under Australia, Northern Territory)
The Geology of Indonesia, by R. W. Van Bemmelen (2 vols., 2nd edn.,
Nijhoff, 1970)
'Geological Results of the Exploration for Oil in Netherlands New Guinea',
by W. A. Visser and J. J. Hermes (*Verhandelingen van het Koninklijk
Nederlandsch Geologisch Mijnbouwkundig Genootschap Geologische Serie,*
20, special issue, 1962)
'A Geological History of Eastern New Guinea', by J. E. Thompson (*Jour-
nal of the Australian Petroleum Exploration Association,* **7**, 83–93, 1967)
JOURNAL: *Nova Guinea Geology* (1960–64)

NEW ZEALAND
BOOKS AND REVIEWS: *Handbuch der Regionalen Geologie.* VII. pt. I. *New
Zealand and Adjacent Islands,* by P. Marshall (Winter, 1911)
Geology of New Zealand, by P. Marshall (New Zealand Geological Sur-
vey, 1912)
The Geology of the British Empire, by F. R. C. Reed (2nd edn., Arnold,
1949)
The Geological Evolution of Australia and New Zealand, by D. A. Brown
et al. (Pergamon, 1968)
'The Mesozoic of New Zealand: Chapters in the History of the Circum-
Pacific Mobile Belt', by C. A. Fleming (*Quarterly Journal of the Geo-
logical Society of London,* **125**, 125–170, 1970)
BIBLIOGRAPHIES: 'A Bibliography of New Zealand Geology to 1950', by
G. L. Adkin and B. W. Collins (*Bulletin of the New Zealand Geological
Survey N.S.,* **65**, 1967). This includes details of other compilations and is
supplemented for the years 1949–55 in *New Zealand Journal of Science
and Technology* and then in *New Zealand Journal of Geology and Geo-
physics.*
JOURNALS: *Bulletin of the Geological Survey of New Zealand* (1906–)
Earth Science Journal, Waikato Geological Society (1967–)
Journal of the Royal Society of New Zealand, Geology (1970–)
Memoirs of the Geological Survey of New Zealand (1928–)

New Zealand Journal of Geology and Geophysics (1958–)
New Zealand Journal of Marine and Freshwater Research (1967–)
Newsletter of the Geological Society of New Zealand (1960–)

NORTH AMERICA

BOOKS: *Geology of North America*, by R. Ruedemann and R. Balk (Borntraeger, 1939)
Historical Geology: The Geologic History of North America, by R. C. Hussey (McGraw-Hill, 1947)
Atlas of Paleogeographic Maps of North America, by C. Schuchert (Wiley, 1955)
The Evolution of North America, by P. B. King (Princeton University Press, 1959)
The Geological Evolution of North America; A Regional Approach to Historical Geology, by T. H. Clark and C. W. Stearn (Ronald, 1960)
Geochronology of North America (National Academy of Sciences, Washington, 1965)

The following works deal with broad areas that transcend national boundaries:

Geology of the Great Lakes, by J. L. Hough (University of Illinois, 1958)
Geology of the Atlantic and Gulf Coast Province of North America, by G. E. Murray (Harper, 1961)
'Rocky Mountain Sedimentary Basins' (*Bulletin of the American Association of Petroleum Geologists*, **49** (11), 1965)
The Geologic Atlas of the Rocky Mountain Region (Rocky Mountain Association of Geologists, 1972)
'Correlation of the North American Silurian Rocks', by W. B. N. Berry and A. J. Boucot (*Special Paper of the Geological Society of America*, **102**, 1970)
BIBLIOGRAPHIES: *Abstracts of North American Geology* (see Chapter 5)
Bibliography of North American Geology (see Chapter 5)
Selected Guides for Geologic Field Study in Canada and the United States of America, by D. H. Lokke (Earth Science Curriculum Project Reference Series RS-9, Prentice-Hall)

CANADA
BOOKS: *Geology and Economic Minerals of Canada*, by R. J. W. Douglas (2 vols., 5th edn., Geological Survey of Canada, 1970)

Major areas of Canada are dealt with in:

Geology of Nova Scotia (Nova Scotia Department of Mines)
Geology of Quebec, by J. Dressner and T. C. Denis (3 vols., Quebec Department of Mines, 1941–49)
Western Canada Sedimentary Basin. A Symposium, edited by L. M. Clark (American Association of Petroleum Geologists, 1954)
Jurassic and Carboniferous of Western Canada. A Symposium, edited by A. J. Goodman (American Association of Petroleum Geologists, 1958)
Geology and Mineral Resources of Manitoba, by J. F. Davies *et al.* (Mines Branch, Manitoba, 1962)

Geological History of Western Canada: An Atlas and Text (2nd edn., (Alberta Society of Petroleum Geologists, 1966)
Beiträge zur Regionalen Geologie der Erde. 5. Rocky Mountains, by D. H. Roeder (Borntraeger, 1967)

BIBLIOGRAPHIES: The Geological Survey of Canada has produced useful catalogues and indexes of its own publications. The *Canadian Index to Geoscience Data* is a computer-based co-ordinate index which identifies sources on the data of the geology of Canada. The first edition issued in 1970 includes about 20 000 document titles, mainly from the Geological Survey of Canada but also from the Mineral Resources Branch, Quebec Department of Natural Resources and Ontario Department of Energy and Resources Management. Individual parts of the index are available for the Provinces and Territories. A full review of this new and ambitious venture appeared in *Geoscience Documentation*, **2**, 134–136 (1970). Other relevant bibliographies are:

'Bibliography of Geology, Palaeontology, Industrial Minerals and Fuels in the Post-Cambrian Regions of Manitoba to 1950', by L. B. Kerr; '1950–1957', by B. A. Mills; '1958–1965', by B. B. Bannatyne. These were issued as *Publications of the Department of Mines and Natural Resources, Manitoba*, **51–2, 57–4, 66–1** (1951–66)
'Cross-index to the Geological Illustrations of Canada', by C. Faessler *Contributions, Laval Université, Géologie et Minéralogie*, **75, 117, 118, 127**, 1956–57). Further volumes are planned
Annotated Bibliography of Geology of the Sedimentary Basin of Alberta and of Adjacent Parts of British Columbia and North Territories, compiled by R. G. McCrossan *et al.* (Alberta Society of Petroleum Geologists, 1958)
Annotated Bibliography of Saskatchewan Geology (1823–1958), compiled by W. O. Kupsch (Saskatchewan Department of Mineral Resources, 1959). Supplements have been issued
Lexicon of Geologic Names in the Western Canada Sedimentary Basin and Arctic Archipelago (Alberta Society of Petroleum Geologists, 1960)
Lexicon of Paleozoic Names in South-Eastern Ontario, by C. G. Winder (University of Toronto, 1961)
'Bibliography of New Brunswick Geology', edited by D. Abbott (*Report of the New Brunswick Research and Productivity Council*, **2**, part C, 1965)
'Bibliography of the Geology of Newfoundland and Labrador, 1814 through 1968', by S. Butler and G. Bartlett (*Bulletin of the Mineral Resources Division, Department of Mines, Agriculture and Resources, Newfoundland*, **38**, 1970)

JOURNALS: *Bulletin of Canadian Petroleum Geology* (1963–)
Bulletin of the Department of Mines and Petroleum Resources, British Columbia (1960–)
Bulletin of the Department of Mines, Agriculture and Resources, Newfoundland (1960–)
Bulletin (Geology) of the Research Council of Alberta (1958–)
Bulletin of the Geological Survey of Canada (1945–)
Bulletin of the Ontario Department of Mines (1896–)
Bulletin of the Saskatchewan Geological Survey (1948–)
Canadian Journal of Earth Sciences (1964–)
Geological Circular of the Ontario Department of Mines (1955–)
Geological Report of the Department of Mines, Quebec (1939–)

Geological Report of the Ontario Department of Mines (1960–)
Memoir of the Geological Survey of Canada (1910–)
Memoir (Geology) of the Research Council of Alberta (1959–)
Miscellaneous Reports of the Geological Survey of Canada (1962–)
Papers of the Geological Survey of Canada (1939–)
PreCambrian Geology Series, Reports of the Saskatchewan Geological Survey (1948–)
Proceedings of the Geological Association of Canada (1947–)
Report of the Department of Mines and Natural Resources (1961–)
Report (Geology) of the Research Council of Alberta (1966–)
Transactions of the Royal Society of Canada, Section IV (1882–)

MEXICO
Good accounts of Mexican stratigraphy are to be found in the field guides produced for the 20th International Geological Congress, Mexico City, 1956, and for the Annual Meeting of the Geological Society of America, Mexico City, 1968.
BOOK: *Geología de México*, by V. R. Garfias and T. C. Chapin (Editorial Jus, 1949)
BIBLIOGRAPHIES: 'Bibliografía Geológica y Minera de la República Mexicana', compiled by R. Aguilar y Santillán (*Boletín del Instituto Geológico de México*, **10**, 1898, 'Supplements', **17**, 1908; and in *Boletín Minero, Mexico*, 1918)
Bibliografía Geológica y Minera de la République Mexicana Correspondiente a los Anos 1919–1930, by R. Aguilar y Santillan (Talleres Gráficos de la Nasion, 1936)
JOURNALS: *Anales del Instituto Geológico de México* (1917–)
Boletín de la Asociación Mexicana de Geólogos Petroleros (1949–)
Boletín del Instituto Geológico de México (1895–)
Boletín de la Sociedad Geológica Mexicana (1905–)

UNITED STATES OF AMERICA
The United States Geological Survey (USGS) is sponsored by the US Government. However, most states maintain (or have maintained, in the past) independent geological surveys. The USGS therefore operates primarily in states without independent surveys; however, it quite often collaborates with the state surveys on particular projects or its officers may be invited to undertake particular projects within these states.
So far as the subjects of this chapter are concerned the principal publications of the USGS are the *Bulletin* (1883–), the *Circular* (1933–) and the *Professional Papers* (1902–). Details of these and other Survey publications can be found in *Publications of the Geological Survey 1879–1961* (United States Government Printing Office, 1964), which is supplemented annually from 1962. This annual publication itself cumulates the monthly *New Publications of the Geological Survey*. Work by the officers of the Survey is fully recorded for certain years in the *Professional Papers* —for example, No. 650A, 1969. Much of the information will be included in a new journal published by the USGS, *The Journal of Research* (1973–).
Still with the central authority the value of Congressional publications to the geologists is reviewed in 'Congressional Geology', by H. R. Pestana and B. D. Bonta (*Bulletin of the Geological Society of America*, **81**, 899–904, 1970)
State geological surveys range in size from the very small to the very large. Some, such as the California, Illinois, Oklahoma and Texas surveys,

produce long series of publications of admirable quality. Their activity has, in many cases, been markedly discontinuous. Even now some states do not have active geological surveys, while others have combined geological and natural history surveys.

The publications of the state surveys are too numerous and changes of name too frequent for ready listing. However, a reasonably comprehensive listing is available, and it is of particular use in searching out the earlier publications. This is *An Index of State Geological Survey Publications Issued in Series*, by J. B. Corbin (Scarecrow Press, 1965). Also of use is *A Bibliography of Bibliographies of Geology of the States of the United States*, by H. K. Long (American Geological Institute, 1971). In addition, most state surveys will readily provide up-to-date lists of their publications.

Besides the surveys a number of US universities and museums either produce geological publications or feature geological papers in journals with a wider coverage. To help locate these there is a good bibliography, which forms a companion to Corbin, mentioned above. This is *Museum Publications: A Classified List and Index of Books, Pamphlets, and other Monographs, and of Serial Reprints,* Part 2. *Publications in the Biological and Earth Sciences*, by J. Clapp (Scarecrow Press, 1962)

Each state also has its geological societies. The publications of these are the bane of the librarian. Many are most useful but their bibliographic control leaves much to be desired. This is not to say that attempts have not been made. Many addresses, for instance, are listed in the issues of the *Bulletin of the American Association of Petroleum Geologists*. Some of the publications themselves are listed in *Index of Geological Publications for 1966* (American Association of Petroleum Geologists, 1969) and in *Geologic Field Trip Guidebooks of North America. A Union List Incorporating Monographic Titles*, by the Guidebook and Ephemeral Materials Committee of the Geoscience Information Society (Wilson, 1969) and a later edition.

It would take more than a book to record all the relevant geological books, series and bibliographies for the USA, but some of the more important examples of the literature are given below.

BOOKS: *Handbuch der Regionalen Geologie.* 8, pt. 2. *United States of North America*, by E. Blackweider (Winter, 1912). This is now considerably out of date.

Particular geologic periods are dealt with in:

The Quaternary of the United States, edited by H. E. Wright and D. G. Frey (Princeton University Press, 1965)
Pennsylvanian System in the United States: A Symposium, edited by C. C. Branson (American Association of Petroleum Geologists, 1962)

Particular regions and states are covered by:

Geology of California, by R. D. Reed (American Association of Petroleum Geologists, 1933)
Miocene Stratigraphy of California, by R. M. Kleinpell (American Association of Petroleum Geologists, 1938)
Geology of the Atlantic and Gulf Coastal Province of North America, by G. E. Murray (Harper, 1961)
A Guide to the National Parks. Their Landscape and Geology. Vol. 1:

The Western Parks. Vol. 2: *The Eastern Parks,* by W. H. Matthews (Natural History Press, 1968)
Geology of the Northern Channel Islands, edited by D. W. Weaver (Pacific Section, American Association of Petroleum Geologists/Society of Economic Paleontologists and Mineralogists, 1969)
'Geology of the American Mediterranean. 19th Annual Meeting of the Gulf Coast Association of Geological Societies and Regional AAPG Meeting', edited by S. S. Winters (*Transactions of the Gulf Coast Association of Geological Societies,* **19**, 1969)
Studies of Appalachian Geology: Northern and Maritime, edited by E-An Zen *et al.,* and *Central and Southern,* edited by G. W. Fisher *et al.* (Interscience, 1968, 1970)
The Tectonics of the Appalachians, by J. Rodgers (Wiley/Interscience, 1970)

BIBLIOGRAPHIES: *Bibliography of North American Geology* and *Abstracts of North American Geology* have been discussed previously, as has the bibliography of state bibliographies by Long. Of a somewhat more specialised nature are:

Geological Literature on the Cook Inlet Basin and Vicinity, Alaska, by J. C. Maher and W. M. Trollman (Department of Natural Resources, Alaska, 1969)
Bibliography of Gulf Coast Geology, edited by J. Braunstein (Gulf Coast Association of Geological Societies, 1970)
Geologic Literature on the North Slope of Alaska, by J. C. Maher and W. M. Trollman (American Association of Petroleum Geologists, 1970)

JOURNALS: There are, of course, many relevant journals produced in the USA and the list that follows is no more than an indication of what exists. Some of the journals have come to be regarded more as international publications than ones devoted solely to North America.

American Journal of Science (1818–)
Baylor Geological Studies (1961–)
Brigham Young University Geology Studies (1954–)
Bulletin of the American Association of Petroleum Geologists (1917–)
Bulletin of the Geological Society of America (1890–)
Contributions to Geology (1962–)
Earth Science Bulletin (1968–)
Guidebook to the Geology of Utah (1946–)
Journal of Geology (1893–)
Memoirs of the Geological Society of America (1934–)
Quarterly of the Colorado School of Mines (1906–)
Southeastern Geology (1965–)
Special Papers of the Geological Society of America (1934–)
Transactions of the Gulf Coast Association of Geological Societies (1953–)

THE CARIBBEAN
BOOKS AND REVIEWS: *Historical Geology of the Antillean-Caribbean Region,* by C. Schuchert (Wiley, 1935)
The Geology of the British Empire, by F. R. C. Reed (2nd edn., Arnold, 1949)

La Constitution Géologique et la Structure des Antilles, by J. Butterlin
(CNRS, Paris, 1956)
'Caribbean Geological Investigations', edited by H. H. Hess (*Memoir of
the Geological Society of America*, **98**, 1966)
Beiträge zur Regionalen Geologie der Erde. 4. Geologie der Antillen, by
R. Weyl (Borntraeger, 1966)
'A Summary of the Geology of the Lesser Antilles', by P. H. A. M. Kaye
(*Overseas Geology and Mineral Resources*, **10**, 172–206, 1969)
'Geological History of the Caribbean', by H. J. MacGillavry (*Proceedings.
Koninklijke Nederlandse Akademie van Wetenschappen B*, **73**, 64–96, 1970)
BIBLIOGRAPHIES: *Bibliography of West Indian Geology*, by L. M. R. Rutten
(Oosthoek, Utrecht, 1938): '1938–1955', by R. C. Mitchell and J. Butterlin
(20th International Geological Congress, 1956)
Regional Bibliography of Caribbean Geology, by L. K. Fink (Institute
of Marine Science, Miami, 1964)
Selected Bibliography of Geology and Geophysics for the Gulf of Mexico,
by H. R. Eensminger and W. T. Morton (Naval Oceanographic Office,
Washington, 1968)
JOURNALS: *Status of Geological Research in the Caribbean* (1959–)
Transactions of the Caribbean Geological Conference (1958–)

BAHAMAS
REVIEW: 'Geology of the Bahamas', by R. M. Field *et al.* (*Bulletin of
the Geological Society of America*, **42**, 759–784, 1931)

BARBADOS
BOOK: Reference may be made to the relevant sections of *Beiträge zur
Regionalen Geologie der Erde. 4. Geologie von Antillen*, by R. Weyl
(Borntraeger, 1966)

CARRIACOU
REVIEW: 'Geology and Fossils of Carriacou, W.I.', by C. T. Trechman
(*Geological Magazine*, **72**, 529–555, 1935)

CAYMAN ISLANDS
REVIEW: 'Geology of the Cayman Islands (British West Indies) and their
relation to the Bartlett Trough', by C. A. Matley (*Quarterly Journal of the
Geological Society of London*, **82**, 352–386, 1926)

CUBA
BOOKS AND REVIEW: 'Outline of the Geology of Cuba', by R. A. Palmer
(*Journal of Geology*, **53**, 1–34, 1945)
Las Formaciónes Geológicas de Cuba, by P. J. Bermudez (2nd edn.,
Instituto Cubano de Recursos Minerales, 1963)
Geología de Cuba, by G. Furrazola-Bermúdez (2 vols., Consejo Nacional
de Universidades, 1964)
BIBLIOGRAPHY: *Bibliografía Geológica Cubana*, by P. J. Bermudez (Publi-
caciónes de la Revista 'Universidad de la Habana', 1938)
JOURNALS: *Memórias de la Facultad de Ciencias, Universidade de la
Habana* (1963–)
Publicaciónes Especial, Departamento Científico de Geología (1964–)

DOMINICAN REPUBLIC
REVIEWS: 'A Geological Reconnaissance of the Dominican Republic', by

T. W. Vaughan *et al.* (*Memoir US Geological Survey of the Dominican Republic*, **1**, 1921)
'Geology of the Central Dominican Republic (A Case History of Part of an Island Arc)', by C. O. Bowin, in H. H. Hess (editor): 'Caribbean Geological Investigations' (*Memoir of the Geological Society of America*, **98**, 11–84, 1966)
JOURNAL: *Publicaciónes del Instituto Geográfico y Geológico de la Universidad de Santo Domingo* (1946–50)

HAITI
REVIEW: 'Géologie Générale et Régionale de la République d'Haiti', by J. Butterlin (*Travaux et Mémoires, Institut des Hautes Etudes de l'Amérique Latine*, **6**, 1960)

JAMAICA
REVIEW: 'Synopsis of the Geology of Jamaica', by V. A. Zans *et al.* (*Bulletin of the Geological Survey Department of Jamaica*, **4**, 1963)
JOURNALS: *Bulletin of the Geological Survey Department of Jamaica* (1951–)
Bulletin of the Institute of Jamaica, Science Series (1940–)
Journal of the Geological Society of Jamaica (1964–)

MONTSERRAT, ST. KITTS, NEVIS AND ANGUILLA
BOOK AND REVIEW: *Report on the Geology of St. Kitts, Nevis and Anguilla*, by K. W. Earle (Colonial Office, London, 1924)
'The Geology of Saba and St. Eustatius, with Notes on the Geology of St. Kitts, Nevis and Montserrat (Lesser Antilles)', by J. H. Westermann and H. Kiel (*Uitgaven van de Natuurwetenschappelijke Studiekring voor Suriname en de Nederlandse Antillen*, **24**, 1961)

PUERTO RICO
BOOK: 'Geology of Puerto Rico', by H. A. Meyerhoff (*Monographs of the University of Puerto Rico, B*, **1**, 1933)
BIBLIOGRAPHY: *Bibliography and Index of the Geology of Puerto Rico and Vicinity 1866–1968*, by M. Hooker (Geological Society of Puerto Rico, 1969)

SABA, ST. EUSTATIUS, ETC.
BOOK: See under Montserrat, St. Kitts, Nevis and Anguilla.
BIBLIOGRAPHY: *De Geologische Literatur over van belang voor Nederlandsch-Guyana (Suriname) en de Nederlandsche Westindische Eilandern (Antillen)* by J. F. Steenhuis (Moulton, 1934); *1934–50* (1950)

TRINIDAD AND TOBAGO
BOOKS AND REVIEW: *The Geology of Venezuela and Trinidad*, by R. A. Liddle (2nd edn., Palaeontological Research Institution, Ithaca, 1946)
'Geology of Tobago, British West Indies', by J. C. Maxwell (*Bulletin of the Geological Society of America*, **59**, 801–854, 1948)
The General and Economic Geology of Trinidad, British West Indies, by H. H. Suter (HMSO, 1960)

VIRGIN ISLANDS
BOOK AND REVIEWS: 'Geology of the Virgin Islands', by K. W. Earle (*Geological Magazine*, **61**, 339–351, 1924)

'Geology and Groundwater Resources of St. Croix, Virgin Islands', by D. J. Cederstrom (*Water Supply Paper, United States Geological Survey*, No. 1067, 1950)
'Geology of St. Thomas and St. John, US Virgin Islands', by T. W. Donnelly; and 'Geology of St. Croix, US Virgin Islands', by J. T. Whetten, in H. H. Hess (editor): 'Caribbean Geological Investigations' (*Memoir of the Geological Society of America*, **98**, 85–176, 177–239, 1966)

CENTRAL AND SOUTH AMERICA

BOOKS: *Geologie Südamerikas*, by H. Gerth (3 vols., Borntraeger, 1932–41)
Handbuch der Regionalen Geologie. VIII, pt. 4a. *Mittelamerika*, by K. Sapper (Winter, 1937)
'Handbook of South American Geology: An Explanation of the Geologic Map of South America', edited by W. F. Jenks (*Memoir of the Geological Society of America*, **65**, 1956)
Regionalen Geologie der Erde. I. *Die Geologie Mittelamerikas*, by R. Weyl (Borntraeger, 1961)
A Résumé of the Geology of South America, by R. P. Morrison (Institute of Earth Sciences, Toronto University, 1962)
Estructura Geológica Historia Tectonica y Morfología de America Central (Instituto Centroamericano de Investigación y Tecnología Industrial (ICAITI), Guatemala, 1968)
BIBLIOGRAPHIES: 'Geological and Paleontological Bibliography of Central America', by M. Maldonado-Koerdell (*Publication of the Pan-American Institute of Geography and History*, **204**, 1958)
'Selected Bibliography of South American Geology', by H. R. Cramer (*Tulsa Geological Society Digest*, **31**, 213–239, 1963)
Selected Geologic Bibliography. Regional Descriptions and Maps. 3. South America (CGMW, Paris, 1964)
Bibliographie und Dokumentation. 7. Wirtschaftsgeologische Literatur über Iberoamerika, by G. Bischoff and F. Renger (Institut für Iberoamerika-Kunde, Hamburg, 1966)
'Sources of Geological Information in Latin America', by H. W. Smith (*Working Paper, 14th Seminar on the Acquisition of Latin American Library Materials (Pan American Union)*, **4**, 1969)
JOURNALS: An excellent list of serials will be found in *Directorio Latinoamericano de Ciencias Geológicas* (UNESCO, Montevideo, 1968), which also includes details of institutions and individual geologists.
Informe de la . . . Reunion de Geológos de America Central (Guatemala) 1965–)
Publicaciones Geológicas del ICAITI (Instituto Centroamericano de Investigación y Tecnología Industrial) (1966–)

ARGENTINA

BOOKS: *Geología Argentina*, by A. Windhausen (Casa Jacabo Peuser, 1931)
Descripción Geológica de la Patagonia, by E. Feruglio (3 vols., Dirección General de Yacimientos Petrolíferos, 1949–50)
Geografía de la Republica Argentina (Sociedad Argentina de Estudios Geograficos, 1957). Volumes are devoted to the various geological eras. The work is not geographical
Indice Bibliografico de Estratigrafía Argentina, by A. V. Borrello (Comision de Investigación Científico, Buenos Aires, 1965)
BIBLIOGRAPHY: Bibliographic coverage is provided by the 'Bibliografía de la Geología, Mineralogía y Paleontología de la Republica Argentina (Incluso

de la Antartica Americana)' in *Boletín de la Academia Nacional de Ciencias en Córdoba* (1914–)
JOURNALS: *Acta Geologica Lilloana* (1956–)
Anales del Instituto Nacional de Geología y Minería (1947–)
Anales de la Jornadas Geológicas Argentinas (1962–)
Anales de la Universidad de la Patagonia 'San Juan Bosco' Ciencias Geológicas (1966–)
Boletín del Instituto Nacional de Ciencias en Cordoba (1874–)
Boletín del Instituto Nacional de Geología y Minería (1913–)
Contribuciones Científicas de la Facultad de Ciencias Exactas Físicas y Naturales Universidad de Buenos Aires. Ser. E. Geología (1956–)
Publicaciones de la Dirección Nacional de Geología y Minería (1958–)
Publicaciones del Instituto de Fisiografía e Geología Universidad Nacional del Litoral (1937–)
Revista de la Asociación Geológica Argentina (1948–)
Revista de la Facultad de Ciencias Naturales de Salta (1959–)
Revista del Instituto Nacional de Geología y Minería (1965–)
Revista del Museo Argentina de Ciencias Naturales 'Bernardino Rivadavia' e Instituto Nacional de Investigación de las Ciencias Naturales, Ciencias Geológicas (1957–)
Revista del Museo de La Plata, Sección Geología (1936–)

BOLIVIA
BOOK: *Geología de Bolivia*, by F. Ahlfeld and L. Branisa (Instituto Boliviano del Petroleo, 1960)
BIBLIOGRAPHY: 'Bibliografía Geológica, Mineralógica y Paleontológica de Bolivia', compiled by J. M. Reyes (*Boletín del Servicio Geológico de Bolivia*, **4**, 1962)
JOURNALS: *Boletín del Instituto Boliviano del Petroleo* (1959–)
Boletín del Servicio Geológico de Bolivia (1961–)

BRAZIL
BOOKS: *International Field Institute, Brazil* (American Geological Institute, 1966)
Problems in Brazilian Gondwana Geology, edited by J. J. Bigarella *et al.* (Instituto de Geologia, Universidade Federal do Paraná, 1967)
Beiträge zur Regionalen Geologie der Erde. 9. Geologie von Brasilien, by K. Beurlen (Borntraeger, 1970)
BIBLIOGRAPHIES: A series 'Bibliografia e Indice da Geologia do Brasil' is published at irregular intervals in *Boletim da Divisão de Geologia e Mineralogia, Brasil*, an example being 1962–63 as **244**, 1969. Other compilations include:

'Bibliografia da Geologia, Mineralogia, Petrografia e Paleontologia do Estado de Sao Paulo', by S. Mezzalira and A. Wohlers (*Boletim do Instituto Geográfico e Geológico do Estado de São Paulo*, **33**, 1952)
'Bibliografia Comentada é Indice de Geologia da Bahia', by P. R. Cruz (*Boletim do Serviço Geológico e Mineralógico do Brasil*, **242**, 1968)
'Bibliografia e Indice da Geologia da Amazonia Legal Brasileira 1641–1946', by P. Loewenstein *et al.* (*Publicações Avulsos do Museu Paranaense Emilo Goeldi*, **11**, 1969)

JOURNALS: *Anais da Academia Brasileira de Ciências* (1929–)

Archivos do Museu Paranaense, Geologia (1954–)
Boletim da Divisão de Geologia e Mineralogia, Brasil (1940–)
Boletim da Faculdade de Filosofia, Ciências e Letras Universidad de São Paulo, Série Geologia (1945–)
Boletim Geologia da Universidade Federal do Rio de Janeiro (1967–)
Boletim Paranaense de Geociências (1967–)
Boletim da Sociedade Brasileira de Geologia (1952–)
Boletim da Universidade Federal do Paraná, Geologia (1959–)
Iheringia, Museum Rio-Grandense de Ciências Naturais, Série Geologia (1967–)
Monografias do Divisão Geológico e Mineralógico, Ministera da Agricultura, Brasil (1913–)
Notas Preliminares e Estudos, Divisão de Geologia e Mineralogia do Brasil (1939–)
Publicações Avulsas da Universidade de Bahia Escola de Geologia (1964–)
Relatório Anual da Divisão de Geologia e Mineralogia do Brasil (1939–)

BRITISH HONDURAS
'Geology of British Honduras', by L. A. Ower (*Journal of Geology*, **36**, 494–509, 1928)
'Geology of Northern Honduras', by G. Flores (*Bulletin of the American Association of Petroleum Geologists*, **36**, 404–409, 1952)
Notes on the Geology of Southern British Honduras, by C. G. Dixon (Government Printer, Belize, 1957)

CHILE
BOOKS: *The Geology and Metal Deposits of Chile*, by J. M. Little (Branwell, N.Y., 1926)
Beiträge zur Regionalen Geologie der Erde. 3. Geologie von Chile, by W. Zeil (Borntraeger, 1964)
Geología y Yacimientos Metalíferos de Chile, by C. Ruiz Fuller *et al.* (Instituto de Investigaciones Geológicas, Chile, 1965)
BIBLIOGRAPHY: 'Bibliografia Geológica de Chile (1927–1953)', by J. Munoz Cristi and J. K. Kokot (*Publicaciones del Instituto de Geología Universidade de Chile*, **5**, 1955)
JOURNALS: *Anales de la Academia Chilena de Ciencias Naturales* (1938–)
Communicaciones de la Escuela Geología, Facultad de Físicas y Matemáticas, Universidad de Chile (1960–)
Publicaciones del Instituto de Geología, Universidad de Chile (1950–)

COLOMBIA
BOOKS: 'Historia Geológica de Columbia', by H. Burgl (*Revista de la Academia Colombiana de Ciencias Exactas, Físicas y Naturales*, **11**, 137–194, 1961). An English translation is available by N. R. Rowlinson: *The Geological History of Columbia* (1962)
BIBLIOGRAPHY: 'Bibliografía de los Informes del Instituto Geológico Nacional', compiled by A. Venegas Leyva (*Compilación de los Estudios Geológicos Oficiales en Colombia*, **9**, 1960: *Suplemento* (1960–64), 1965)
'Bibliografía de la Biblioteca del Instituto Geofísica de los Andes Colombianos sobre Geología y Geofísica de Colombia', compiled by J. E. Ramirez (*Boletín del Instituto Geofísica de los Andes Colombianos, Série C, Geología*, **6**, 1957)

JOURNALS: *Boletín Geológico del Instituto Geológico Nacional* (1952– *Geología Colombiana* (1962–)
(*Guidebooks*) *Colombian Society of Petroleum Geologists* (1959–)

COSTA RICA
BIBLIOGRAPHY: 'Bibliografía de la Geología de Costa Rica', by G. Dengo (*Publicaciones de la Universidad de Costa Rica, Série Ciencias Naturales*, **3**, 1962)

ECUADOR
BOOKS: *Geología del Ecuador*, by W. Sauer (Libreria Cima, 1969)
'The Geology of Ecuador. Explanatory Note for the Geological Map of the Republic of Ecuador (1:500 000)', by P. J. Goosens (*Annales de la Société Géologique de Belgique*, **93**, 255–263, 1970)
Beiträge zur Regionalen Geologie der Erde. 11. *Geologie von Ecuador*, by W. Sauer (Borntraeger, 1971)
BIBLIOGRAPHY: 'Bibliography of the Geology and Geography of Ecuador', by R. B. Colton (*United States Geological Survey, Open File*, **1040**, 1968)

FRENCH GUIANA
BOOK: *Géologie et Pétrographie de la Guyane Française*, by B. Choubert (Office de la Recherche Scientifique, 1949). Reference should also be made to the conferences cited under Guyana.
JOURNAL: *Mémoires du Service de la Carte Géologique Détaillée, Département de la Guyane Française* (1959–)

GUYANA
BOOKS: *The Geology of the British Empire*, by F. R. C. Reed (2nd edn., Arnold, 1949)
'The Geology of British Guiana and the Development of its Mineral Reserves', by G. M. Stockley (*Bulletin of the Geological Survey of British Guiana*, **25**, 1955)

There is also a series of conferences of which the proceedings have been published, e.g. the 6th in *Avulso, Divisão de Geologia e Mineralogia, Brasil*, **41** (1966) and the 7th (1966) in *Verhandelingen van het Koninklijk Nederlandsch Geologisch Mijnbouwkundig Genootschap*, **27** (1969). These proceedings also deal with Surinam and French Guiana.
BIBLIOGRAPHY: 'Bibliography of the Geology and Mining of British Guiana', compiled by C. G. Dixon and H. K. George (*Bulletin of the Geological Survey of British Guiana*, **32**, 1964)
JOURNAL: *Bulletin of the Geological Survey of Guyana* (1933–)

HONDURAS
BOOK: 'Volcanic History of Honduras', by H. Williams and A. R. McBirney (*University of California Publications, Geological Sciences*, **85**, 1969)

NICARAGUA
JOURNAL: *Boletín del Servicio Geológico Nacional de Nicaragua* (1957–)

PANAMA (AND CANAL ZONE)
BOOKS: 'A Geological Reconnaissance of Panama', by R. A. Terry (*Occasional Paper of the California Academy of Science*, **23**, 1956)

'Geology and Paleontology of the Canal Zone and Adjoining Areas' (*Professional Paper, United States Geological Survey*, No. 306, 1957–64)

PARAGUAY
BOOKS: 'Geology and Mineral Resources of Paraguay—A Reconnaissance', by E. B. Eckel (*Professional Paper, United States Geological Survey*, No. 327, 1959)
Beiträge zur Regionalen Geologie der Erde. 2. Geologie von Paraguay, by H. Putzer (Borntraeger, 1962)

PERU
BOOKS: *Geology of the Tertiary and Quaternary Periods in the North-West Part of Peru*, by T. O. Bosworth (Macmillan, 1922)
Geologie von Peru, by G. Steinmann (Winter, 1929)
'Memoria Explicativa del Mapa Geológico del Perú', by B. E. Bellido and F. S. Simons (*Boletín de la Sociedad Geológica del Perú*, **31**, 1951)
BIBLIOGRAPHIES: *Bibliografía Geológica del Perú*, by L. B. Castro (the author, Lima, 1960)
'1000 Selected References to the Geography, Oceanography, Geology, Ecology and Archaeology of Coastal Peru and Adjacent Areas', by N. P. Psuty *et al.* (*The Paracas Papers*, **1**, 1968)
JOURNALS: *Boletín del Servicio de Geología y Minería* (1967–   )
Boletín de la Sociedad Geológica del Perú (1925–   )

SURINAM
BOOKS: *Outline of the Geology and Petrology of Surinam (Dutch Guiana)*, by R. Ijzerman (Klemink en Zoon, Utrecht, 1931)
Geologie en Mijnbouw, **15** (6) (1953) is devoted to Surinam stratigraphy

Reference should also be made to the conferences listed under Guyana.
BIBLIOGRAPHY: *De Geologische Literatuur over of van belang voor Neder-landsch-Guyana (Suriname) en de Nederlandsche Westindische Eilandern (Antillen)*, by J. F. Steenhuis (Moulton, 1934); *1934–50* (1951)
JOURNAL: *Uitgaven van de Natuurwetenschappelijke Studiekring voor Suri-name en de Nederlandse Antillen* (1941–   )

URUGUAY
BOOK: *Geología del Uruguay*, by J. Bossi (Universidad de la Republica, Montevideo, 1966)
BIBLIOGRAPHY: 'Bibliographie Géologique de la République Orientale de l'Uruguay', compiled by R. Lambert (*Boletín del Instituto Geológico del Uruguay*, **26**, 1939)
JOURNAL: *Boletín del Instituto Geológico del Uruguay* (1938–   )

VENEZUELA
BOOK AND REVIEW: *The Geology of Venezuela and Trinidad*, by R. A. Liddle (2nd edn., Paleontological Research Institution, Ithaca, N.Y., 1946)
'Geology of Venezuela and its Oilfields', by E. Mencher *et al.* (*Bulletin of the American Association of Petroleum Geologists*, **37**, 690–777, 1953)
BIBLIOGRAPHIES: 'Bibliografía e Indice de Geología Minería y Petróleo de Venezuela 1813–1944', by H. D. Hedberg (*Revista de Fomento*, **7**, 1945); '1950–1958', by B. Korol and C. J. Forjonel (*Boletín de Geología, Dirección de Geología Venezuela*, **5**, 121–211, 1959)

'Stratigraphic Lexicon of Venezuela (English edition)' (*Boletín de Geologia, Publicación Especial*, **1**, 1956)

'Internal Publications on Geology of Venezuela, 1958–mid-1965', by R. M. Stainforth (*Bulletin of the American Association of Petroleum Geologists*, **49**, 2289–2294, 1965)

JOURNALS: *Boletín de Geología, Dirección de Geología, Venezuela* (1951–)
Boletín Informativo de la Asociación Venezolana de Geología Minería y Petróleo (1958–)
Boletín de la Sociedad Venezolana de Geólogos (1965–)
Geos (1959–)

ARCTIC REGIONS

Mention should first be made of the *Polar Record* (1931–), which contains a bibliographic section 'Recent Polar Literature'.

THE ARCTIC

BOOKS: *The Geology of the Arctic*, edited by G. O. Raasch (2 vols., University of Toronto Press, 1961). This is a symposium volume. A second symposium has been held and its *Proceedings* were published in 1972 by Oslo University

The Arctic Basin, edited by J. E. Slater (Arctic Institute of North America, 1963)

BIBLIOGRAPHY: *Arctic Bibliography* (1953–). See also Chapter 15

JOURNALS: *Årbok Norsk Polarinstitutt* (1960–)
Meddelelser om Norsk Polarinstitutt (1949–)

GREENLAND

BOOKS: *Handbuch der Regionalen Geologie*. 4, pt. 2. *Grönland*, by O. B. Bøggild (Winter, 1917)

Geologie von Grönland, by L. Koch (Borntraeger, 1935)

BIBLIOGRAPHY: See under Denmark.

JOURNALS: *Meddelelser om Grønland* (1879–)
Miscellaneous Reports of the Geological Survey of Greenland (1964–)
Reports of the Geological Survey of Greenland (1964–)

ICELAND (AND SURTSEY)

On the Geology and Geophysics of Iceland (21st International Geological Congress, Guidebook A2, 1960). Papers relating to the geological survey of Iceland are published in *Náttúrufroeðingurinn* (1931–), *Surtsey Research Progress Report* (1965–) and *Proceedings of the Surtsey Research Conference, Reykjavik, June 25–28, 1967* (1967)

SPITZBERGEN

BOOK AND REVIEW: 'The Geology of Spitzbergen', by D. L. Allen (*Proceedings of the Liverpool Geological Society*, **18**, 37–56, 1941)

Geology of Spitzbergen, edited by V. N. Sokolov (2 vols., National Lending Library for Science and Technology, 1970). Translation of Russian original dated 1965

BIBLIOGRAPHIES: 'Littérature Géologique Concernant la Norvège avec Svalbard et la Territoire des Recherches de la Mer Glaciale et la Mer Norvégienne et Groenlandienne', by H. Rosendahl (*Norsk Geologisk Tidsskrift*, **18**, 129–301, 1934)

'Bibliography of the Literature about Geology, Physical Geography, Useful Minerals and Mining of Svalbard', by A. Orvin (*Norges Svalbard-og Ishavs-Undersökelser Skrifter*, **89**, 1947)

Many works dealing with the Canadian and Soviet Arctic also exist and the reader is referred to the appropriate country sections. An example of one Canadian serial publication is *Axel Heiberg Island Research Reports, McGill University* (1959–   )

THE ANTARCTIC
General accounts of the geology of Antarctica become rapidly out of date, because of the rate of progress of knowledge; attention is therefore confined to works published since 1958.
BOOKS: *El Conocimiento Geológico de la Antártica*, by I. R. Cordini (Instituto Antártico Argentino, 1959)
'Antarctic Research (The Mathew Fontaine Maury Memorial Symposium). Presented to the 10th Pacific Science Congress' (*Geophysical Monographs, American Geophysical Union*, **7**, 1962)
Géologie de l'Antarctique, by A. Cailleux (Centre de Documentation Universitaire, Paris, 1963)
Antarctic Geology: Proceedings of the First International Symposium, Cape Town, 1963, edited by R. J. Adie (Interscience, 1964)
Antarctic Research: A Review of British Scientific Achievement in Antarctica, edited by R. E. Priestley (Butterworths, 1964)
'Geology and Paleontology of the Antarctic', edited by J. B. Hadley (*Antarctic Research Study, American Geophysical Union*, **6**, 1965)

The geology of the British Antarctic territories is reviewed in *The Geology of the British Empire*, by F. R. C. Reed (2nd edn., Arnold, 1949).
The *New Zealand Journal of Geology and Geophysics* has had, so far, five special issues devoted to Antarctic geology, viz: **5** (5) 1962, **6** (3) 1963, **8** (2) 1965, **10** (2) 1967 and **11** (4) 1968.
BIBLIOGRAPHIES: *Antarctic Bibliography* (1965–   ). See also Chapter 15
'Abstracting and Indexing Service for Current Antarctic Literature', by G. A. Doumani (*Antarctic Journal of the United States*, **1**, 226–227, 1966)
'Antarctic Geological Literature', by J. F. Splettstoesser (*Geotimes*, **11**, 33–35, 1966)
'Bibliography of Reports on Geology, Geomorphology and Glacial Geology Resulting from Australian Work in Antarctica', by I. R. McLeod (*Report of the Bureau of Mineral Resources, Geology and Geophysics*, **146**, 1970)
JOURNALS: *Antarctic Journal of the United States* (1966–   )
Antarctic Record, Reports of the Japanese Antarctic Research Expedition (1957–   )
Antarctic Research (1964–   )
Boletín del Instituto Antártico Argentino (1957–   )
Bulletin of the British Antarctic Survey (1963–   )
CNFRA Publications du Comité Français des Recherches Antarctiques (1962–   )
Publicaciones del Instituto Antártico Argentino (1955–   )
Publicaciones del Instituto Antártico Chileno (1964–   )
Scientific Reports of the British Antarctic Survey (1962–   )

ATLANTIC OCEAN ISLANDS

Works dealing with more than one island or island group include:

Handbuch der Regionalen Geologie, 3, pt. 10. *Die Mittelatlantischen Vulkaninseln*, by C. Gagel (Winter, 1910)
Handbuch der Regionalen Geologie. 4, pt. 24. *Die Nordatlantischen Polarinseln*, by O. Nordenskjöld (Winter, 1921)
'Géologie des Iles Atlantides', by J. Bourcart (*Mémoires de la Société de Biogéographie*, 1946)
Iceland and Mid-Ocean Ridges. Report of a Symposium, Reykjavik 1967, edited by S. Bjornsson (Visindafelag Islendinga No. 38: Prentsmidjan Leiftur, Reykjavik, 1967)
'North Atlantic Geology and Continental Drift', edited by M. Kay (*Memoir of the American Association of Petroleum Geologists*, 12, 1969)
Beiträge zur Regionalen Geologie der Erde. 10. *Geology of the South Atlantic Islands*, by R. C. Mitchell-Thomé (Borntraeger, 1970). This valuable work is extremely comprehensive and covers the following islands: Ascension, St. Helena, Tristan da Cunha, Gough, Falkland, Saint Paul Rocks, Fernando de Noronha, Rocas, Trinidad, Martin Vas Archipelago, Sao Thomé, Fernando Poo, Principe and Annoban

Some of the more important studies on individual islands are given below but the major source, *Geology of the South Atlantic Islands* (see above), should be remembered, as should *The Geology of the British Empire*, by F. R. C. Reed (2nd edn., Arnold, 1949). Neither of these is cited again in this section.

BERMUDA
'Bermuda a Partially Drowned Late Mature Pleistocene Karst', by J. H. Bretz (*Bulletin of the Geological Society of America*, 71, 1729–1754, 1960)

CANARY ISLANDS
'On the Geology of Fuerteventura (Canary Islands)', by H. Hausen (*Commentationes Physico-Mathematicae Societas Scientiarum Fennica*, 22 (1), 1958)
'On the Geology of Lanzarote, Graciosa and the Isletas (Canarian Archipelago)', by H. Hausen (*Commentationes Physico-Mathematicae Societas Scientiarum Fennica*, 23 (4), 1959)
'New Contributions to the Geology of Grand Canary', by H. Hausen (*Commentationes Physico-Mathematicae Societas Scientiarum Fennica*, 27 (1), 1962)

CAPE VERDE ISLANDS
'Geologia da Provincia de Cabo Verde', by C. Torre de Assunção, in *Curso de Geologia do Ultramar*. 1 (Junta de Investigações do Ultramar, 1968, pp. 1–52)
Geologia da Ilha de Maio (*Cabo Verde*), by A. Serralheiro (Junta de Investigações do Ultramar, 1970)

FAEROE ISLANDS
'Geology of the Faeroe Islands (Pre-Quaternary)' by J. Rasmussen and

A. Noe-Nygaard (*Danmarks Geologiske Undersøgelse, I Raekke*, **25**, 1970). Reference might also be made to the junior author's paper 'The Geology of the Faeroes' (*Quarterly Journal of the Geological Society of London*, **118**, 375–383, 1962)

FALKLAND ISLANDS (AND SCOTIA SEA ISLANDS)

Reference should be made to the general Atlantic Ocean Islands section for major references to the Falkland Islands themselves. They originally lent their name to a major series of works, *Scientific Reports of the Falkland Islands Dependencies Survey* (**1–35**), which then became *Scientific Reports of the British Antarctic Survey* (**36–**). Major issues are listed below:

7, 19. 'Geology of South Georgia' I, II, by A. F. Trendall (1953, 1959)
25, 62. 'The Geology of South Orkney Island' I by D. H. Matthews and D. H. Maling; II by J. W. Thomson (1967, 1968)
26, 27, 44, 47. 'The Geology of South Shetland' I, II, by D. D. Hawkes; III by L. M. Barton; IV by G. J. Hobbs (1961, 1961, 1965, 1968)

TRISTAN DA CUNHA

'The Volcanological Report of the Royal Society Expedition to Tristan da Cunha, 1962', by P. E. Baker *et al.* (*Philosophical Transactions of the Royal Society of London A*, **1075**, 439–578, 1964)

INDIAN OCEAN ISLANDS

No general account of the geology of the Indian Ocean is available; however, geological works figure in a useful series of bibliographies:

A Partial Bibliography of the Indian Ocean, compiled by A. E. Yentsch (Woods Hole Oceanographic Institution, 1962)
'Bibliography of the Indian Ocean 1900–1930', by K. Alagarswami *et al.* (*Bulletin of the Central Marine Fisheries Research Institute*, **4**, 1–117, 1968)
'Bibliography of the Indian Ocean 1931–1961. A supplement to the Partial Bibliography', by R. S. Lal Mohan *et al.* (*Bulletin of the Central Marine Fisheries Research Institute*, **11**, 1969). There have been further supplements on a continuing basis.

The geology of the present and former British colonies is reviewed in *The Geology of the British Empire*, by F. R. C. Reed (2nd edn., Arnold, 1949). Reference might also be made to *The Geology of Indonesia*, by R. W. Van Bemmelen (2 vols., 2nd edn., Nijhoff, 1970)

ALDABRA

'Geomorphology of Aldabra Atoll', by D. R. Stoddart (*Philosophical Transactions of the Royal Society of London B*, **260**, 31–66, 1971)

ANDAMAN AND NICOBAR ISLANDS, CHRISTMAS ISLAND

'Geology of the Andaman and Nicobar Islands', by E. R. Gee (*Records of the Geological Survey of India*, **59**, 208–232, 1926)
The Geology of Indonesia, by R. W. Van Bemmelen (2 vols., 2nd edn., Nijhoff, 1970)

'Geology of South Andaman Island', by C. Karuinakaran *et al.* (*Report of the 22nd International Geological Congress,* **11**, 79–100, 1964)

THE CHAGOS ARCHIPELAGO AND OTHER CORAL REEFS
'The Reefs of the Western Indian Ocean', by J. S. Gardiner (*Transactions of the Linnean Society of London,* **19**, 393–436, 1936)

FARSAN AND KAMARAN ISLANDS
'Geology of the Farsan Islands, Gizan and Kamaran', by W. A. Macfadyen (*Geological Magazine,* **67**, 310–315, 1930)

THE MASCARENE ISLANDS (MAURITIUS, REUNION, RODRIGUEZ)
'Une Excursion Géologique à la Réunion et à l'Ile Maurice', by E. A. Aubert de la Rue (*Revue de Géographie Physique et Géologie Dynamique,* **4**, 201–221, 1931)
'The Geology and Mineral Resources of Mauritius', by E. S. W. Simpson (*Colonial Geology and Mineral Resources,* **1**, 217–238, 1950)
'Etude Géologique de l'Ile de la Réunion', by P. Bussière (*Travaux du Bureau de Géologie, Madagascar,* **84**, 1958)
Geology of Africa, by R. Furon (Oliver and Boyd, 1963)

PERIM
'Notes on the Geology of Perim Island', by C. A. Raisin (*Geological Magazine,* **49**, 206–210, 1902)

ST. JOHNS ISLAND
Preliminary Geological Report on St. Johns (Zeberged) Island in the Red Sea, by F. W. Moon (Geological Survey of Egypt, 1923)

SHADWAN ISLAND
'The Geology of Shadwan Island, Northern Red Sea', by N. M. Shukri (*Bulletin de la Société Géographique d'Egypte,* **27**, 83–92, 1954)

SOCOTRA ISLAND
'The Geology of Socotra Island, Gulf of Aden', by Z. R. Beydoun and H. R. Bichan (*Quarterly Journal of the Geological Society of London,* **125**, 413–446, 1970)

PACIFIC OCEAN AND ITS ISLANDS
BOOKS: The most useful book on the submarine geology of the Pacific is *Marine Geology of the Pacific,* by H. W. Menard (McGraw-Hill, 1964). However, in view of the rapid developments in this field, the work should be regarded more as a progress report. Dealing with a more limited area is the definitive *Marine Geology of the Gulf of California,* by T. H. Van Andel and G. G. Shor (American Association of Petroleum Geologists, 1964). The most relevant general accounts, although only the first is comprehensive, are:

Handbuch der Regionalen Geologie. 7, pt. 2. *Oceania,* by P. Marshall (Winter, 1911)
The Geology of the British Empire, by F. R. C. Reed (2nd edn., Arnold, 1949), This covers the geology of all British and formerly British islands

Recherche Géologique et Minérale en Polynésie Française (Inspection Générale des Mines et de la Géologie, Paris, 1959)
'Geology of the Solomon and New Hebrides Islands as Part of the Melanesian Re-entrant, Southwest Pacific', by P. J. Coleman (*Pacific Science,* **24**, 289–314, 1970)

BIBLIOGRAPHIES: *Annotated Bibliography of Geologic and Soils Literature of Western North Pacific Islands,* compiled by H. L. Foster (US Army Intelligence Division, Far East, Tokyo, 1956)
Catalog of Translations of Japanese Geologic, Soils and Allied Literature of the Pacific Islands through 31 December 1958 (US Department of the Army, Pacific Division, 1959)
A Bibliography of Fiji, Tonga and Rotuma; Preliminary Working Edition, by P. A. Snow (Australian National University Press, 1969)
JOURNALS: The *New Zealand Journal of Geology and Geophysics* has so far devoted three issues to the topic of Pacific geology. Other relevant journals are:
Bulletin of the Bernice P. Bishop Museum (1922–)
Occasional Papers, Bernice P. Bishop Museum (1898–)
Pacific Geology (1968–)
Pacific Science (1947–)
Proceedings of the Pacific Science Congress (1929–)

The following accounts of the geology of particular islands or island groups are available:

AUSTRAL ISLANDS
'Geology of the Austral Islands', by L. J. Chubb and W. Campbell Smith (*Quarterly Journal of the Geological Society of London,* **83**, 291–316, 1927)

CAROLINE ISLANDS
Military Geology of Yap Islands, Caroline Islands, by C. G. Johnson (US Army, 1960)

COOK ISLANDS
'Geology of Mangaia', by P. Marshall (*Bulletin of the Bernice P. Bishop Museum,* **36**, 1927)
'The Geology of Rarotonga and Atiu', by P. Marshall (*Bulletin of the Bernice P. Bishop Museum,* **75**, 1930)
'The Geology of the Cook Islands', by B. L. Wood (*New Zealand Journal of Geology and Geophysics,* **10**, 1429–1445, 1967)

FIJI ISLANDS
REVIEWS: 'Geology of Vitilevu', by H. S. Ladd (*Bulletin of the Bernice P. Bishop Museum,* **119**, 1934; reprinted 1970)
'Geology of Lau, Fiji', by H. S. Ladd and J. E. Hoffmeister (*Bulletin of the Bernice P. Bishop Museum,* **181**, 1945; reprinted 1970)
'Outline of the Geology of Vitilevu', by P. Rodda (*New Zealand Journal of Geology and Geophysics,* **10**, 1260–1273, 1967)
BIBLIOGRAPHY: *Bibliography of the Geology of Fiji up to November 1968,* by R. F. Duberal and P. Rodda (1969)
JOURNALS: *Bulletin of the Geological Survey Department of Fiji* (1958–)
Memoir of the Geological Survey of Fiji (1964–)
Report of the Geological Survey Department of Fiji (1958–)

THE GALAPAGOS ISLANDS, COCOS ISLAND AND EASTER ISLAND
BOOKS AND REVIEW: 'Geology of Galapagos, Cocos and Easter Islands', by
L. J. Chubb (*Bulletin of the Bernice P. Bishop Museum*, **110**, 1933)
'Geology and Petrology of Easter Island', by M. C. Bandy (*Bulletin of
the Geological Society of America*, **48**, 1589–1609, 1937)
'Geology and Petrology of the Galapagos Islands', by A. R. McBirney
and H. Williams (*Memoir of the Geological Society of America*, **118**, 1969)
JOURNAL: *Noticias de Galapagos* (1963–)

HAWAII
BOOKS: *Geology of the State of Hawaii*, by H. T. Stearns (Pacific Books,
Palo Alto, California, 1966)
Road Guide to Points of Geologic Interest in the Hawaiian Islands, by
H. S. Stearns (Pacific Books, Palo Alto, California, 1966)
Volcanoes in the Sea. The Geology of Hawaii, by G. A. MacDonald and
A. T. Abbott (University of Hawaii Press, 1970)
BIBLIOGRAPHY: 'Bibliography of the Geology and Water Resources of the
Island of Hawaii', by G. A. MacDonald (*Bulletin of the Division of Hydro-
graphy, Hawaii*, **10**, 1947)
JOURNAL: *Bulletin of the Division of Hydrography, Hawaii* (1935–)

JUAN FERNANDEZ ISLANDS
'Die Geologie der Juan Fernandez Inseln', by P. D. Quensel (*Bulletin
of the Geological Institution of the University of Upsala*, **11**, 252–290, 1912)
'Additional Comments on the Geology of the Juan Fernandez Islands',
by P. Quensel, in *Natural History of Juan Fernandez and Easter Island*
(Almqvist and Wiksell, 1952)

MACQUARIE ISLAND
Macquarie Island, its Geology and Geography, by D. Mawson (Government Printing Office, Sydney, 1943)

MARIANA ISLANDS
'Geology of Saipan, Mariana Islands', by P. E. Cloud Jr. (*Professional
Paper, United States Geological Survey*, No. 280A, 1956)
'General Geology of Guam', by J. I. Tracey *et al.* (*Professional Paper, United
States Geological Survey*, No. 403A, 1964)

MARQUESA ISLANDS
'Geology of the Marquesa Islands', by L. J. Chubb (*Bulletin of the Bernice
P. Bishop Museum*, **68**, 1930)

MARSHALL ISLANDS
'Geology of Bikini and Nearby Atolls, Marshall Islands', by K. O. Emery
et al. (*Professional Paper, United States Geological Survey*, No. 260, 1954)

NEW CALEDONIA
REVIEW: 'The Geology of New Caledonia', by A. R. Little and R. N.
Brothers (*New Zealand Journal of Geology and Geophysics*, **13**, 145–183,
1970)
BIBLIOGRAPHY: 'Geological and Geophysical Publications of New Cale-
donia', by F. Baltzer *et al.* (*New Zealand Journal of Geology and Geo-
physics*, **10**, 1275–1279, 1967)
JOURNAL: *Bulletin Géologique de la Nouvelle-Caledonie* (1958–)

NEW HEBRIDES
REVIEW: 'Geology of the New Hebrides', by D. Mawson (*Proceedings of the Linnean Society of New South Wales*, **30**, 400–484, 1905)
BIBLIOGRAPHY: 'Geological and Geophysical Publications of the New Hebrides', by A. H. G. Mitchell (*New Zealand Journal of Geology and Geophysics*, **10**, 1274, 1967)
JOURNAL: *Progress Report of the Geological Survey, New Hebrides Anglo-French Condominium* (1959–). The Geological Survey also produce a series of regional guides (1964–)

PITCAIRN ISLAND
'The Geology of Pitcairn Island, South Pacific Ocean', by R. M. Carter (*Bulletin of the Bernice P. Bishop Museum*, **231**, 1–38, 1967)

SAMOA AND WESTERN SAMOA
'The Geology of American Samoa', by R. A. Daly (*Publication of the Carnegie Institution, Washington*, **340**, 1924)
'Geology of the Samoan Islands', by H. O. Stearns (*Bulletin of the Geological Society of America*, **55**, 1279–1331, 1944)
'Geological Notes on Western Samoa', by D. Kear (*New Zealand Journal of Geology and Geophysics*, **10**, 1446-1451, 1967)

SAN FELIX AND SAN AMBROSIO
'San Felix and San Ambrosio; Their Geology and Petrology', by B. Willis and H. S. Washington (*Bulletin of the Geological Society of America*, **35**, 365–384, 1924)

SOCIETY ISLANDS
'Geology of Borabora, Society Islands', by J. T. Stark and A. L. Howland (*Bulletin of the Bernice P. Bishop Museum*, **169**, 1941)
'Geology of Tahiti, Moorea and Maiao', by H. Williams (*Bulletin of the Bernice P. Bishop Museum*, **105**, 1933)

SOLOMON ISLANDS
REVIEWS: 'Geology of the British Solomon Islands', by R. J. A. W. Lever (*Geological Magazine*, **74**, 271–277, 1937)
'A Brief History of Geological and Geophysical Investigations in the British Solomon Islands 1881–1961', by J. C. Grover (*British Solomon Islands Geological Record*, **2**, 9–14, 1959–62)
JOURNALS: *British Solomon Islands Geological Record* (1957–)
Memoir of the Geological Survey of the British Solomon Islands (1955–)

TONGA ISLANDS
'Geology of Eua, Tonga', by J. E. Hoffmeister (*Bulletin of the Bernice P. Bishop Museum*, **96**, 1932)
'The Geology of Tofua Island Tonga', by G. R. Bauer (*Pacific Science*, **24**, 333–350, 1960)
'Notes on the Geology of the Tongan Islands', by J. C. Schofield (*New Zealand Journal of Geology and Geophysics*, **10**, 1424–1428, 1967)

10

Palaeontology

W. A. S. Sarjeant and A. P. Harvey

Palaeontology may be defined as *the study of the preserved remains of organisms and their activity*. There is no implicit starting or finishing date for the period of attention (in the words of a geology students' song: 'You count as a fossil as soon as you've died'). It is today conventional to exclude from attention all artefacts and artificial inhumations; but there remains a broad zone of overlap with archaeology and anthropology, since the preserved bones of early man and his immediate ancestors are certainly considered within the province of the palaeontologist.

The palaeontologist, then, is concerned with *fossils* (the preserved remains of organisms) and with *trace fossils* (the remains of the activity of organisms). In his endeavours to gain a more complete picture of the nature and mode of life of the organisms he is studying he also uses the techniques of stratigraphy and sedimentology and, to a lesser extent, the other geological disciplines.

His involvement with the literature will be relatively slight if his interest is merely cultural or even if he is taking a course of study in this discipline. If he is interested as a collector, concerned only to identify material found on excursions, then he may be fortunate enough to have a handy guide in which about 80 per cent of the fossils he finds will be illustrated. The identification of those remaining will, however, prove a real headache. If he is undertaking serious research, he will be faced with a task which appears impossible, i.e. the necessity of reading and digesting the whole of the world's literature on palaeontology and stratigraphy, to be sure that he is able to name his fossils correctly, to know their geographic and stratigraphic ranges, and to comprehend, as fully as possible, their way of life. However, there are several factors which serve to greatly reduce this burden. These are the general availability of summary texts which, although inevitably out of date, provide a sound starting point; the availability, in most fields, of bibliographic works; the willingness of fellow workers in his particular field to send him offprints of their papers, on an exchange basis; and the fact that he will be normally receiving guidance, during the earliest stages, from an experienced worker in his field. However, no research palaeontologist can ever be truly confident, when he publishes a paper, that it is free from error caused by unawareness of a previously published work.

Besides the formal literature there is a considerable 'underground' literature in palaeontology, various groups producing newsletters. These are not normally available to libraries and the like, although they contain much valuable information. There is a useful list of such newsletters in *Palaeontological Association Circular*, **64**, 14–15 (1971), but certainly they require greater bibliographic control than is at present the case. One of us (A.P.H.) is at present studying geological newsletters as a source of information.

Future development of computer techniques may solve some of the problems concerning current awareness, etc., but these developments seem some way off and for some time mastery of the literature is likely to remain the most formidable task which confronts the palaeontologist.

The observation of fossils is on record throughout written history. Recognition that they were the remains of formerly living organisms can be traced back to the Greek philosopher Xenophanes (560–480 B.C.) and it is probable that the dragon myths of China and myths of giants in Europe were attempts to rationalise observations of gigantic fossil mammalian bones. The beginnings of serious study were in the late eighteenth and early nineteenth centuries, principally in France through the work of Lamarck (1744–1829) on vertebrates, Adolphe Brongniart (1801–79) on fossil plants and Ehrenberg (1795–1876) on microscopic fossils. Some relevant studies on the history of palaeontology are given in Chapter 17.

Fossils have great intrinsic interest as evidence of the changes that the life of the earth has undergone and of the evolutionary process that has produced the present biological environment. They also possess a considerable extrinsic value as indices of conditions and as means of correlation between sequences of sediments, permitting of relative dating, the determination of structure and the identification of levels of especial economic significance. For these purposes, attention was originally focused especially on the invertebrates. In the present era of subsurface sampling by boreholes, etc., the much greater relative abundance of microfossils has caused commercial attention to be focused on the larger microfossils (foraminifera and ostracods) and, most recently, on the truly microscopic groups (spores, pollen, dinoflagellates and coccoliths). The commercial palaeontologist of today most commonly has a chemical laboratory at his command and works with a high-powered microscope.

Literature has, however, lagged behind these developments. Many so-called 'general texts' on palaeontology are devoted principally to invertebrates, with vertebrates receiving scant treatment and plants and microfossils scarcely mentioned—a procedure which is exactly paralleled in the courses still given in many European and North American universities.

An example of such unbalanced procedure is *An Introduction to Palaeontology*, by A. Morley Davies (3rd edn., revised by C. J. Stubblefield, Murby, 1961). In contrast, more balanced and very readable introductory works include *Fossils*, by H. H. Swinnerton (Collins, 1960); *Life of the Past: An Introduction to Paleontology*, by G. G. Simpson (2nd edn., Yale University Press, 1961); and *The History of Life*, by A. L. McAlester (Prentice-Hall, 1968).

A refreshing and readable work, which deals with the ideas of palaeontology and avoids description of particular groups is *General Palaeontology*, by A. Brouwer (Oliver and Boyd, 1968). An example of a textbook for the new style of palaeontological teaching is *Principles of Paleontology*, by D. M. Raup and S. M. Stanley (Freeman, 1971).

It is worth mentioning a few other general texts in view of the numerous enquiries which librarians receive from adults and children for reading material about fossils:

Prehistoric Animals, by J. Augusta and Z. Burian (Spring Books, 1958)
The Fossil Book, by C. L. Fenton and M. A. Fenton (Doubleday, 1958)
Prehistoric Life on Earth, by K. Petersen (Methuen, 1963)
Fossils. A Guide to Prehistoric Life, by F. H. T. Rhodes *et al.* (Hamlyn, 1965)
Fossils in Colour, by J. F. Kirkaldy (Blandford, 1967)
The Procession of Life, by A. S. Romer (Weidenfeld and Nicolson, 1968)
Evolution of Life, by M. S. Randhawa *et al.* (Publications and Information Directorate, New Delhi, 1969)
The Wonderful World of Prehistoric Animals, by W. E. Swinton (Macdonald, 1969)
The Elements of Palaeontology, by R. M. Black (Cambridge University Press, 1970)

To cover thoroughly all aspects of palaeontology is impossible within the restricted compass of a single volume. What is required is a series produced by a team of authors. The most comprehensive attempt is that by the palaeontologists of the USSR under the general editorship of Yu. A. Orlov, *Osnovy Paleontologii* (Princciples of Palaeontology) (15 vols., Akademiya Nauk SSSR, 1959–64). Translations of this series are being prepared by the Israel Program for Scientific Translations and are readily available in Europe and North America. Because of the multifarious authorship, the quality of the volumes is inevitably variable, as is the degree of awareness of non-Russian texts.

A comprehensive survey of the whole field of biology is provided by several series published by Masson, Paris. These include *Traité de Zoologie*, edited by P. Grassé (Masson, 1952–), in which fossil occurrences of animals are summarised; *Traité de Paléontologie*, edited by J. Priveteau (7 vols., Masson, 1952–69), which deals in much greater detail with fossil animals and stresses the vertebrates; *Traité de Botanique (Systématique)*, edited by M. Chadefaud and L. Emberger (Masson, 1960–) and *Traité de Paléobotanique*, edited by E. Boureau (9 vols. planned, Masson, 1964–) treat the plant kingdom in a similar fashion. Also well known and used is A. H. Müller's *Lehrbuch der Paläozoologie* (7 vols., Fischer, 1957–70). In addition, the stratigraphical applications of fossils are stressed in *Paléontologie Stratigraphique*, by H. and G. Termier (Masson, 1960), and *Stratigraphical Palaeontology*, by E. Neaverson (Clarendon Press, 1955).

Micro-organisms and invertebrates are dealt with in *Treatise on Invertebrate Paleontology*, edited by R. C. Moore (Geological Society of America and University of Kansas Press, 1953– ; 2nd edn., 1970–). Because of its multiple authorship, this, again, is of variable quality. However, a standardised presentation and a very high standard of illustration render the series of special value.

In briefer compass and still useful, though many parts are now outdated, is the classic work *Textbook of Palaeontology*, by K. A. Zittel (1925–37; reprinted by Stechert-Hafner, 1964). The first volume is devoted to invertebrates, the second and third to the vertebrates. Fossil plants are omitted from consideration.

A series of works which attempts a comprehensive index to published generic and specific names of fossils is the *Fossilium Catalogus*, published as two series—*I. Animalia* and *II. Plantae* (Junk, 1913– and 1914– , respectively). Despite forbidding Latin titles, these works are readily comprehensible and simple to use. The series has been subject to reprinting and is therefore still available.

There have been a number of important conferences over the years devoted to palaeontology; a recent one is the North American Paleontological Convention (1969), whose *Proceedings* are being published by Allen Press Inc., Kansas. The *Proceedings* are in 12 parts, a number of them providing material not readily available before.

There are three abstracting/indexing services of importance to the whole field of palaeontology. In addition, a number of general secondary services (see Chapter 5) are also useful.

Bulletin Signalétique, 227; *Bibliographie des Sciences de la Terre*. H. *Paléontologie* (Centre National de la Recherche Scientifique: Bureau de Recherches Géologiques et Minières, 1972–). Issued monthly, each part being classified. There are author, subject and geographic indexes.

Biological Abstracts (Biosciences Information Service of Biological Abstracts, 1926–). This is issued in 24 parts per annum, each being arranged by subject. There are author, biosystematic and subject indexes. A keyword-in-context subject index is included with each issue. There are many other services offered by BIOSIS and a useful paper about the range of activities is 'Biosciences Information Service,' by P. V. Parkins, in *Encyclopedia of Library and Information Science*, Vol. 2, pp. 603–621 (Marcel Dekker, 1969).

Zoological Record (Zoological Society of London, 1865– , covering the literature from 1864). Basically an annual publication, this is one of the principal sources of information for the palaeontologist. Its usefulness, however, is impaired by substantial publishing delays. The *Record* is divided into 20 separately issued sections: 1. *Comprehensive Zoology*; 2. *Protozoa*; 3. *Porifera*, which includes Archaeocyatha; 4. *Coelenterata;* 5. *Echinodermata;* 6. *Vermes* (including conodonts and scolecodonts); 7. *Brachiopoda*; 8. *Bryozoa*; 9. *Mollusca*; 10. *Crustacea*; 11. *Trilobita*; 12. *Arachnida*; 13. *Insecta*; 14. *Protochordata* (including graptolites); 15. *Pisces*; 16. *Amphibia*; 17. *Reptilia*; 18. *Aves*; 19. *Mammalia*; 20. *List of New Genera and Subgenera* (this forms an updating of S. A. Neave's *Nomenclator Zoologicus*). Each part consists of an author list and subject and systematic indexes. A useful paper in conjunction with the *Record* is 'The Zoological Record. A Centenary Appraisal', by G. D. R. Bridson (*Journal of the Society for the Bibliography of Natural History*, **5**, 23–34, 1968).

Besides these continuing bibliographies, two bibliographies of considerable reference value are 'Treatise on Marine Ecology and Paleoecology. Vol. II', edited by H. S. Ladd (*Memoir of the Geological Society of America*, **67**, 1957), which contains excellent bibliographies on many fossil groups; and *Handbook of Paleontological Techniques*, edited by B. Kummel and D. Raup (Freeman, 1965), which contains a section by B. Kummel, 'Compilation of Bibliographies of Use to Paleontologists and Stratigraphers'.

It is also appropriate to mention here one other work which is of great value in locating the work of the earlier palaeontologists: 'Palaeontologi. Catalogus Bio-bibliographicus', by K. Lambrecht and W. and A. Quenstedt (*Fossilium Catalogus I*, **72**, 1938). This book lists palaeontologists alphabetically and provides brief biographical data together with a list of biographies and obituaries. It indicates whether or not portraits and bibliographies of individuals are available.

REGIONAL PALAEONTOLOGY

Adequately illustrated works summarising the most typical fossils of a country or those that are of especial stratigraphic importance are growing in numbers. Regional palaeontological bibliographies are also being issued with greater regularity. Below are a few random examples of these types of work arranged by country.

ARGENTINA
Bibliography of Argentine Palaeontology (1960–   )

AUSTRALASIA
'Bibliography of the Mesozoic Palaeontology of Australia and Eastern New Guinea', by S. K. Shwarko (*Bulletin of the Bureau of Mineral Resources Geology and Geophysics, Australia*, **108**, 237–279, 1970)

AUSTRIA
Catalogus Fossilium Austriae (1965–   )

BULGARIA
Fosilite na Bŭlgariya (1958–   )

HUNGARY
Catalogus Originalium Fossilium Hungariae. Pars Zoologica, by J. Boda (M. All. Földtani Intézet, 1964)

JAPAN
'Bibliography of Japanese Palaeontology and Related Sciences, 1941–1950', by R. Endo; '1951–1960', by F. Takai (*Special Papers, Palaeontological Society of Japan*, **1**, **9**, 1951, 1962)
Catalogue of Type-Specimens of Fossils in Japan, by S. Hanzawa *et al.* (Palaeontological Society of Japan, Anniversary Volume, 1961)

MALAGASY REPUBLIC
Atlas des Fossiles Caractéristiques de Madagascar, by M. Collignon (Service Géologique, Tananarive, 1958–   )

MALAYSIA
'Bibliography of Palaeontological Literature on Sarawak, Brunei and Sabah, 1945–1964', by A. J. Keij (*Annual Report of the Geological Survey of British Borneo*, 1964–65)

PHILIPPINES
'Palaeontology of the Philippines', by W. Hashimoto (*Geology and Palaeontology of Southeast Asia*, **6**, 293–329, 1969)

UNITED KINGDOM
British Palaeozoic Fossils (3rd edn., British Museum (Natural History), 1969)
British Mesozoic Fossils (4th edn., British Museum (Natural History), 1972)
British Cainozoic Fossils (4th edn., British Museum (Natural History), 1971)

UNITED STATES OF AMERICA
Index Fossils of North America, by H. W. Shimer and R. R. Shrock (Wiley, 1967)

Long series of works treating in detail with particular fossil groups or with particular stratigraphic levels are the *Palaeontographical Society Monographs* (1847–   ) and *Paleontologiya SSSR* (1935–   ). Although originally based on specimens from one country, these have become of international importance to palaeontologists.

PALAEONTOLOGICAL TECHNIQUES

Techniques of collection, preparation, display and cataloguing of fossils are dealt with in the following works:

Methods in Paleontology, by C. L. Camp and G. D. Hanna (University of California Press, 1937)
Handbook of Paleontological Techniques, edited by B. Kummel and D. Raup (Freeman, 1965)
The Preservation of Natural History Specimens, Vol. 2, by R. Wagstaffe and J. H. Fidler (Witherby, 1968)
Instructions for Collectors, No. 11, Fossils, Minerals and Rocks (9th edn., British Museum (Natural History), 1970)
'Field Methods in Paleontology', by S. J. Nelson (*Bulletin of Canadian Petroleum Geology*, **13**, 1–138, 1965)

PALAEOECOLOGY

Ecology may be defined as *the study of the relation between living organisms and their environment at the present day*. Palaeoecology is the extension of this study into the past—an attempt to elucidate the life conditions of organisms formerly living and, by this means, to reconstruct the environments of the past. General texts on this topic are:

'Treatise on Marine Ecology and Palaeoecology. I. Ecology', edited by J. W. Hedgpeth; 'II. Paleoecology', edited by H. S. Ladd (*Memoir of the Geological Society of America*, **67**, 1957)
Principles of Paleoecology, by D. V. Ager (McGraw-Hill, 1963)
Approaches to Paleoecology, edited by J. Imbrie and N. D. Newall (Wiley, 1964)

Introduction to Paleoecology, by R. F. Hecker, translated and edited by M. K. Elias and R. C. Moore (Elsevier, 1965)
Ancient Environments, by L. F. Laporte (Prentice-Hall, 1968)

The palaeoecology of the latest phase in geological history is treated in two symposium volumes: *Quaternary Paleoecology*, edited by E. J. Cushing and H. E. Wright (Yale University Press, 1967); and Vol. 6 of the Proceedings of the VII INQUA Congress, *Pleistocene Extinctions: The Search for a Cause*, edited by P. S. Martin and H. E. Wright (Yale University Press, 1967).

TAXONOMY

Taxonomy is the science of classification of living organisms, in which the palaeontologist works alongside the botanist and zoologist. Two generally agreed systems of nomenclature exist: the *International Code of Botanical Nomenclature*, of which the latest edition is that *adopted by the Tenth International Botanical Congress, Edinburgh, August 1964* (International Association for Plant Taxonomy, Utrecht, 1966); and the *International Code of Zoological Nomenclature*, of which the most recent revision was that *adopted by the XV International Congress of Zoology* (2nd edn., International Trust for Zoological Nomenclature, London, 1964). In addition, there are two works which serve as most useful guides to the field. These are *Methods and Principles of Systematic Zoology*, by E. Mayr *et al.* (2nd edn., McGraw-Hill, 1953) and *Taxonomy*, by R. E. Blackwelder (Wiley, 1967).

Discussions of problems of botanical nomenclature, and points of general interest, are to be found in the journal *Taxon* (1951–   ) and in some volumes of *Regnum Vegetabile* (1952–   ). Of interest to the palaeozoologist is *Systematic Zoology* (1952–   ). Proposals for changes and clarifications in zoological nomenclature are put forward, in severely legalistic fashion, in the *Bulletin of Zoological Nomenclature* (1953–   ).

The idea of having a list of all the genera and species of animals and plants, both recent and fossil, together with a reference to the original description has long been a dream of the palaeontologist and zoologist. It will most certainly remain one. However, this is not to say that attempts have not been made. Natural history's most famous bibliographer, Charles Davies Sherborn (1861–1942), made a splendid attempt with his *Index Animalium* (Cambridge University Press and British Museum (Natural History), 1902–33) but this only covered the literature up to 1850. The *Nomen-*

clator Zoologicus, by S. A. Neave (Zoological Society of London, 1939–40), lists genera and subgenera from the 10th edition of Linnaeus, 1758 to 1935. Supplements have covered the periods *1936–1945* (1950) and *1956–1965* (1966). This is of course continued on an annual basis by Section 20 of the *Zoological Record* and a further supplement is in active preparation.

In order to help the taxonomist, many depositories of type specimens have produced catalogues. They are too numerous to mention in any detail in this brief account but it is worth recording *Une Liste Préliminaire de Catalogues des Specimens Types en Zoologie et Paléontologie*, edited by A. W. F. Banfield (Comité d'Etat pour la Culture et l'Art, for ICOM, 1968).

PALAEOBIOCHEMISTRY

The application of biochemical techniques to palaeontology, as a field of research, is steadily assuming importance, despite the inherent difficulties introduced by partial preservation and by diagenetic effects and other post-depositional changes to which the fossils have been subjected. The following list includes most major studies currently available:

'Paleobiochemistry', by P. H. Abelson (*Scientific American*, **195**, 83–89, 1956)

Organic Geochemistry, edited by I. A. Breger (Pergamon, 1963)

'Collagen in Fossil Teeth and Bones', by J. M. Shackeford and R. W. G. Wyckoff (*Journal of Ultrastructure Research*, **11**, 173–180, 1964)

'The Amino-Acid Composition of Bone and Tooth Proteins in Late Pleistocene Mammals', by T. Y. Ho (*Proceedings of the National Academy of Sciences*, **54**, 26–31, 1965)

'Amino-Acid Components in Fossil Calcified Tissues', by W. G. Armstrong and L. B. Halstead Tarlo (*Nature*, **210**, 481–482, 1966)

'Chemical Fossils', by C. Eglinton and M. Calvin (*Scientific American*, **216**, 32–43, 1967)

'Biochemical Evolution and the Fossil Record', by L. B. Halstead Tarlo, in *The Fossil Record*, edited by W. B. Harland *et al.* (Geological Society of London, 1967)

Chemical Evolution, by M. Calvin (Oxford University Press, 1969)

THE ORIGIN OF LIFE

The origin of life has always been a subject of interest and much has been written. Some of the writings are mere speculation but below are given the major scientific references to this fascinating topic:

Proceedings of the 1st International Symposium on the Origin of Life on the Earth, held at Moscow, 19–24 August 1957, edited by A. I. Oparin (Pergamon, 1959)
The Geological Aspects of the Origin of Life on Earth, by M. G. Rutten (Elsevier, 1962)
The Chemical Origin of Life, by A. I. Oparin (Thomas, 1964)
The Origin of Life, by J. D. Bernal (Weidenfeld and Nicolson, 1967)
Genesis and Evolutionary Development of Life, by A. I. Oparin (Academic Press, 1968)
Chemical Evolution: Molecular Evolution. Towards the Origin of Living Systems on the Earth and Elsewhere, by M. Calvin (Oxford University Press, 1969)
Chemical Evolution and the Origin of Life, edited by R. Buvet and C. Ponnamperuma (North-Holland, 1971)
Biochemical Evolution and the Origin of Life, edited by E. Schoffeniels (North-Holland, 1971)

Important bibliographies are:

'Organic Evolution, Selections from the Literature. IV. Cosmic Evolution and the Origin of Life', by E. B. Ehrle and H. J. Birx (*Bioscience*, **20**, 834–835, 1970)
'Chemical Evolution and the Origin of Life: A Comprehensive Bibliography', by M. W. West and C. Ponnamperuma (*Space-Life Sciences*, **2**, 225–295, 1970)

EXTRA-TERRESTRIAL LIFE: THE FOSSIL EVIDENCE

As fascinating as the origin of life is the idea that there may be life-forms on other planets or in other parts of the universe. Below are listed a few of the more important references and reviews to supposed life forms in meteorites.

'A Microbiological Examination of some Carbonaceous Chondrites', by G. Claus and B. Nagy (*Nature*, **192**, 594–596, 1961)
'Microscopic "Life-Forms" in Meteorites. A Review of Some Recent Literature', by F. W. Price (*Journal of the Queckett Microscopical Club*, **29**, 68–71, 1962)
'Organic Remains in Meteorites', by F. L. Staplin, in *Current Aspects of Exobiology*, edited by G. Mamikunian and M. H. Briggs (Pergamon, 1965)
A Review of Evidence for Biological Material in Meteorites, by H. C. Urey (Cospar, 1965)
'Microfossil-like Objects in Meteorites', by A. A. Manten (*Earth-Science Reviews*, **1**, 337–341, 1966)
'Investigations of the Orgueil Carbonaceous Meteorite', by B. Nagy (*Geologiska Föreningens i Stockholm Förhandlingar*, **88**, 235–272, 1966)
'Observations on Fossil-like Objects in the Orgueil Meteorite', by

G. O. W. Kremp (*Journal of the British Interplanetary Society*, **21**, 99–112, 1968)
'The Mineralogy and Petrology of Chrondritic Meteorites', by W. R. Van Schmus (*Earth-Science Reviews*, **5**, 145–184, 1969)
'Review of Organic Matter in the Orgueil Meteorite', by B. L. Baker (*Space-Life Sciences*, **2**, 472–497, 1971)

MICROPALAEONTOLOGY

Micropalaeontology is *the palaeontology of groups of organisms, or discrete parts of higher organisms, whose study necessitates the use of a microscope.* It has recently become customary to distinguish between palynology and micropalaeontology *sensu stricto.* The definition of palynology is to some degree a matter for argument, but it is most generally defined as *the study of the so-called 'acid-insoluble' microfossils, having walls made up of organic substances that are not affected by hydrochloric and hydrofluoric acids,* i.e. principally spores, pollen, dinoflagellates, acritarchs and chitinozoans. On this basis, micropalaeontology becomes the study of microfossils whose walls are composed of silica, calcium carbonate or strontium sulphate, or of mixtures of these substances with organic substances. With certain exceptions (e.g. coccoliths) the fossils which are the concern of the micropalaeontologist tend to be larger than those which concern the palynologist—indeed, the largest foraminifers and ostracods are measurable in centimetres and can be collected and readily examined with the unaided eye.

During the last 50 years, micropalaeontology has been intensively employed in industry and its literature is voluminous. Catalogues of, and indexes to, a number of groups are available but no satisfactory single-volume textbook on this field is available. The most comprehensive is *Introduction to Microfossils*, by D. J. Jones (Harper, 1956); unfortunately, no recent revision of this work has been attempted. Other works include *Principles of Zoological Micropalaeontology*, by V. Pokorný, translated by J. W. Neale (2 vols., Pergamon, 1963, 1965); *Principles of Micropalaeontology*, by M. F. Glaessner (2nd edn., Hafner, 1965), which lays heavy stress on foraminifera and ostracods; and *Leitfossilien der Mikropaläontologie*, by H. Bartenstein *et al.* (Borntraeger, 1962), which deals with microfossils of especial stratigraphic importance and is not by intent comprehensive.

A *Bibliography and Index of Micropalaeontology* started publication in 1972. Produced by the American Museum of Natural History in co-operation with the American Geological Institute, it is issued monthly and lists about 3000–4000 papers per year. References are arranged by microfossil group and monthly author and annual author and subject indexes are provided.

Journals in the field include *Micropaleontology* (1955–), *Revista Española di Micropaleontología* (1969–), *Revue de Micropaléontologie* (1958–), *Utrecht Micropalaeontological Bulletin* (1969–) and *Voprosy Mikropaleontologii* (1956–).

The following is a list of the more important papers, books and bibliographies dealing with the various groups of microfossils:

ALCYONARIAN SCLERITES

'A Classification of Fossil Alcyonarian Sclerites', by M. Deflandre-Rigaud (*Micropaleontology*, **3**, 357–366, 1957)

BACTERIA

Geologic Activity of Micro-organisms, edited by S. I. Kuznetsov (Consultants Bureau, 1962)

Introduction to Geological Microbiology, by S. I. Kuznetsov *et al.* (McGraw-Hill, 1963)

BIBLIOGRAPHY: 'Annotated Bibliography, Bacteria', by C. E. ZoBell, in 'Treatise on Marine Ecology and Paleoecology. II', edited by H. S. Ladd (*Memoir of the Geological Society of America*, **67**, 693–697, 1957)

CALCAREOUS NANNOPLANKTON (INCLUDING COCCOLITHS)

Coccoliths, by M. Black (*Endeavour*, **24**, 131–137, 1965)

Sur les Coccolithes du Jurassique Européen et d'Afrique du Nord. Essai de Classification des Coccolithes Fossiles, by D. Noel (CNRS, Paris, 1965)

Fichier Micropaléontologique, Général Séries 17–18, by G. Deflandre and M. Deflandre-Rigaud (CNRS, Paris, 1967)

Catalogue of Calcareous Nannofossils, by A. Farinacci (Edizioni Tecnoscienza, 1969–)

'State of Knowledge and Geological Significance of Coccoliths', by S. I. Shumenko (*International Geology Review*, **11**, 1290–1297, 1969)

BIBLIOGRAPHY: 'Annotated Index and Bibliography of the Calcareous Nannoplankton', by A. R. Loeblich Jr. and H. Tappan (*Phycologia*, **5**, 1966). Supplements have been issued regularly in the *Journal of Paleontology*

CHAROPHYTES

'Charophyten und Halophyten', by K. Mädler (*Fortschritte der Geologie von Rheinland und Westfalen*, **10**, 121–128, 1963)

BIBLIOGRAPHY: 'Charophyta', by J. Groves (*Fossilium Catalogus II*, **19**, 1933)

CONODONTS

Treatise on Invertebrate Paleontology. W. Miscellanea (*Conodonts, etc.*), by W. W. Hass *et al.* (Geological Society of America, University of Kansas Press, 1962). This was supplemented by F. H. T. Rhodes *et al.* in the form of *University of Kansas Paleontological Contributions*, **9** (1966)

Conodonts, by M. Lindström (Elsevier, 1964)

'The Nature of Conodonts', by Y. Globensky (*Naturaliste Canadien*, **97**, 213–228, 1970)

'A Suprageneric Taxonomy of the Conodonts', by M. Lindström (*Lethaia*, **3**, 427–445, 1970)
BIBLIOGRAPHIES: 'Catalogue of Conodonts', by R. O. Fay (*University of Kansas Paleontological Contributions*, **3**, 1952). This is supplemented by compilations by S. R. Ash—'Bibliography and Index of Conodonts, 1949–1958' and '1959–1963'—in, respectively, *Micropaleontology*, **7**, 213–244 (1961) and *Brigham Young University Studies in Geology*, **10**, 3–50 (1963)
'Annotated Bibliography and Index of Conodonts', by S. P. Ellison (*Publication, Bureau of Economic Geology, Texas*, No. 6210, 1962), supplemented by 'Supplement to Annotated Bibliography and Index of Conodonts', by S. P. Ellison (*Texas Journal of Science*, **15**, 50–67, 1963)

DIATOMS
Publications on diatoms are especially numerous and diffuse, in part because they have attracted much study by amateur microscopists. A convenient introductory work, with a chapter on diatom literature is: *Notes on Diatoms. An Introduction to the Study of the Diatomaceae*, by F. B. Taylor (Privately published, Bournemouth, 1929). Other works include:

A Treatise on the Diatomaceae, by H. Von Huerck, translated by W. E. Baxter (reprint of 1896 edition, Junk, 1963)
Diatomeenschalen im Elektronmikroscopischen Bild, by J. G. Helmcke and W. Krieger (Cramer, 1954–). Seven volumes have been issued so far
'Diatomaceae', by B. J. Cholnoky *et al.* (*Nova Hedwigia Beihefte*, **21**, **31**, 1966, 1970)
Catalogue of the Fossil and Recent Genera and Species of Diatoms and their Synonyms (*A Revision of F. W. Mills, 'An Index to the Genera and Species of the Diatomaceae and their Synonyms 1933–1935'*), by S. L. Van Landingham (Cramer, 1967–). Three volumes have so far been issued

EBRIDIANS
BIBLIOGRAPHY: 'Annotated Index of Fossil and Recent Silicoflagellates and Ebridians, with Descriptions and Illustrations of Validly Proposed Taxa', by A. R. Loeblich III *et al.* (*Memoir of the Geological Society of America*, **106**, 1968)

FORAMINIFERA
Since foraminifera are the most economically important group of fossils, the literature is vast and bibliographic aids are especially needful. The American Museum of Natural History is responsible for the following catalogues:

Catalogue of Foraminifera, by B. F. Ellis and A. R. Messina (1940–)
Catalogue of Index Foraminifera, by B. F. Ellis and A. R. Messina (3 vols., 1965–67)
Catalogue of Index Smaller Foraminifera, by B. F. Ellis *et al.* (3 vols., 1968–69)

General texts on the Foraminifera include Pokorny's *Principles of Zoological Micropalaeontology* and Glaessner's *Principles of Micropalaeontology*, both of which have already been discussed.
The first volume of Piveteau's *Traité de Paléontologie* and the first volume of the *Osnovy Paleontologii*, which has been translated as *Fundamentals*

of Palaeontology: Protozoa-Foraminifera and Radiolaria (Israel Program for Scientific Translations, 1965), are relevant.

Other works include:

Foraminifera, their Classification and Economic Use, by J. A. Cushman (Harvard University Press, 1948)
Ecology and Distribution of Recent Foraminifera, edited by F. B. Phleger (Johns Hopkins University Press, 1960)
Evolutionary Trends in Foraminifera, edited by G. H. R. Von Koenigswald (Elsevier, 1963)
'History and Classification of Fossil Foraminifera', by F. G. Leavill and J. Lintz (*Special Paper, Geological Society of America*, No. 76, 1964)
Treatise on Invertebrate Paleontology. C. Sarcodina, Chiefly 'Thecamoebians and Foraminifera', by A. R. Loeblich Jr. and H. Tappan (2 vols., Geological Society of America and University of Kansas Press, 1964)
Short-course Lecture Notes [on] Paleontology, by O. L. Bandy *et al.* (American Geological Institute, 1967). Almost entirely concerned with foraminifera
Manuel de Micropaléontologie des Foraminifères, by M. Neumann (Gauthier-Villars, 1967–)
Fossil Foraminifera of the USSR. Globigerinidae, Hantkeninidae and Globorotaliidae, by N. N. Subbotina (Collet's, 1971). Translation of *Trudy VNIGRI*, **76** (1953)

BIBLIOGRAPHIES: *A Bibliography of the Foraminifera, Recent and Fossil Forms 1565–1888*, by C. D. Sherborn (London, 1888)
'A Bibliography of the Family Fusulinidae', by D. F. Toomey (*Journal of Paleontology*, **28**, 465–484, 1954). Supplemented in the same journal, mainly by G. A. Sanderson
'Annotated Bibliography of Paleozoic Nonfusulinid Foraminifera', by D. F. Toomey (*Contributions from the Cushman Foundation for Foraminiferal Research*, **10**, 1959). Supplements appear in later issues of the same journal
'Annotated Bibliography of Pre-Carboniferous Foraminifera', by D. F. Toomey (*Contributions from the Cushman Foundation for Foraminiferal Research*, **12**, 33–46, 1961)
'A Bibliography of Foraminiferal Ecology and Paleoecology (with Annotations)', by H. C. Skinner and H. C. Eppert (*Transactions of the Gulf Coast Association of Geological Societies*, **16**, 355–372, 1966)

Relevant parts of the *Fossilium Catalogus I* are:

Parts **49, 59, 60**. 'Bibliographia Foraminiferum Recentium et Fossilium', by A. Liebus and H. Thalmann (1931, 1933)
Parts **111–114**. 'Fusulinidae (Foraminifera)', by F. and G. Kahler (1966–67)

Reference might also be made to:

An Index to the Genera and Species of the Foraminifera 1890–1950, by H. E. Thalmann (Stanford University, George Vanderbilt Foundation, 1960). This up-dates 'An Index to the Genera and Species of the Foraminifera', by C. D. Sherborn (*Smithsonian Miscellaneous Collections*, **37**, No. 856, 1893, and No. 1031, 1896)

JOURNALS: *Contributions from the Cushman Foundation for Foraminiferal Research* (1950–70). This includes a section 'Recent Literature on the Foraminifera'
Journal of Foraminiferal Research (1971–)
Special Publications of the Cushman Foundation for Foraminiferal Research (1952–)

HOLOTHURIAN SCLERITES

'Monograph of Fossil Holothurian Sclerites', by D. L. Frizzell and H. Exline (*Technical Series, University of Missouri School of Mines and Metallurgy*, **89**, 1955)
'Contribution à la Connaissance des Sclérites d'Holothuridés Fossiles', by M. Deflandre-Rigaud (*Mémoires du Muséum d'Histoire Naturelle N.S. Sér. C.*, **11**, 1962)

OSTRACODS

A *Catalogue of Ostracoda* is produced by the American Museum of Natural History (1952–), though the coverage is less comprehensive than that on the Foraminifera. A second catalogue, *Stereo-atlas of Ostracod Shells*, edited by P. C. Sylvester-Bradley (Leicester University Press), illustrated by scanning electromicrographs, is to commence publication in the near future.
Treatise on Invertebrate Paleontology. Q. Arthropoda 3 (*Crustacea and Ostracoda*), by R. H. Benson *et al.* (Geological Society of America and University of Kansas Press, 1961)
Ostracod Taxonomy, by H. V. Howe (Louisiana State University Press, 1962)
Post-Paleozoic Ostracoda, their Morphology, Taxonomy, and Economic Use, by F. P. C. M. van Morkhoven (Elsevier, 1962–63)
BIBLIOGRAPHY: *Bibliographic Index and Classification of the Mesozoic Ostracoda*, by H. N. Coryell (2 vols., University of Dayton Press, 1963)
Additional bibliographies appear in *The Ostracodologist* (1963–), a newsletter for those concerned with ostracod research.

OTOLITHS

These calcareous particles from fishes' ears are of great interest in tracing evolutionary lineages. No comprehensive general account is available, but two bibliographic works are: 'Otolithi Piscium', by O. Posthumous (*Fossilium Catalogus I*, **24**, 1924), and its supplement by H. Weiler (**117**, 1968).

PHYTOLITHS

These minute particles of silica laid down during physiological processes by some land-plants are described in 'Fossil Opal-phytoliths', by G. Baker (*Micropaleontology*, **6**, 79–85, 1960) and 'Les Phytolithaires (Ehrenberg): Nature et Signification Micropaléontologique, Pédologique et Géologique', by G. Deflandre (*Protoplasma*, **57**, 234–259, 1963)

RADIOLARIA

Fichier Micropaléontologique Général, Série 9, Ciliès, Tintinnoidea et Ciliatae Incertae by G. Deflandre and M. Deflandre-Rigaud (2nd edn., CNRS, Paris, 1965)
'An Introduction to the Study of Radiolaria', by A. S. Campbell (*The Micropaleontologist*, **6**, 29–44, 1952)
Treatise on Invertebrate Paleontology. D. Protista 3 (*Chiefly Radiolaria and Tintinnia*), by A. S. Campbell and R. C. Moore (Geological Society of America and University of Kansas Press, 1954)

Fundamentals of Palaeontology: Protozoa-Foraminifera and Radiolaria (Israel Program for Scientific Translations, 1965). Translation *Osnovy Paleontologii*, **1**
BIBLIOGRAPHY: 'Annotated Bibliography—Radiolaria', by A. S. Campbell and E. A. Holm, in 'Treatise on Marine Ecology and Paleoecology II', edited by H. S. Ladd (*Memoir of the Geological Society of America*, **67**, 737–743, 1957)

SCOLECODONTS
'Polychaete Jaw Apparatuses from the Ordovician and Silurian of Poland, and a Comparison With Modern Forms', by Z. Kielan-Jawarowska (*Palaeontologica Polonica*, **16**, 1966)

SILICOFLAGELLATES
'Annotated Index of Fossil and Recent Silicoflagellates and Ebridians, with Descriptions and Illustrations of Validly Proposed Taxa', by A. R. Loeblich III *et al.* (*Memoir of the Geological Society of America*, **106**, 1968)

SPONGE SPICULES
See 'Invertebrate Palaeontology'.

THECAMOEBIANS
Treatise on Invertebrate Paleontology. C. Protista 2, by A. R. Loeblich Jr. and H. Tappan (Geological Society of America and University of Kansas Press, 1964)

TINTINNIDS
Fichier Micropaléontologique Général, Série 9, Ciliès, Tintinnoidea et Ciliatae Incertae (2nd edn., CNRS, Paris, 1965)
Treatise on Invertebrate Paleontology. D. Protista 3 (*Chiefly Radiolaria and Tintinnia*), by A. S. Campbell and R. C. Moore (Geological Society of America and University of Kansas Press, 1954)
'Annotated Index to Genera, Subgenera and Suprageneric Taxa of the Ciliate Order Tintinnida', by A. R. Loeblich Jr. and H. Tappan (*Journal of Protozoology*, **15**, 185–192, 1968)

PALYNOLOGY

Texts on palynology which by title would appear to cover the whole field very often prove to be concerned exclusively with pollen and spores. Exceptions to this are *Handbook of Palynology. Morphology-Taxonomy-Ecology*, by G. Erdtman, with *Appendix on Groups Other than Spores and Pollen*, by W. A. S. Sarjeant (Munksgaard, 1969); and *Aspects of Palynology*, edited by R. H. Tschudy and R. A. Scott (Wiley/Interscience, 1969).

Techniques are discussed in *Palynological Techniques*, by C. A. Brown (Privately published, 1960), and in the section by J. Grey, 'Techniques in Palynology', in *Handbook of Paleontological Techniques*, edited by B. Kummel and D. Raup (Freeman, 1965).

The most useful bibliography is 'Bibliography of Palaeopalynology 1836–1966', by A. A. Manten (*Review of Palaeobotany and Palynology*, **8**, 1969); and the best current journals in the field are *Grana* (formerly *Grana Palynologica*) (1954–), *Journal of Palynology* (1965–) and *Review of Palaeobotany and Palynology* (1967–).

Papers will, of course, be found in other palaeobotanical journals of a more general nature and in palaeontological and botanical journals. Important recent works covering the various fossil groups are:

ANELLOTUBULATES
'Rätselhäfte Mikrofossilien des Oberlias (ε). Neue Funde von "Anellotubulaten" O. WE. 1957', by O. Wetzel (*Neues Jahrbuch für Geologie und Paläontologie, Abhandlungen*, **128**, 341–352, 1967)

CHITINOZOA
Fichier Micropaléontologique Général, Série 7, Chitinozoaires, by G. Deflandre and M. Deflandre-Rigaud (2nd edn., CNRS, Paris, 1965)
Microfossiles Organiques du Paléozoique; les Chitinozoaires, edited by A. Combaz (2 vols., CNRS, Paris, 1967). This incorporates a comprehensive bibliography
'Chitinozoa', by W. A. M. Jenkins, in 'Proceedings of the First Annual Meeting, American Association of Stratigraphic Palynologists', edited by B. F. Perkins (*Geoscience and Man*, **1**, 1–21, 1970)

DINOFLAGELLATES AND ACRITARCHS
The most comprehensive general accounts of these economically important groups are:

'Xanthidia, Palinospheres and "Hystrix". A Review of the Study of Unicellular Microplankton with Organic Cell Walls', by W. A. S. Sarjeant (*Microscopy*, **31**, 221–253, 1970)
'Dinoflagellates—a Selective Review', by W. R. Evitt, in 'Proceedings of the First Annual Meeting, American Association of Stratigraphic Palynologists' (*Geoscience and Man*, **1**, 29–45, 1970)

Two catalogue series cover these groups: the card-file *Fichier Micropaléontologique Général, Séries 11–16*, by G. Deflandre and M. Deflandre-Rigaud (CNRS, Paris, 1962–), and the *Katalog der Fossilien Dinoflagellaten, Hystrichosphären und verwandten Mikrofossilen*, edited by A. Eisenack *et al.* (Schweizerbart'sche, 1964–)
BIBLIOGRAPHIES: 'Bibliography and Index of Fossil Dinoflagellates and Acritarchs', compiled by C. Downie and W. A. S. Sarjeant (*Memoir of the Geological Society of America*, **94**, 1964)
'A Descriptive Index of Genera of Fossil Dinophyceae and Acritarcha', by G. Norris and W. A. S. Sarjeant (*Palaeontological Bulletin of the New Zealand Geological Survey*, **40**, 1965)
'Index to the Genera, Subgenera and Species of the Pyrrhophyta', compiled by A. R. Loeblich Jr. and A. R. Loeblich III (*University of Miami Studies in Tropical Oceanography*, **3**, 1966). Supplements, on an approximately annual basis, are issued in *Journal of Paleontology*
'An Annotated Bibliography of the Tasmanaceae and of Related Living

Forms (Algae: Prasinophyceae)', by M. D. Muir and W. A. S. Sarjeant, in *Microfossiles Organiques du Paléozoique.* No. 3. *Les Acritarches* (CNRS, 1971)

MELANOSCLERITES
'Melanoskleriten aus anstehenden Sedimenten und aus Geschieben', by A. Eisenack (*Paläontologische Zeitschrift*, **37**, 122–134, 1963)

OPHIOBOLIDS
'The Cretaceous Microfossil *Ophiobolus lapidaris* O. Wetzel and its Flagellum-like Filaments', by W. R. Evitt (*Stanford University Publications, Geological Sciences*, **12** (3), 1968)

POLLEN AND SPORES
These are now rivalling the foraminifera as the group of microfossils of greatest economic significance. General texts concerned exclusively, or almost so, with pollen and spores are:

An Introduction to Palynology, by G. Erdtman (3 vols., Chronica Botanica, 1952–66)
An Introduction to Pollen Analysis, by G. Erdtman (2nd edn., Chronica Botanica, 1954)
Pollen Grains, by R. P. Wodehouse (Hafner, 1959)
Textbook of Pollen Analysis, by K. Faegri and J. Iverson (2nd edn., Munksgaard, 1964)
Morphologic Encyclopedia of Palynology, by G. O. W. Kremp (2nd edn., University of Arizona Press, 1969)
Preparation Procedures for Fossil Pollen and Spores Currently Used in the Pollen and Spore Laboratory, Paleontology and Stratigraphy Branch, United States Geological Survey Denver, by L. I. Doher (1965)

A *Catalogue of Fossil Spores and Pollen* has been produced by the University of Pennsylvania since 1957. Since 1961 an annual translations volume has been appended. In addition, a computer system for the processing of data on these groups has been launched by the University of Oklahoma (see 'General Information Processing System. Permian Palynology of North America and Some Associated Problems', by L. R. Wilson *et al.*, *University of Oklahoma Information Science Series Monograph*, **2**, 1969)
BIBLIOGRAPHY: 'Bibliography of Palaeopalynology 1936–1966', by A. A. Manten (*Review of Palaeobotany and Palynology*, **8**, 1969). A bibliographic supplement is also issued with the journal *Pollen et Spores* (1959–)

TASMANITIDS
'An Annotated Bibliography of the Tasmanaceae and of Related Living Forms (Algae: Prasinophyceae)', by M. D. Muir and W. A. S. Sarjeant, in *Microfossiles Organiques du Paléozoique*. No. 3. *Les Acritarches* (CNRS, 1971)

INVERTEBRATE PALAEONTOLOGY

The palaeontology of the invertebrate animals is sometimes considered to include unicellular organisms behaving like animals but most often it is now interpreted as dealing with multicellular animals only. Nonetheless, it has a broad overlap with micropalaeontology in that a number of animal groups decay on death so completely that only particulate skeletal elements of microscopic size survive (e.g. holothurians and many sponges). Textbooks on invertebrate palaeontology tend not to treat, or to treat only partially, with these groups. Texts currently available include:

Invertebrate Fossils, by R. C. Moore *et al.* (McGraw-Hill, 1952)
Principles of Invertebrate Paleontology, by R. R. Shrock and W. H. Twenhofel (McGraw-Hill, 1953)
Invertebrate Paleontology, by W. H. Easton (Harper, 1960)
Palaeontology, Invertebrate, by H. Woods (reprint of 8th edn., Cambridge University Press, 1967)
Die Vorzeitlichen Wirbellösen. System and Evolution, by O. Kuhn (Oeben, 1966)

The main series of volumes is *Treatise on Invertebrate Paleontology* already discussed, and cited as appropriate in the following sections.

PORIFERA AND ARCHAEOCYATHIDS
Treatise on Invertebrate Paleontology. E. Archaeocyatha, Porifera, by V. J. Okulitch (Geological Society of America and University of Kansas, 1955). Volume 1 of the 2nd edition, covering only the Archaeocyatha, by D. Hill, was issued in 1972
Osnovy Paleontologii. 2. Gubki, Arkeotsiaty, Kishechnopolostnye, Chervi, edited by B. S. Sokolov (Izdatel'stvo Akademii Nauk SSSR, 1962)
'The Phylum Archaeocyatha', by D. Hill (*Biological Reviews*, **39**, 232–258, 1964)
'Archaeocyatha from Antarctica and a Review of the Phylum', by D. Hill (*Scientific Reports, Trans-Antarctic Expedition, Geology*, **3**, 1965)
'A Revision of Australian Genera of Archaeocyatha', by F. Debrenne (*Transactions of the Royal Society of South Australia*, **94**, 21–50, 1970)
BIBLIOGRAPHIES: *Bibliography of Sponges 1551–1913*, by G. C. J. Vosmaer (Cambridge University Press, 1928)
'Annotated Bibliography. Sponges of the Paleozoic', by V. J. Okulitch and S. J. Nelson, and 'Annotated Bibliography. Sponges of the Post Paleozoic', by M. W. de Laubenfels, in 'Treatise on Marine Ecology and Paleoecology. II', edited by H. S. Ladd (*Memoir of the Geological Society of America*, **67**, 763–769, 771–772, 1957)
'Bibliographic Index of North American Archaeocyathids', by M. H. Nitecki (*Fieldiana Geology*, **17**, 1962)

COELENTERATES (INCLUDING STROMATOPOROIDEA AND CONULARIDA AND EXCLUDING ALCYONARIAN SCLERITES)
Treatise on Invertebrate Paleontology. F. Coelenterata, by F. M. Bayer *et al.* (Geological Society of America and University of Kansas Press, 1963)
Osnovy Paleontologii. 2. Gubki Arkeotsiaty, Kishechnopolostnye, Chervi,

edited by B. S. Sokolov (Izdatel'stvo Akademii Nauk SSSR, 1962)
'The Cnidaria and their Evolution', edited by W. J. Rees (*Symposia, Zoological Society of London*, **16**, 1966)
BIBLIOGRAPHIES: The following parts of the *Fossilium Catalogus I* are relevant to coelenterates:

Pt. 5. *Anthozoa Palaeocretacea*, by J. Felix (1914)
Pt. 6. *Anthozoa Cenomanica*, by J. Felix (1914)
Pt. 7. *Anthozoa Neocretacea*, by J. Felix (1914)
Pt. 13. *Cnidaria Triadica*, by C. Diener (1921)
Pt. 26. *Medusae Fossiles*, by A. Kieslinger (1924)
Pt. 28. *Anthozoa Eocaenica et Oligocaenica*, by J. Felix (1925)
Pt. 35. *Anthozoa Miocaenica*, by J. Felix (1927)
Pt. 36. *Hydrozoa*, by O. Kuhn (1928)
Pt. 44. *Anthozoa Pliocaenica et Pleistocaenica* (*cum indice ad omnes partes Anthozoarum Tertiarium et Quaternarium*), by J. Felix (1929)
Pts. 115–116. *Stromatoporoidea* (*Hydrozoa Palaeozoica*), by E. Flügel and E. Flügel-Kahler (1968)

Other bibliographies include:

'A Bibliography of the Conularida', by G. W. Sinclair and E. S. Richardson (*Bulletin of American Paleontology*, **34**, No. 145, 1954)
'A Bibliography of the Order Stromatoporoidea', by J. J. Galloway and J. St. Jean Jr. (*Journal of Paleontology*, **30**, 170-185, 1956)
'Zur Bibliographie der Stromatoparen', by E. Flügel (*Mitteilungen des Naturwissenschaftlichen Vereins für Steiermark*, **86**, 26–31, 1956)
'Annotated Bibliography. Corals', by J. W. Wells, in 'Treatise on Marine Ecology and Paleoecology. II', edited by H. S. Ladd (*Memoir of the Geological Society of America*, **67**, 773–782, 1957)

BRYOZOA
Treatise on Invertebrate Paleontology. G. Bryozoa, by R. S. Bassler (Geological Society of America and University of Kansas Press, 1953)
Osnovy Paleontologii. 7. Mshanki, Brakhiopody, Prilozhenie, Foronidy, edited by T. G. Sarycheva (Izdatel'stvo Akademii Nauk SSSR, 1960)
'Proceedings of the First International Conference on Bryozoa', edited by E. Annoscia (*Atti della Società Italiana di Scienze Naturali e del Museo Civile di Storia Naturale di Milano*, **108**, 1968). The second conference was held in 1971 and publication of the proceedings is in the hands of Academic Press
'Skeletal Growth, Intracolony Variation, and Evolution in Bryozoa; A Review', by R. S. Boardman and A. H. Cheetham (*Journal of Paleontology*, **43**, 205-244, 1969)
BIBLIOGRAPHIES: 'Bryozoa (genera)', by R. S. Bassler (*Fossilium Catalogus I*, **67**, 1934)
'Annotated Bibliography. Bryozoa', by H. Duncan, in 'Treatise on Marine Ecology and Paleoecology. II', edited by H. S. Ladd (*Memoir of the Geological Society of America*, **67**, 783–800, 1957)

BRACHIOPODS
Treatise on Invertebrate Paleontology. H. Brachiopoda, by A. Williams *et al.* (2 vols., Geological Society of America and University of Kansas Press, 1965)

Osnovy Paleontologii. 7. Mshanki, Brakhiopody. Prilozhenie Foronidy, edited by T. G. Sarycheva (Izdatel'stvo Akademii Nauk SSSR, 1960)
'Brachiopod Palaeoecology', by D. V. Ager (*Earth-Science Reviews*, **3**, 157–179, 1967)
Living and Fossil Brachiopods, by M. J. S. Rudwick (Hutchinson, 1970)
BIBLIOGRAPHIES: 'Brachiopoda Triadica', by C. Diener (*Fossilium Catalogus I*, **10**, 1920)
'Brachiopoda (general)', by C. Schuchert and C. H. Le Vene (*Fossilium Catalogus I*, **2**, 1929)
'Bibliography and Index of North American Carboniferous Brachiopods', by J. L. and R. C. Carter (*Memoir of the Geological Society of America*, **128**, 1970)

MOLLUSCS

Treatise on Invertebrate Paleontology. I. Mollusca 1 (Mollusca General Features, Scaphopoda, Amphineura, Monoplacophora, Gastropoda General Features, Archaeogastropoda, Mainly Paleozoic Caenogastropoda and Opisthobranchia), by J. B. Knight *et al.* (Geological Society of America and University of Kansas Press, 1964)
Treatise on Invertebrate Paleontology. K. Mollusca 3 (Cephalopoda General Features, Endoceratoidea, Actinoceratoidea, Nautiloidea, Bactrito idea), by C. Teichert *et al.* (Geological Society of America and University of Kansas Press, 1964)
Treatise on Invertebrate Paleontology. L. Mollusca 4 (Ammonoidea), by W. J. Arkell *et al.* (Geological Society of America and University of Kansas Press, 1957)
Treatise on Invertebrate Paleontology. N. Mollusca 6 (Bivalvia), by L. R. Cox *et al.* (Geological Society of America and University of Kansas Press, 1969). The other two parts of the sections dealing with mollusca will be J. *Mollusca 2 (Gastropoda, Streptoneura Exclusive of Archaeogastropoda, Euthyneura)* and M. *Mollusca 5 (Coleoidea)*
Osnovy Paleontologii. 3. Mollyuski—Pantsirnye, Dvustvorchatye, Lopatonogie, edited by A. G. Eberzin (Izdatel'stvo Akademii Nauk SSSR, 1960)
Osnovy Paleontologii. 4. Mollyuski-Bryukhonogie, edited by V. F. Pchelintsev and I. A. Korobkov (Izdatel'stvo Akademii Nauk SSSR, 1960)
Osnovy Paleontologii. 5. Mollyuski-golovonogie I. Nautiloidei, Endotseratoidei, Aktinotseratoidei, Baktritoidei, Ammonoidei (Agoniaty, Goniatity, Klimenii), edited by V. E. Ruzhentsev (Izdatel'stvo Akademii Nauk SSSR, 1962)
Osnovy Paleontologii. 6. Mollyuski-golovonogie II. Ammonoidei (Tseratity i Ammonity) Vnutrennerakovinnye Prilozhenie Konikonkhii, edited by N. P. Lunnov and V. C. Drushchii (Izdatel'stvo Akademii Nauk SSSR, 1958)
BIBLIOGRAPHIES: The following parts of *Fossilium Catalogus I* are relevant:

Pt. 1. *Ammoneae Devonicae (Clymeniidae, Aphyllitidae, Gephyroceratidae Cheiloceratidae)*, by F. Frech (1913)
Pts. 2, 15. *Lamellibranchiata Tertiaria 'Anisomyaria'*, by W. Teppner (1914, 1922)
Pts. 8, 56. *Cephalopoda Triadica*, by C. Diener and A. Kutassy (1915, 1933)
Pt. 11. *Cephalopoda Dibranchiata*, by E. von Bulow-Trummer (1930)
Pt. 14. *Ammonoidea Permiana*, by C. Diener (1921)
Pts. 17, 18, 20–23, 32, 38, 40, 43, 46. *Gastropoda Extra Marina Tertiaria*, by W. Wenz (1923–30)

Pts. 19, 51. *Lamellibranchiata Triadica*, by C. Diener and A. Kutassy (1923, 1931)
Pt. 29. *Ammonoidea Neocretacea*, by C. Diener (1925)
Pt. 30. *Trigoniidae Mesozoicae* (*Myophoriis Exclusis*), by W. Deecke (1925)
Pt. 31. *Gastropoda Mesozoica; Familia Nerineidae*, by W. O. Dietrich (1925)
Pts. 34, 81. *Glossophora Triadica*, by C. Diener and A. Kutassy (1926, 1940)
Pts. 53, 65. *Gastropoda, Amphineura et Scaphopoda Jurassica*, by G. Haber (1932, 1934)
Pt. 54. *Rudistae*, by O. Kuhn (1932)
Pt. 55. *Cypraeacea*, by F. A. Schilder (1932)
Pt. 68. *Pachyodontia Mesozoica* (*Rudistis Exclusis*), by A. Kutassy (1934)
Pt. 91. *Lamellibranchiata Infracarbonica*, by H. Paul (1941)
Pt. 95. *Teredinidae*, by F. Moll (1941)
Pt. 96. *Hecticoceratinae* (*Ammonoidea Jurassica*), by A. Zeiss (1959)

Other relevant bibliographic compilations are:

'Recent Literature on Mesozoic Ammonites', by O. Haas, in *Journal of Paleontology* from 1957
'Genera of the Bivalvia: A Systematic and Bibliographic Catalogue', by H. E. Vokes (*Bulletin of American Paleontology*, **51**, No. 232, 1967)
'A Bibliographic Index of North American Late-Paleozoic Hyolitha Amphineura, Scaphopoda and Gastropoda', by E. L. Yochelson and B. W. Saunders (*Bulletin of the United States Geological Survey*, No. 1210, 1967)

VERMES

Treatise on Invertebrate Paleontology. W. Miscellanea, by W. H. Hass *et al.* (Geological Society of America and University of Kansas Press, 1962)
Osnovy Paleontologii. 2. Gubki, Arkheotsiaty, Kishechnopolostyne Chervi, edited by B. S. Sokolov (Izdatel'stvo Akademii Nauk SSSR, 1962)
'Annotated Bibliography. Vermes', by B. F. Howell, in 'Treatise on Marine Ecology and Paleoecology II', edited by H. S. Ladd (*Memoir of the Geological Society of America*, **67**, 805–816, 1957)

ARTHROPODS (EXCEPT OSTRACODS)

Treatise on Invertebrate Paleontology. O. Arthropoda 1 (*Arthropoda General Features, Protarthropoda, Euarthropoda General Features, Trilobitomorpha*), by H. J. Harring *et al.* (Geological Society of America, University of Kansas Press, 1964)
Treatise on Invertebrate Paleontology. P. Arthropoda 2 (*Chelicerata with Sections on Pycnogonida and Palaeoisopus*), by L. Størmer *et al.* (Geological Society of America and University of Kansas Press, 1964)
Treatise on Invertebrate Paleontology. R. Arthropoda 4. Crustacea (*except Ostracoda, Myriapoda, Hexapoda*), edited by H. K. Brooks *et al.* (Geological Society of America and University of Kansas Press, 1969)
Osnovy Paleontologii. 8. Chlenistonogie. Trilobitoobraznye i Rakoobraznye, edited by N. E. Chernysheva (Izdatel'stvo Akademii Nauk SSSR, 1960)
Osnovy Paleontologii. 9. Trakheiny i Khelitserovye, edited by B. B. Rodendorf (Izdatel'stvo Akademii Nauk SSSR, 1962)
'Paleozoic and Mesozoic Arachnida of Europe', by A. Petrunkevitch (*Memoir of the Geological Society of America*, **53**, 1953)
Introduction to Entomology, by R. G. Jeannel (2nd edn., Hutchinson, 1960)

'Studies on Fossil Aphids (Homoptera: Aphidoidea)', by O. E. Heie (*Spolia Zoologica Musei Nauniensis*, **26**, 1–274, 1967)
'Remarks on the Appendages of Trilobites', by J. Bergström (*Lethaia*, **2**, 395–414, 1969)
Die Stammesgeschichte der Insekten, by W. Hennig (Kramer, 1969)
BIBLIOGRAPHIES: The following parts of the *Fossilium Catalogus I* are relevant:

Pt. 16. *Insecta Palaeozoica*, by A. Handlirsch (1922)
Pt. 25. *Eurypterida*, by C. Diener (1924)
Pt. 37. *Trilobitae Neodevonici*, by R. and E. Richter (1928)
Pt. 41. *Crustacea Decapoda*, by M. F. Glaessner (1929)
Pt. 48. *Crustacea Eumalacostraca* (*Crustaceis Decapodis Exclusis*), by V. van Straelen (1931)
Pt. 64. *Crustacea Phyllocarida* (*Archaeostraca*), by V. Van Straelen and G. Schmitz (1934)
Pt. 118. *Trilobitae Carbonici et Permici. I.* (*Brachymetopidae, Otarionidae, Proetidae, Proetinae, Dechenellinae, Drevermanniinae, Cyrtosymbolinae*), by G. and R. Hahn (1969)
Pt. 119. *Trilobitae Carbonici et Permici. II.* (*Proetidae, Griffithidinae*), by G. and R. Hahn (1970)

'A Bibliography of Paleozoic Crustacea from 1698 to 1892, to which is Added a Catalogue of North American Species', by A. W. Vogdes (*Occasional Paper, California Academy of Sciences*, **4**, 1893). Supplements in *Proceedings of the California Academy of Sciences*, **5** (2) (1895) and *Transactions of the San Diego Society of Natural History*, **3** (1917), **4** (1925)

ECHINODERMS (EXCEPT HOLOTHURIANS)
Treatise on Invertebrate Paleontology. S. Echinodermata 1. (*Echinodermata General Features, Homalozoa, Crinozoa, Exclusive of Crinoidea*), by H. H. Beaver *et al.* (2 vols., Geological Society of America and University of Kansas Press, 1968)
Treatise on Invertebrate Paleontology. U. Echinodermata 3. (*Asterozoans, Echinozoans*), by J. W. Durham *et al.* (2 vols., Geological Society of America and University of Kansas Press, 1966)
Osnovy Paleontologii. 10. Iglokozhie, Gemikhordovye, Pogonofory i Shchetinkochelyustnye, edited by R. F. Gekker (Izdatel'stvo Akademii Nauk SSSR, 1964)
'Classification of Echinoids', by G. M. Philip (*Journal of Paleontology*, **39**, 45–62, 1965)
'Evolution among the Echinoidea', by J. W. Durham (*Biological Reviews*, **41**, 368–391, 1966)
'Echinoderm Biology', edited by N. M. Holt (*Symposia of the Zoological Society of London*, **20**, 1967)
BIBLIOGRAPHIES: The following parts of the *Fossilium Catalogus I* are relevant:

Pt. 3. *Stelleroidea Palaeozoica*, by C. Schuchert (1914)
Pt. 39. *Echinoidea Jurassica*, by W. Deecke (1928)
Pt. 66. *Crinoidea Triadica*, by W. Biese (1934)
Pt. 70, 73, 76. *Crinoidea Jurassica*, by W. Biese (1935–37)
Pt. 77. *Crinoidea Cretacea*, by W. Biese and H. Sieverts-Doreck (1939)
Pt. 80. *Crinoidea Caenozoica*, by W. Biese and H. Sieverts-Doreck (1939)

Pt. 83. *Pelmatazoa Palaeozoica*, by R. S. Bassler (1939)
Pt. 88. *Supplementum ad: Crinoidea Triadica, Jurassica, Cretacea and Caenozoica*, by H. Sieverts-Doreck (1939)

'Bibliography of Cenozoic Echinoidea, Including Some Mesozoic and Paleozoic Titles', by N. E. Weisbord (*Bulletin of American Paleontology*, **59**, No. 263, 1971)

GRAPTOLITES
Treatise on Invertebrate Paleontology. V. Graptolithina, by O. M. B. Bulman (2nd edn., Geological Society of America and University of Kansas Press, 1970)
Osnovy Paleontologii. 10. Iglokozhie, Gemikhordovye, Pogonofory i Shchetinkochelyustnye, edited by R. F. Gekker (Izdatel'stvo Akademii Nauk SSSR, 1964)

VERTEBRATE PALAEONTOLOGY

Vertebrate palaeontology is generally taken to comprise not only the vertebrates but also the other groups of chordates that are their near relatives. The romance associated with the ancestry of man and the familiar spectacular reconstructions of many extinct vertebrates have produced a market for, and therefore an ample supply of, popular works. Many of these, however, are of doubtful quality. The following general accounts of vertebrate palaeontology are recommended:

The Life of Vertebrates, by J. Z. Young (Clarendon Press, 1950)
Paläontologie und Phylogenie der Niederen Tetrapoden, by F. von Huene (Fischer, 1956)
Studies on Fossil Vertebrates, edited by T. S. Westoll (Athlone Press, 1958)
L'Evolution des Vertébrés Inférieurs, by J. P. Lehman (Dunod, 1959)
Théories de l'Evolution des Vertébrés, by E. Jarvik (Masson, 1960)
The Vertebrate Story, by A. S. Romer (4th edn., University of Chicago Press, 1961)
Structure and Habit in Vertebrate Evolution, by G. S. Carter (Sidgwick and Jackson, 1967)
Chordate Morphology, by M. Jollie (Reinhold, 1968)
The Origin of Terrestrial Vertebrates, by I. I. Schmalhausen (Academic Press, 1968)
Evolution of the Vertebrates, by E. C. Colbert (2nd edn., Wiley, 1969)
The Pattern of Vertebrate Evolution, by L. B. Halstead (Oliver and Boyd, 1969)
Morphologie Evolutive des Chordés, by P. Pirlot (Université de Montréal, 1969)
Evolution des Vertébrés de leur Origine à l'Homme, by G. Vandebroek (Masson, 1969)
Vertebrate Paleozoology, by E. C. Olson (Wiley, 1971)

The most comprehensive single-volume work, including an outline classification and featuring an extensive bibliography arranged by groups, is *Vertebrate Paleontology*, by A. S. Romer (3rd edn., University of Chicago Press, 1966). To this may be added the same author's *Notes and Comments on Vertebrate Paleontology* (Chicago University Press, 1968).

A series of books characterised by superb illustrations, which amply repay examination, have been produced by two Czech authors, J. Augusta and Z. Burian. Among their works are *Prehistoric Animals* (Spring Books, 1958), *A Book of Mammoths* (Hamlyn, 1962), *Prehistoric Sea Monsters* (Hamlyn, 1964) and *The Age of Monsters* (Hamlyn, 1966).

Of course, some of the textbooks on palaeontology cover vertebrates but the two major comprehensive works, of value to vertebrate palaeontologists, are:

Traité de Paléontologie. IV, V, VI, VII, edited by J. Piveteau (Masson, 1955–69). Individual volumes and parts are dealt with in the text as appropriate
Osnovy Paleontologii. 11, 12, 13, edited by Y. A. Orlov (Izdatel'stvo Akademii Nauk SSSR, 1962–64). Again, individual parts are mentioned in the text as appropriate.

The vertebrates are particularly well served by bibliographies. The *Fossilium Catalogus I* contains many works on particular vertebrate groups. The following are, however, more comprehensive:

'Bibliography and Catalogue of the Fossil Vertebrata of North America', by O. P. Hay (*Bulletin of the United States Geological Survey*, No. 179, 1902). This was followed by 'Second Bibliography and Catalogue of the Fossil Vertebrata of North America', by O. P. Hay (*Publication, Carnegie Institution, Washington*, No. 390 (2 vols.), 1929–30)
Covering world literature is 'Bibliography of Fossil Vertebrates', which has been issued as *Special Papers, Geological Society of America*, **27** and **42**, and *Memoirs of the Geological Society of America*, **37, 57, 84, 92** and **117**. These update the Hay bibliographies
'Bibliography of Fossil Vertebrates Exclusive of North America, 1509–1927', by A. S. Romer *et al.* (*Memoir of the Geological Society of America*, **87**, 1962)
Bibliography of Vertebrate Paleontology and Related Subjects (1945–71) (Society of Vertebrate Paleontology, 1947–72)

The following are the principal journals concerned with vertebrate palaeontology: *Fossil Vertebrates of Africa* (Academic Press, 1969–), which continues in part *Fossil Mammals of Africa*, produced by the British Museum (Natural History); *News Bulletin*

Society of Vertebrate Paleontology (1941–); *Palaeovertebrata* (1967–); and *Vertebrata Palasiatica* (1957–).

Papers on fossil vertebrates are frequently found in zoological journals and therefore the field of literature to be searched is very wide.

PRIMITIVE CHORDATES

The relevant parts of the comprehensive works are *Osnovy Paleontologii.* 10, edited by R. F. Gekker (Izdatel'stvo Akademii Nauk SSSR, 1964), and *Traité de Paléontologie.* IV. *L'Origine des Vertébrés, leur Expansion dans les Eaux Douces et le Milieu Marin,* Part 1, edited by J. Piveteau (Masson, 1964)

The following two books are also of value:

The Origin of the Vertebrates, by N. J. Berrill (Clarendon Press, 1955)
The Biology of Hemichordata and Protochordata, by E. J. W. Barrington (Oliver and Boyd, 1965)

FISH

The reader's attention is drawn to the introduction to the section on vertebrate palaeontology and also to the following comprehensive accounts:

Osnovy Paleontologii. 11. *Beschelyustnye, ryby,* edited by D. V. Obruchev (Izdatel'stvo Akademii Nauk SSSR, 1964). This volume has been translated by the Israel Program for Scientific Translations (1967)
Traité de Paléontologie. IV. *L'Origine des Vertébrés, leur Expansion dans les Eaux Douces et le Milieu Marin.* Part 1. *Vertébrés (Généralités) Agnathes;* Part 2. *Gnathostomes, Acanthodiens, Placodermes, Elasmobranches;* Part 3. *Actinoptérygiens, Crossoptérygiens, Dipneustes,* edited by J. Piveteau (Masson, 1964–69)

These two works are of the utmost importance. Others of value are:

Catalogue of the Fossil Fishes in the British Museum (Natural History), by A. S. Woodward (British Museum (Natural History), 1889–1901)
System der Rezenten und Fossilien Fischartigen und Fische, by L. S. Berg (Veb Deutscher Verlag der Wissenschaften, 1958). The original was issued, in Russian and English, as *Trudy Zoologicheskogo Instituta,* **5** (2), (1940) and this in turn was reprinted by Edwards, Ann Arbor, in 1947 and the English text again by Thai National Documentation Centre, 1965
'Recent Advances in the Palaeontology of Fishes', by T. S. Westoll (*Liverpool and Manchester Geological Journal,* **2**, 586–596, 1960)
Palaeozoic Fishes, by J. A. Moy-Thomas, revised by R. S. Miles (2nd edn., Chapman and Hall, 1971). This work is of major importance and contains excellent bibliographies.

BIBLIOGRAPHY: *Bibliography of Fishes,* by B. Dean (3 vols., American Museum of Natural History, 1916–23). The American Museum of Natural History has also just announced, and produced the first part of, a new computer-based bibliography entitled *Dean Bibliography of Fishes.* The first volume covers the literature of fossil and recent fish for the year 1968. Volumes for later years are listed as being forthcoming in the near future.

Relevant sections of *Fossilium Catalogus I* include:

Pt. 33. *Pisces Triadici*, by D. Deeke (1926)
Pt. 57. *Antiarchi*, by W. Gross (1932)

AMPHIBIANS
Of the major treatises the following are relevant:

Osnovy Paleontologii. 12. Zemnovodyne, Presmykayushchiesya i Ptitsy, edited by A. K. Rozhdestvenskii and L. P. Tatarinov (Izdatel'stvo Akademii Nauk SSSR, 1964)
Traité de Paléontologie. V. La Sortie des Eaux, Naissance de la Tétrapodie. L'Exubérance de la Vie Végétative. Le Conquête de l'Air. Amphibiens, Reptiles, Oiseaux, edited by J. Piveteau (Masson, 1955)

Among the more important books on amphibians are:

Die Familien der Rezenten und Fossilen Amphibien und Reptilien, by O. Kuhn (Meisenbach, 1961)
Die Amphibien: System und Stammesgeschichte, by O. Kuhn (Oeben, 1965)
Amphibien und Reptilien. Catalog der Subfamilien und Höheren Taxa mit Nachwies des Ersten Auftretens, by O. Kuhn (Fischer, 1967)
Handbuch der Paläoherpetologie. Encyclopedia of Paleoherpetology, edited by O. Kuhn (Fischer, 1969–). This is scheduled to be published in 19 parts, six covering the amphibians, of which so far only Part 5. *Batrachosauria. A. Anthracosauria*, by A. L. Panchen (1970), has been published

BIBLIOGRAPHIES: *Amphibia*, by O. Kuhn (*Fossilium Catalogus I*, **61**, **84**, 1933, 1938)
Amphibia (Supplementum), by O. Kuhn (*Fossilium Catalogus I*, **97**, 1960)
'Amphibians and Reptiles from the Pliocene and Pleistocene of North America: A Chronological Summary and Selected Bibliography', by F. R. Gehlbach (*Texas Journal of Science*, **17**, 56–70, 1965)

REPTILES (AND MAMMAL-LIKE REPTILES)
The major treatises are:

Osnovy Paleontologii. 12. Zemnovodyne, Presmykayushchiesya i Ptitsy, edited by A. K. Rozhdestvenskii and L. P. Tatarinov (Izdatel'stvo Akademii Nauk SSSR, 1964)
Traité de Paléontologie. V. La Sortie des Eaux. Naissance de la Tétrapodie. L'Exubérance de la Vie Végétative. La Conquête de l'Air. Amphibiens, Reptiles, Oiseaux, edited by J. Piveteau (Masson, 1955)
Handbuch der Paläoherpetologie. Encyclopedia of Paleoherpetology, edited by O. Kuhn (Fischer, 1969–). To be published in 19 volumes, of which 14 will deal with various groups of the reptilia. Those so far published include:

6. *Cotylosauria*, by O. Kuhn (1969)
9. *Proganosauria, Bolosauria, Placodontia, Araescelidia, Trilophosauria, Weigeltisauria, Millerosauria, Rhynchocephalia, Protorosauria*, by O. Kuhn (1969)
14. *Saurischia*, by R. Steel (1970)
15. *Ornithischia*, by R. Steel (1969)

Books with a general coverage include:

Paläontologie und Phylogenie der Niederen Tetrapoden, by F. von Huene (Fischer, 1956), with supplement, *Nachträge und Ergänzungen* (Fischer, 1959)
Osteology of the Reptiles, by A. S. Romer (University of Chicago Press, 1956). Deals with both fossil and recent forms
Die Familien der Rezenten und Fossilien Amphibien und Reptilien, by O. Kuhn (Meisenbach, 1961)
The Age of Reptiles, by E. H. Colbert (Weidenfeld and Nicolson, 1965)
Die Reptilien. System und Stammesgeschichte, by O. Kuhn (Oeben, 1966)
Amphibien und Reptilien. Catalog der Subfamilien und Höheren Taxa mit Nachwies des Ersten Auftretens, by O. Kuhn (Fischer, 1967)
Die Deutschen Saurier, by O. Kuhn (Oeben, 1968)
The Life of Reptiles, by A. d'A. Bellairs (2 vols., Weidenfeld and Nicolson, 1969)
Biology of the Reptilia, edited by C. Gans (Academic Press, 1969)

Books dealing with particular reptilian groups include:

The Mammal-like Reptiles of South Africa and the Origin of the Mammals, by R. Broom (Witherby, 1932)
Les Iguanodons de Bernissart, by E. Casier (Editions du Patrimoine de l'Institut Royal des Sciences Naturelles de Belgique, 1960)
Dinosaurs; their Discovery and their World, by E. H. Colbert (Hutchinson, 1962)
Prehistoric Sea Monsters, by J. Augusta and Z. Burian (Hamlyn, 1964). This is a lavishly illustrated account of the fossil marine reptiles
Die fossile Wirteltierklasse Pterosauria, by O. Kuhn (Oeben, 1967)
Die Vorzeitlichen Krokodile, by O. Kuhn (Oeben, 1968)
The Age of the Dinosaurs, by B. Kurtén (Weidenfeld and Nicolson, 1968)
Dinosaurs, by W. E. Swinton (4th edn., British Museum (Natural History), 1969)
The Dinosaurs, by W. E. Swinton (2nd edn., Allen and Unwin, 1970)
The Last of the Ruling Reptiles. Alligators, Crocodiles and their Kin, by W. T. Neill (Columbia University Press, 1971)

BIBLIOGRAPHIES: The following bibliographies, all part of the series *Fossilium Catalogus I* are important:

Pt. 4. *Saurischia et Ornithischia Triadica*, by F. von Huene (1914)
Pt. 9. *Stegosauria*, by E. Hennig (1915)
Pts. 27, 50. *Osteologia Reptilium Fossilium et Recentium* (*cum Appendice*), by F. von Nopcsa (1926, 1931)
Pt. 45. *Pterosauria*, by F. Plieninger (1929)
Pt. 52. *Tryionychia Fossilia*, by K. Hummel (1932)
Pt. 58. *Thecodontia*, by O. Kuhn (1932)
Pt. 62. *Placodontia*, by O. Kuhn (1933)
Pt. 63. *Icthyosauria, with Thecodontia: Supplementum*, by O. Kuhn (1934)
Pt. 69. *Sauropterygia*, by O. Kuhn (1934)
Pt. 71. *Rhynchocephalia (Eosuchia)*, by O. Kuhn (1935)
Pt. 75. *Crocodilia*, by O. Kuhn (1936)
Pt. 78. *Ornithischia (Stegosauriis exclusis)*, by O. Kuhn (1936)

Pt. 79. *Cotylosauria et Theromorpha*, by O. Kuhn (1937)
Pt. 85. *Protosauria, Mesosauria*, by O. Kuhn (1939)
Pt. 86. *Squamata: Lacertilia et Ophidia*, by O. Kuhn (1939)
Pt. 87. *Saurischia*, by O. Kuhn (1939)
Pt. 94. *Testudinata Triadica*, by O. Kuhn (1941)
Pts. 98–99. *Reptilia: Supplementum*, by O. Kuhn (1961). Deals mostly with pre-Jurassic groups
Pt. 103. *Serpentes (Supplementum 1)*, by O. Kuhn (1963)
Pt. 104. *Sauria (Supplementum 1)*, by O. Kuhn (1963)
Pt. 105. *Ornithischia (Supplementum 1)*, by O. Kuhn (1963)
Pt. 106. *Sauropterygia (Supplementum 1)*, by O. Kuhn (1964)
Pt. 107. *Testudines*, by O. Kuhn (1964)
Pt. 109. *Saurischia (Supplementum 1)*, by O. Kuhn (1965)
Pt. 110. *Therapsida (Supplementum 1)*, by O. Kuhn (1965)

BIRDS

The relevant parts of the major treatises are: *Osnovy Paleontologii. 12. Zemnovodnye, Presmykayushchiesya i Ptitsy*, edited by A. K. Rozhdestvenskii and L. P. Tatarinov (Izdatel'stvo Akademii Nauk SSSR, 1964); and *Traité de Paléontologie. V. La Sortie des Eaux. Naissance de la Tétrapodie. L'Exubérance de la Vie Végétative. La Conquête de l'Air. Amphibiens, Reptiles, Oiseaux*, edited by J. Piveteau (Masson, 1955)

The following are the more important books on fossil birds:

The Origin of Birds, by G. Heilmann (Witherby, 1926)
Handbuch der Paläornithologie, by K. Lambrecht (Borntraeger, 1933; reprinted 1964)
Prehistoric Reptiles and Birds, by J. Augusta and Z. Burian (Hamlyn, 1961). Excellent illustrations of the Mesozoic birds and their presumed reptilian relations. Not as comprehensive as the title suggests
Die Fossilen Vögel: Osteologie, Stammesgeschichte und System der 42 Ordnungen, by O. Kuhn (Oeben, 1965)
Fossil Birds, by W. E. Swinton (2nd edn., British Museum (Natural History), 1965)

BIBLIOGRAPHIES: 'Catalogue of Fossil Birds', by P. Brodkorb (*Bulletin of the Florida State Museum, Biological Sciences*, **7** (4); **8** (3); **11** (3); **15** (4) (1963–71)
'Zur Archaeopteryx-Bibliographie', by B. v. Freyberg (*Geologische Blätter für Nordost-Bayern und Angrenzende Gebiete*, **14**, 122–123, 1964)

MAMMALS (OTHER THAN PRIMATES)

The relevant volumes of the major treatises are: *Osnovy Paleontologii. 13. Mlekopitatushchie*, edited by V. I. Gromova (Izdatel'stvo Akademii Nauk SSSR, 1962; translation by the Israel Program for Scientific Translations, (1968); and *Traité de Paléontologie VI. L'Origine des Mammifères et les Aspects Fondamentaux de leur Evolution. I. Mammifères. Origine Reptilienne Evolution. II. Mammifères, Evolution*, edited by J. Piveteau (Masson, 1961, 1958)

Books with a general coverage include:

A Catalogue of the Mesozoic Mammalia in the Geological Department of the British Museum (Natural History), by G. G. Simpson (British Museum (Natural History), 1928)

A History of Land Mammals in the Western Hemisphere, by W. B. Scott (2nd edn., Macmillan, 1937). This has recently been reprinted
Bones for the Archaeologist, by I. W. Cornwall (Phoenix House, 1956). Excellent treatment of British Pleistocene and post-Pleistocene mammalian osteology
The Life of Mammals, by J. Z. Young (Clarendon Press, 1957)
International Colloquium on the Evolution of Lower and Non-Specialized Mammals, edited by G. Vandebroek (2 vols., Koninklijke Vlaamse Academie voor Wetenschappen, Letteren en Schone Kunsten, Klasse der Wetenschappen, Brussels, 1961)
A History of Domesticated Animals, by F. E. Zeuner (Hutchinson, 1963)
Animals and Plants of the Cenozoic Era, by R. Pearson (Butterworths, 1964)
The Age of Monsters, Prehistoric and Legendary, by J. Augusta and Z. Burian (Hamlyn, 1966). A well-illustrated account of the giant Tertiary mammals
Pleistocene Mammals of Europe, by B. Kurtén (Weidenfeld and Nicolson, 1968)
The Age of Mammals, by B. Kurtén (Weidenfeld and Nicolson, 1971)

Books dealing with particular mammalian groups include:

Proboscidae, by H. F. Osborn (2 vols., American Museum of Natural History, 1936, 1942)
A Review of Fossil and Recent Bears of the Old World, by D. P. Erdbrink (Drukkerij Jan de Lange, 1953)
Elephants. A Short Account of their Natural History, Evolution and Influence on Mankind, by R. Carrington (Chatto and Windus, 1958)
Horses. The Story of the Horse Family in the Modern World and Through 60 Million Years of History, by G. G. Simpson (reprint of the 1951 edition, Doubleday, 1961)
Whales, by E. J. Slijper (Hutchinson, 1962). Discusses their fossil record
A Book of Mammoths, by J. Augusta and Z. Burian (Hamlyn, 1963). Includes excellent illustrations of most major elephant types

BIBLIOGRAPHIES: *Fauna Fossilis Cavernarum* (*Homino excluso*), by B. Wolf (*Fossilium Catalogus I*, **82**, **89**, **92**, 1938–41)
'An Annotated Bibliography of the Fossil Mammals of Africa (1742–1950)', by A. T. Hopwood and J. P. Hollyfield (*Fossil Mammals of Africa*, **8**, 1954). This has been updated by 'Bibliography of Quaternary African Palaeontology', by R. G. Welbourne, in *Palaeontologia Africana*, **12**, 151–202 (1969)
'An Annotated Bibliography on the Origin and Descent of Domestic Mammals, 1900–1955', by S. Angress and C. A. Reed (*Fieldiana Anthropology*, **54**, 1962)
JOURNALS: *Fossil Mammals of Africa* (1951–1967). The last five issues of this valuable series were issued as parts of the *Bulletin of the British Museum (Natural History) (Geology)*. It is, in part, continued by *Fossil Vertebrates of Africa* (1969–) published by Academic Press

PRIMATES (INCLUDING MAN)
The ancestry and development of Man is a topic treated in a vast number of works, scientific and popular. The works listed below emphasise the

geological and palaeontological aspects rather than the archaeological or ethnographic approach.

Traité de Paléontologie. VII. Vers la Forme Humaine. Le Problème Bio-logique de l'Homme. Les Epoques de l'Intelligence. Primates. Paléontologie Humaine, edited by J. Piveteau (Masson, 1957)
Fossil Men: A Textbook of Human Palaeontology, by M. Boule and H. V. Vallois (translation of the 4th French edition, Thames and Hudson, 1957)
Prehistoric Man, by J. Augusta and Z. Burian (Hamlyn, 1960). A high-quality picture book
Evolution after Darwin. II. The Evolution of Man, edited by S. Tax (University of Chicago Press, 1960)
The Progress and Evolution of Man in Africa, by L. S. B. Leakey (Oxford University Press, 1961)
Ideas on Human Evolution, by W. W. Howells (Harvard University Press, 1962)
Evolutionary and Genetic Biology of Primates, edited by J. Buettner-Janusch (2 vols., Academic Press, 1963)
The Antecedents of Man, by W. E. Le Gros Clark (reprint of 1963 edition, Harper and Row, 1969)
African Ecology and Human Evolution, edited by F. C. Howell and F. F. Bourlière (Aldine Publishing Company, 1963). Viking Fund Publications in Anthrology, No. 36
Classification and Human Evolution, edited by S. L. Washburn (Aldine Publishing Company, 1963). Viking Fund Publications in Anthropology, No. 37
The Fossil Evidence for Human Evolution, by W. E. Le Gros Clark (2nd edn., University of Chicago Press, 1964)
History of the Primates. An Introduction to the Study of Fossil Man, by W. E. Le Gros Clark (10th edn., British Museum (Natural History), 1970)
Guide to Fossil Man. A Handbook of Human Palaeontology, by M. H. Day (Cassell, 1965)
Human Evolution. An Introduction to Man's Adaptations, by B. G. Campbell (Heinemann, 1967)
Man-Apes or Ape-Men? The Story of Discoveries in Africa, by W. E. Le Gros Clark (Holt, Rinehart and Winston, 1967)
Frameworks for Dating Fossil Man, by K. P. Oakley (3rd edn., Weidenfeld and Nicolson, 1969)
Unveiling Man's Origins, by L. S. B. Leakey and V. M. Goodall (Methuen, 1970)
The Emergence of Man, by J. E. Pfeiffer (Nelson, 1970)
The Evolution of Man, by D. Pilbeam (Thames and Hudson, 1970)

Reference should be made to the recent and current issues of *Nature*, for progress reports on the development of primate research, especially on the Leakeys' work at Olduvai Gorge and elsewhere. There are also review papers in such journals as *Scientific American*.
BIBLIOGRAPHIES: *Hominidae Fossiles*, by W. and A. Quenstedt (*Fossilium Catalogus I*, **74**, 1936)
A Bibliography of Early Man. Pleistocene Studies and Palaeolithic Archaeology in Southern Equatorial and Eastern Africa, by H. L. Movius, Jr., and D. F. Jordan (Wenner-Gren Foundation, 1954)
The Australopithecinae. Bibliography, by R. Musiker (School of Librarian-

ship, Bibliographical Series, University of Cape Town, 1954)
A Bibliography of Fossil Man, by G. E. Fay (Southern State College, Magnolia, Arkansas, 1959)
Catalogue of Fossil Hominids. I. *Africa*, edited by K. P. Oakley and B. G. Campbell; II. *Europe*, edited by K. P. Oakley *et al*. (British Museum (Natural History), 1967–71). Other volumes are planned
JOURNALS: The principal journals covering human evolution are: *American Journal of Physical Anthropology* (1918–), *Current Anthropology* (1960–), *Folia Primatologia* (1963–), *Journal of Human Evolution* (1972–), *Perspectives on Human Evolution* (1968–) and *Primates* (1957–).

PALAEOICHNOLOGY

The study of the traces left in sediments and rocks by the life processes of animals and plants (palaeoichnology) is profoundly important in palaeoecology and is considered in all textbooks on that subject. There exists, however, a separate literature on the topic:

'Triassic Life of the Connecticut Valley', by R. S. Lull (*Bulletin of the Connecticut Geological and Natural History Survey*, **81**, 1953). This contains a classic study of vertebrate footprints and their interpretation
Die Fährten der Vorzeitlichen Amphibien und Reptilien, by O. Kuhn (Meisenbach, 1958)
Treatise on Invertebrate Paleontology. W. Miscellanea, by W. H. Hass *et al*. (Geological Society of America and University of Kansas Press, 1962). Coverage is confined to invertebrate traces
'Fossil Behaviour', by A. Seilacher (*Scientific American*, **217**, 72–80, 1967)
'Fossiele Levenssporen I–II', by A. P. Schuddebeurs (*Grondboor en Hamer*, 125–216, 1969)
Trace Fossils, edited by T. P. Crimes and J. C. Harper (Seel House Press, 1970). *Geological Journal*, Special Issue, No. 3
Handbuch der Paläoherpetologie. 18. Ichnia Amphibiorum et Reptiliorum Fossilium, by H. Haubold (Fischer, 1971)
BIBLIOGRAPHIES: *Ichnia Tetrapodorum*, by O. Kuhn (*Fossilium Catalogus I*, **101**, 1963). Unusually for this series, this contains a diagrammatic key to types
Vestigia Invertebratorum et Problematica, by W. Häntzschel (*Fossilium Catalogus I*, **108**, 1965)
'Coprolites: An Annotated Bibliography', by W. Häntzschel (*Memoir of the Geological Society of America*, **108**, 1968)

PALAEOBOTANY

Sometimes treated as a separate discipline distinct from palaeontology, the study of fossil plants has a broad zone of overlap with the fields of micropalaeontology and palynology and is best treated as dealing with the macroscopic fossils of plants, i.e. those large enough to be capable of recognition and classification without the use of a microscope. It is well served by general texts, of which the first two in the following list are major works:

Osnovy Paleontologii. 14. *Vodorosli, Mokhoobraznye, etc.* 15. *Golose-mennye i Pokrytosemennye*, edited by V. A. Vakhrameev *et al.* (Izdatel-'stvo Akademii Nauk SSSR, 1963)
Traité de Paléobotanique, edited by E. Boureau (Masson, 1964–   ). Nine volumes are planned, of which three have been published to date
An Introduction to the Study of Fossil Plants, by J. Walton (Black, 1953)
Principles of Paleobotany, by W. C. Darrah (2nd edn., Ronald Press, 1960)
Studies in Palaeobotany, by H. N. Andrews (Wiley, 1961)
Morphology and Evolution of Fossil Plants, by T. Delevoryas (Holt, Rinehart and Winston, 1962)
Manuel de Paléontologie Végétale, by L. Moret (3rd edn., Masson, 1964)
An Evolutionary Survey of the Plant Kingdom, by R. F. Scagel *et al.* (Wadsworth Publishing Company, 1966)
Plant Life through the Ages: A Geological and Botanical Retrospect, by A. C. Seward (reprint of the 2nd edn., Harper, 1966)
'Studies on Fossil Plants', edited by K. L. Alvin (*Journal of the Linnean Society of London* (*Botany*), **61** (384), 1968)

Books with a narrower coverage include:

Devonian Floras. A Study of the Origin of Cormophyta, by E. A. Newell Arber (Cambridge University Press, 1921)
Coal Measure Plants, by R. Crookall (Arnold, 1929)
The Book of Amber, by G. C. Williamson (Benn, 1932)
Primitive Land Plants, Also Known as the Archegoniatae, by F. O. Bower (Hafner, 1959)
Animals and Plants of the Cenozoic Era, by R. Pearson (Butterworths, 1964)
A Critical Review of the Upper Pennsylvanian Floras of Eastern United States, by W. C. Darrah (privately published, Gettysburg, Pa., 1969)
Studies in the Vegetational History of the British Isles. Essays in Honour of Harry Goodwin, edited by D. Walker and R. G. West (Cambridge University Press, 1970)
Phytochemical Phylogeny. Proceedings of the Phytochemical Society Symposium, Bristol, April 1969, edited by J. B. Harborne (Academic Press, 1970)
'Symposium on Major Evolutionary Events and Geological Record of Plants' (*Biological Reviews*, **45** (3), 1970)

BIBLIOGRAPHIES: 'World Report on Palaeobotany' (1956–   ) is a major source of information. It is issued as numbers of the series *Regnum Vegetabile*. Published so far are: I—Vol. 7, 1956; II—Vol. 11, 1958; III—Vol. 19, 1960; IV—Vol. 24, 1962; V—Vol. 35, 1964; VI—Vol. 42, 1966; VII—Vol. 57, 1968; VIII—Vol. 78, 1971
The *Fossilium Catalogus II* has been discussed in relation to its series *I*. Series *II* is devoted entirely to the plant kingdom and so far some 78 parts have been issued. The series commenced in 1914
'Index of Generic Names of Fossil Plants 1820–1965', by H. N. Andrews (*Bulletin of the United States Geological Survey*, No. 1300, 1970)

'Literatur über Fossile Gymnospermen-Hölzer (1949–1960)', by J. Schultze-Motel (*Geologie*, **11**, 604–619, 1962)
Bibliography of American Paleobotany (1965–)
'Bibliografía Paleobotanica de America del Sur', by C. A. Menendez (*Revista del Museo Argentino de Ciencias Naturales 'Bernardino Rivadavia' Instituto Nacional de Investigación de las Ciencias Naturales. Paleontología*, **1**, 1968)
A valuable series of papers on algae, many with bibliographies, appears in the *Quarterly of the Colorado School of Mines* between 1956 and 1967
JOURNALS: The more important palaeobotanical journals are *Acta Palaeobotanica* (1960–), *Argumenta Palaeobotanica* (1966–), *The Palaeobotanist* (1952–), *Palaeontographica, Ser. B. Paläophytologie* (1933–), *Paleobotanika* (1956–) and *Review of Palaeobotany and Palynology* (1967–). The last has contained a number of valuable review papers on the palaeobotany and palynology of individual countries—for example, Australia, France, New Zealand and South America. Excellent bibliographies are included. It should also be remembered that many palaeobotanical papers appear in botanical journals.

FOSSILS IN FOLKLORE AND LITERATURE

To end this chapter on an offbeat note, we cite a handful of works that discuss the interaction of fossils with humanity. First of all, two papers which deal with the use of fossils for decorative purposes by ancient people, although the second one does show some modern uses of fossils: 'Folklore of Fossils', by K. P. Oakley (*Antiquity*, **39**, 9–17, 117–125, 1965) and ' "Formed Stones" Folklore and Fossils', by M. G. Bassett (*Amgueddfa*, **7**, 2–17, 1971).

A book which deals with the same topic but also with broader aspects of folk belief about fossils is *Vorzeitliche Tierreste im Deutschen Mythus, Brauchtum und Volksglauben*, by O. Abel (Fischer, 1939). Reference may also be made to various chapters in *The Lungfish, the Dodo and the Unicorn*, by W. Ley (Viking Press, 1948).

The development of the trade in, the ornamental use of, and the understanding of fossils in amber is discussed in *Dragons in Amber*, by W. Ley (Sidgwick and Jackson, 1951).

Two attempts to review the incidence of fossils in literary works are: 'Fossil-lore in Greek and Latin Literature', by E. S. McCartney (*Papers, Michigan Academy of Science, Arts and Letters*, **3**, 23–38, 1923) and 'Palaeontology in Literature', by A. Lamont (*Quarry Managers' Journal*, **30**, 432–441, 542–551, 1947). These two papers are by no means extensive: preparation of a bibliography on this topic has not yet been attempted and ought to prove a fascinating task!

CURRENTLY   PUBLISHED   PALAEONTOLOGICAL   PERIODICALS

Acta Palaeontologica Polonica (1956–   )
Acta Palaeontologica Sinica (1953–   )
Ameghiniana (1957–   )
Annales de Paléontologie (1906–   )
*Bericht der Geologischen Gesellschaft in der Deutschen Demokratischen
   Republik. A. Geologie und Paläontologie* (1967–   )
Bollettino della Società Paleontologica Italiana (1960–   )
Bulletin of the British Museum (Natural History) Geology (1949–   );
   Supplement (1965–   )
Bulletin de la Société Belge de Géologie, de Paléontologie et d'Hydrologie
   (1887–   )
Bulletins of American Paleontology (1895–   )
Col-pa Coloquios de la Catedra de Paleontología. Madrid (1964–   )
*Communicaciones del Museo Argentino de Ciencias Naturales 'Bernardino
   Rivadavia' e Instituto Nacional de Investigación de las Ciencias Naturales
   Paleontología* (1966–   )
Contribuições do Instituto Geobiológico 'La Salle' de Canõas (1951–   )
Contributions, Institute of Geology and Palaeontology Tohoku University
   (1924–   )
Contributions from the Museum of Paleontology University of Michigan
   (1928–   )
Ezhegodnik Vsesoyuznogo Paleontologicheskogo Obshchestva (1931–   )
Fieldiana: Geology (1945–   )
Fieldiana: Geology Memoirs (1947–   )
Folia Quaternaria (1960–   )
*Fossilia; Revista de la Catedra de Paleontología de la Universidád Barce-
   lona* (1965–   )
Fragmenta Mineralogica et Palaeontologica (1969–   )
Geobios (1968–   )
Geologica Hungarica, Series Palaeontologia (1928–   )
Geologica et Palaeontologica (1967–   )
Journal of the Palaeontological Society of India (1956–   )
Journal of Paleontology (1927–   )
Lethaia (1968–   )
*Materialy k Osnovam Paleontologii, Paleontologicheskii Institut, Akademii
   Nauk SSSR* (1957–   )
Memoirs of the Geological Survey of India. Palaeontologia Indica (1861–   )
Memoirs of the Geological Survey of Pakistan. Palaeontologia Pakistanica
   (1956–   )
Memorie Geopalaeontologiche dell'Università di Ferrara (1964–   )
*Mitteilungen aus dem Geologisch Paläontologischen Institut der Universitat
   Hamburg* (1969–   )
*Monograph of the Geological Museum, Geological Survey and Mineral
   Research Department, Palaeontological Series.* Cairo (1962–   )
Neues Jahrbuch für Geologie und Paläontologie. Abhandlungen (1950–   )
Neues Jahrbuch für Geologie und Paläontologie. Monatshefte (1950–   )
Palaeoecology of Africa and of the Surrounding Islands and Antarctica
   (1966–   )

Palaeontographia Italica (1895–)
Palaeontographica (1851–). Since 1933 divided into two series: A. *Paläo-zoologie-Stratigraphie,* B. *Paläophytologie.*
Palaeontographica Americana (1916–)
Palaeontologia Africana (1953–)
Palaeontologia Jugoslavica (1958–)
Palaeontologia Polonica (1929–)
Palaeontological Bulletin, New Zealand Geological Survey (1913–)
Paläontologische Abhandlungen (1961–). A. *Paläozoologie.* B. *Paläo-botanik*
Paläontologische Zeitschrift (1913–)
Palaeontology (1957–)
Paleontología Cubana (1949–)
Paleontología Mexicana (1954–)
Paleontological Contributions, University of Kansas (1947–); *Papers* (1965–)
Paleontological Journal (1967–). A translation of *Paleontologicheskiĭ Zhurnal*
Paleontological Studies, Louisiana Geological Survey (1959–)
Paleontologicheskiĭ Sbornik (1961–)
Paleontologicheskiĭ Zhurnal (1959–)
Paleontologiya i Stratigrafiya Pribaltiki Belorussii (1966–)
Publications in Palaeontology, National Museum of Natural Science, Canada (1970–)
Publications du Service de la Carte Géologique de l'Algérie, Paléontologie; Mémoires (1959–)
Revista Italiana di Paleontologia e Stratigrafia (1947–)
Revista Italiana di Paleontologia e Stratigrafia, Memoria (1952–)
Sbornik Geologickych Véd. Rada P. Paleontologie (1963–)
Schweizerische Paläontologische Abhandlungen (1963–)
Senckenbergiana Lethaea (1954–)
Smithsonian Contributions to Paleobiology (1969–)
Special Papers, Palaeontological Society of Japan (1951–)
Special Papers in Palaeontology (1967–)
Transactions and Proceedings of the Palaeontological Society of Japan (1951–)
Trudy Instituta Paleobiologii, Akademiya Nauk Gruzinskoĭ SSR (1954–)
Trudy Paleontologicheskogo Instituta, Akademiya Nauk SSSR (1937–)
Trudy, Vsesoyuznoe Paleontologicheskoe Obshchestvo (1957–)
Uchenye Zapiski, Nauchno-Issledovatel'skii Institut Geologii Arktiki, Pale-ontologiya i Biostratigrafiya (1966–)
Voprosy Paleontologii (1950–)
Zitteliana. Abhandlungen der Bayerischen Staatssammlung für Paläonto-logie und Historische Geologie (1969–)

In addition, palaeontological papers are to be found in most stratigraphic journals, in many museum and university serial publications, and in the publications of most national and many local geological societies and journals.

11

Mineralogy, petrology, geochemistry and crystallography

Olive R. Bradley

Literature in the fields of mineralogy, petrology, geochemistry and crystallography shows the same general trends as other geological literature, i.e. a considerable recent increase in the amount published and a marked trend towards increasing specialisation. It is only relatively recently that the primary literature has become too bulky to keep track of by conventional methods—that is, by scanning the periodicals and the abstracting journals. Mechanised methods of information storage and retrieval are therefore still in the early stages of development but there will undoubtedly be considerable advances along these lines in the next few years. This is particularly true of the recording of mineralogical and geochemical data, where schemes are already being actively developed. Increasing specialisation, seen in both books and periodicals, is necessary to cope with new developments in petrology, sedimentary petrology and clay minerals, and in geochemistry and chemical geology. Crystallography is such a specialised discipline that it has from early days required its own periodicals and reference books. Mineralogical literature is likewise well-established but the increasingly complex techniques required for research have resulted in new books to describe them and the complexities of mineral structures that are being discovered.

Another feature that mineralogy and petrology share with other geological disciplines is the need to consult foreign literature. It is important to be able to compare rocks and minerals of like character from different localities and for this reason a representative sample of foreign periodicals has been included.

Geology as a subject is only rarely taught in schools and thus there is a need for books suitable for undergraduate teaching. A selection of so-called introductory textbooks is given below, followed by the section on general textbooks, monographs and works of reference.

INTRODUCTORY TEXTBOOKS

It is proposed to consider first those textbooks most suitable for students with little or no knowledge of geology; a background knowledge of some physics and chemistry is usually assumed.

Mineralogy

Many generations of students have been introduced to mineralogy by way of *Rutley's Elements of Mineralogy*, by F. Rutley (25th edn., revised by H. H. Read, Murby, 1962). It has provided the general pattern that many books have followed, i.e. a first half devoted to a discussion of concepts and principles is followed in part two by a detailed account of the occurrence and properties of the most important minerals. Often revised, and with some account of atomic structures now included, the book is still excellent value for money. *Elements of Mineralogy*, by B. Mason and L. G. Berry (Freeman, 1968), is a revised version of the same author's *Mineralogy: Concepts, Descriptions, Definitions* (Freeman, 1959). It follows the same general scheme but goes more fully into atomic structures and crystallography. It includes a general account of geochemistry and covers economic aspects of mineral deposits in more detail than most elementary textbooks. *Mineralogy: A First Course*, by J. Sinkanas (Van Nostrand, 1966), provides a simplified but comprehensive treatment of mineralogical concepts and methods; the excellent drawings and photographs include a 12-page colour section. Details are given about the 250 mineral species most likely to be encountered. *Introduction to Mineralogy*, by C. W. Correns (translated from the German, 2nd edn., Springer, 1969), deals with structural mineralogy and petrology from a physicochemical point of view.

The examination and identification of minerals under the polarising microscope is a new technique to most students. There are a number of books available for instruction, and the student can select one with the degree of specialisation he requires. *Optical Mineralogy*, by P. F. Kerr (3rd edn., McGraw-Hill, 1959), is a well-proven standard textbook for university courses. It provides a non-mathematical but thorough account of the theoretical aspects of optical mineralogy and gives useful descriptions of the main rock-forming minerals. More detail on the variations within mineral groups and species is given by *Microscopic Identification of Minerals*, by E. W. Heinrich (McGraw-Hill, 1965). This book includes information about the identification of crushed fragments and detrital grains, in addition to thin sections. *Optical Crystallography*, by E. E. Wahlstrom (4th edn., Wiley, 1969), is concerned with optical theory and in particular with the behaviour of light as it passes through non-opaque substances under the polarising microscope. The various techniques for examining crystalline substances under the microscope are surveyed in *Methods in Chemical and Mineral Microscopy*, by E. E. El-Hinnawi (Elsevier, 1966). This book assumes some previous experience and, hence, can devote more space to determinative criteria.

Petrology

From the study of minerals the student proceeds naturally to the study of rocks. Petrography is the science of systematic description and classification of rocks, while petrology is the study, by all available methods, of the natural history of rocks, including origin and alteration. It is customary to recognise three main categories of rocks: igneous, sedimentary and metamorphic. Although increasing knowledge has revised earlier ideas about the origin of some rocks, particularly in the igneous and metamorphic sections, the basic sub-division is still a convenient one, particularly for introductory textbooks. One introductory book, remembered with affection by generations of students, is *The Petrology of the Igneous Rocks*, by F. H. Hatch, A. K. Wells and M. K. Wells (12th edn., Allen and Unwin, 1961). It deals with the minerals that form the igneous rocks and with the origin, structure, classification and complex relationships of the rocks themselves. It is well-written, and there are excellent line drawings. A useful summary of igneous activity in the British Isles is included. The companion volume, *The Petrology of the Sedimentary Rocks*, by F. H. Hatch, R. H. Rastall and J. T. Greensmith (4th edn., Allen and Unwin, 1965), has been extensively revised for this last edition, with photomicrographs now replaced by clearer line drawings and with more attention paid to classification and terminology of limestones and sandstones. The book includes an account of the deposits of modern oceans but omits the section, found in earlier editions, on detrital minerals and grain analysis. It is felt that this branch of sediment-ary petrology has developed so much that it requires the fuller treatment of a separate textbook. *Rocks and Mineral Deposits*, by P. Niggli (Freeman, 1954), is a comprehensive textbook suitable for the intending specialist. It discusses first the basic materials (elements and minerals) and then deals with the fabric, i.e. the textural relationships between them. The emphasis is on the physico-chemical principles that underlie both the formation of mineral deposits and the attempts at a systematic classification of minerals.

An introduction to thin-section petrography is provided by *Petrographic Mineralogy*, by E. E. Wahlstrom (Wiley, 1955). This covers a wide field in not too great detail and is suitable for the student who does not intend to specialise in mineralogy. It includes details of classification methods, properties of both minerals and rocks, and an account of the theory of the polarising microscope. *The Study of Rocks in Thin Section*, by W. W. Moorhouse (Harper and Row, 1959), is a laboratory manual suitable for the beginner and the advanced student; it gives detailed information about all the main types of rock and about the component minerals. There is more information about economic minerals, ore specimens and altered rocks than in most books of this type and it would there-fore be a useful manual for the field geologist. *Petrography—An Introduction to the Study of Rocks in Thin Sections*, by H. Williams, F. J. Turner and C. M. Gilbert (Freeman, 1955), pro-vides a detailed study of igneous, metamorphic and sedimentary rocks in thin section, with good clear drawings. It assumes a know-ledge of the principles of optical mineralogy, and devotes the space thus saved to discussion of the modes of origin of the rocks. About half the rocks figured are from the USA, but many speci-mens from European localities are also included.

Geochemistry

Geochemistry is a branch of geology rapidly increasing in impor-tance as its potential value in the search for economic minerals and ore deposits becomes realised. In general, it has been thought sufficient to introduce the beginner to the basic geochemical prin-ciples during his first instruction in petrology. Today the increas-ing demand for a more systematic approach has been met by books such as *Introduction to Geochemistry*, by K. B. Krauskopf

(McGraw-Hill, 1967), which seeks to introduce students to the use of chemistry in solving geologic problems; the current position is reviewed and failures as well as successes are noted. A useful general text on geochemistry is *Principles of Geochemistry*, by B. Mason (3rd edn., Wiley, 1968). This latest edition has been revised to take account of recent laboratory experiments at high pressures. It includes chapters on sedimentary geochemistry and metamorphism and metamorphic cycles. For students of chemistry or those with an interest in this field, a book that can be recommended is *The Chemical Elements in Nature*, by F. H. Day (Harrap, 1963). It is directly concerned with chemical materials, the sources from which they are obtained and the various kinds of transformations they undergo under natural conditions. The main world sources of supply of elements and minerals up to about 1960 are usefully summarised. *Geochemistry of Solids*, by W. S. Fyfe (McGraw-Hill, 1964), treats the subject with particular reference to the atomic structures of minerals; it is equally suited to the student of petrology on the one hand and of chemistry on the other. *Geochemistry of Sediments*, by E. T. Degens (Prentice-Hall, 1965), provides a useful introduction to some selected topics in this field that are at present arousing much interest; these include minerals of low-temperature and aqueous origin, and organic geochemistry in general and carbon isotopes in particular. It is concerned to demonstrate how geochemical data can supplement field research and petrological study.

Crystallography

There is a variety of introductory textbooks about crystallography. Some require a high standard of background knowledge in mathematics and physics, and it is important for the student to select carefully with due regard to the type of course he is going to take. In *An Introduction to Crystal Chemistry*, by R. C. Evans (2nd edn., Cambridge University Press, 1964), the elementary principles of crystal chemistry are clearly explained and there are many illustrative examples. The second edition includes sections on wave mechanics and electron structure and on defects and structural faults in crystals. *An Introduction to Crystallography*, by F. C. Phillips (3rd edn., Longmans, 1963), is equally well suited to mineralogists, chemists or physicists. The first part deals with the external symmetry of the seven crystal systems and the second part with the internal structure of crystals. A book which covers much the same ground and is better suited to geology students without much background knowledge of physics and chemistry is *An Outline of Crystal Morphology*, by A. C. Bishop (Hutchinson, 1967). The explanation and diagrams are clear and easy to follow. *Introduction to Crystallography*, by G. Tunell and J. Murdoch (2nd edn., Freeman, 1964), is a laboratory manual for students of mineralogy and petrology, with the emphasis on practical techniques such as stereographic projection and the calculation of axial angles and ratios. *Elementary Crystallography*, by M. J. Buerger (Wiley, 1956), is not perhaps an easy book for the novice, but provides a thorough treatment of the subject suitable for the serious student.

The diffraction of X-rays by crystals is dealt with in *Elements of X-ray Crystallography*, by L. V. Azaroff (McGraw-Hill, 1968), and in *X-ray Optics*, by A. J. C. Wilson (2nd edn., Methuen, 1961).

ADVANCED TEXTBOOKS AND REFERENCE BOOKS

The books considered in this section are those which the advanced student or research worker would find it useful to own or have available in his departmental library.

Mineralogy

An indispensable work of reference for any mineralogist is M. H. Hey's *An Index of Mineral Species and Varieties* (2nd edn., British Museum (Natural History), 1955) together with the *Appendix to the Second Edition* (1963). This lists all known mineral species together with their varieties and synonyms, and the chemical composition as closely as it is known. No attempt is made to give other properties, but the enquirer is referred to the original literature. In addition to an alphabetical index, the minerals are arranged according to a chemical classification, based first on the anion groups and second on the metals present.

Lists of new minerals and of revised or discredited mineral species are given every three years in *Mineralogical Magazine* and yearly in *Zapiski Vsesoyuznogo Mineralogicheskogo Obschestva*. *American Mineralogist*, **51** (8), (1966) gives an index of new mineral names, discredited minerals and changes in nomenclature that were reported in vols. 1–50.

The most recent full survey of the mineralogical literature is that provided by *Rock-Forming Minerals*, by W. A. Deer, R. A. Howie and J. Zussman (5 vols., Longmans, 1962–63); it covers work up to the end of 1960 for all minerals and up to 1962 for some. The five volumes are arranged on a mainly structural classification and deal with, respectively, ortho- and ring-silicates, chain silicates, sheet silicates, framework silicates and, finally, non-silicates (oxides, sulphates, carbonates and phosphates). Each mineral or mineral group is described under five headings: structure; chemistry; optical and physical properties; distinguishing features (including tests); and paragenesis—the rocks in which the mineral occurs and some typical mineral assemblages. X-ray powder data are not, in general, included. A single-volume condensed edition is available as a students' textbook. In this, doubtful species and many of the chemical analyses have been omitted, as well as much of the bibliographic information.

The classic work of reference for many generations of geologists has been *Dana's System of Mineralogy* and Appendices (6th edn., Wiley, 1892–1915). It still contains a wealth of valuable information and must form part of any reference library, but the preparation of an up to date revised edition has been delayed by difficulties in coping with the vastly increased amount of information available. The aim has always been to make the work as comprehensive and accurate as possible, to include full crystallographic and chemical data and, in the latest edition, a certain amount of X-ray material. Three volumes of the seventh edition, written by C. Palache, H. Berman and C. Frondel, have so far appeared. The classification is chemical. Volume 1 (1944) deals with the elements, sulphides, sulpho-salts and oxides; Vol. 2 (1951) with halides, nitrates, borates, carbonates, sulphates, phosphates, arsenates, tungstates and molybdates; Vol. 3 (1962) is on the polymorphs of silica. The volumes on the silicates are still awaited. *Dana's Manual of Mineralogy*, revised by C. S. Hurlbut, Jr. (18th edn., Wiley, 1971), is another classic work largely rewritten; it now includes more about crystallography and more details of quantitative methods.

A textbook based on a different system of classifying minerals is *Mineralogy*, by I. Kostov (Oliver and Boyd, 1968). The classification is a geochemical–crystallochemical one. The first broad division is into classes according to the principal component anions; after that according to the chief metals, which are grouped in 'geochemical triads' according to the paragenetic relationships and diadochy of the elements. This grouping links together those minerals that tend to occur together and is useful in the consideration of problems of the origin of minerals. The book includes almost all the minerals known at the end of 1966, although only the commonest are described in detail.

After comprehensive texts on general mineralogy, it is appropriate to consider books which deal in fuller detail with selected aspects of determinative mineralogy. *Mineral Tables: Hand Specimen Properties of 1,500 Minerals*, by R. V. Dietrich (McGraw-Hill, 1969), is a useful book designed to assist both mineral collectors and professional mineralogists; the methods described do not need expensive equipment. *Physical Methods in Determinative Mineralogy*, by J. Zussman (Academic Press, 1967), is a most useful addition to the literature since it provides an account of the new techniques recently developed for the more accurate measurement of physical properties and specific gravity. There is also a valuable introduction to the techniques of X-ray fluorescence analysis and electron-probe micro-analysis.

A comprehensive treatment of the optical determination of minerals is given in *Elements of Optical Mineralogy*, by A. N. Winchell and H. Winchell (3 vols., 4th edn., Wiley, 1951). Principles and methods are described in Vol. 1 and minerals in Vol. 2, while Vol. 3 carries determinative tables. Another full compilation of data is provided by W. E. Troger's *Optische Bestimmung der gesteinsbildenden Minerale* (2 vols., Schweizerbart'sche, 1959, 1967), at present available only in the German text. Part 1 carries the determinative tables; Part 2 (edited by O. Braitsch) gives the text and lists references for each group. *Optical Properties of Minerals*, by H. Winchell (Academic Press, 1965), is essentially a series of determinative tables for the identification of minerals by use of optical properties only.

Although not exclusively concerned with minerals, a useful guide to the identification of mineral particles is *The Particle Atlas*, by W. C. McCrone *et al.* (Ann Arbor Science Publishers, 1967). As well as containing a general introduction to microscopy, the book contains over a hundred colour photographs of mineral particles as seen under the polarising microscope.

Other authoritative sources of systematic mineralogy are *Handbuch der Mineralogie*, by C. Hintze, revised by K. F. Chudoba (Walter der Gruyter, 1965–); and *Nouveau Traité de Chimie Minérale*, by P. Pascal (Masson, 1956–). Both are fully documented and are being issued in parts.

Ontogeny of Minerals, by D. P. Grigor'ev (Israel Program for Scientific Translations, 1965), is concerned with the genesis of minerals and, since their properties are held to result from this, the interaction between the mineral and its environment is revealed. The same author's *Fundamentals of the Constitution of Minerals* (Israel Program for Scientific Translations, 1965), an essay in 'higher mineralogy', explains how it should be possible, from knowledge of the composition and structure, to predict the properties as well as the conditions of formation of minerals.

Two useful reference books on applied topics are *Economic Mineral Deposits*, by A. M. Bateman (2nd edn., Chapman and Hall, 1950), a standard textbook, and *Industrial Minerals and Rocks (Non-metallics other than Fuels)*, edited by J. L. Gillson *et al.* (3rd edn., American Society of Mining, Metallurgical and Petroleum Engineers, 1960). The latter book includes useful lists of bibliographic references; the statistics of production relate mainly to the USA.

As the width and magnitude of mineralogical and petrological research increases, it becomes more difficult to compile a comprehensive textbook which is also up to date. Thus there is a tendency to produce books dealing fully with a single group of minerals. The feldspars, an important and ubiquitous series, are, for example, dealt with in: *Feldspars*, by T. F. W. Barth (Wiley, 1969); *The Feldspars: Phase Relations, Optical Properties and Geological Distribution*, by A. S. Marfunin (Israel Program for Scientific Translations, 1966); *Die Optische Orientierung der Plagioclase*, by C. Burri, R. L. Parker and E. Wenk (Birkhauser, 1967); and for those who need to study them in sediments, *The Identification of Detrital Feldspars*, by L. Van der Plas (Elsevier, 1966). Silica itself now merits a 2-volume work in *The Phases of Silica*, by R. B. Sosman (2nd edn., Rutgers University Press, 1965).

In an effort to keep research workers informed about rapidly developing areas of research, the American Geological Institute has started a series of short courses; the discussions are available (at low cost) as mimeographed lecture notes: *Chain Silicates* (1966) and *Layer Silicates* (1967). Further volumes are expected. A similar full discussion of recent work on an important mineral group is given in *Amphiboles*, by W. G. Ernst (Springer, 1968).

Universal stage technique for the measurement of optical properties of minerals is clearly explained in *The Universal Stage*, by R. C. Emmons (Geological Society of America, 1943), and, for more specialised application, in *Determination of Volcanic and Plutonic Plagioclases Using a Three- or Four-Axis Universal Stage*, by D. B. Slemmons (Geological Society of America, 1962).

Gemmology

Suitable authoritative reference books on the study of precious stones are *A Key to Precious Stones*, by L. J. Spencer (2nd edn., Blackie, 1946); *Gemstones*, by G. F. Herbert Smith (13th edn., revised by F. C. Phillips, Methuen, 1958); *Gems, their Source, Description and Identification*, by R. Webster (2nd edn., Butterworths, 1970); and *Dictionary of Gems and Gemology*, by R. M. Shipley (Gemological Institute of America, 1946).

Meteorites

An essential reference book for any work on meteorites is the *Catalogue of Meteorites with Special Reference to Those Represented in the Collection of the British Museum*, by M. H. Hey (3rd edn., British Museum (Natural History), 1966). Also useful, although in a more limited way, is the *Directory of Meteorite Collections and Meteorite Research* (UNESCO: Working Group on Meteorites, Paris, 1968). A survey of the information available on meteorites up to about 1961 is given in *Meteorites*, by B. Mason (Wiley, 1962). In this work the emphasis throughout is on data rather than theory, since it is felt to be too early for any consistent synthesis. A useful list of American falls is given in an appendix. Another less technical book of the same name, *Meteorites*, by F. Heide (Chicago University Press, 1964), includes much information on fall phenomena and many odd and interesting facts.

Petrology

Most books in the field of petrology are devoted to particular groups of rocks or to certain restricted localities. The simple division of rocks into igneous, metamorphic and sedimentary is not as easy as was at one time thought, and thus the books about them cannot be neatly classified either. An essential book for reference purposes is A. Johannsen's *A Descriptive Petrography of the Igneous Rocks* (4 vols., 2nd edn., University of Chicago Press, 1939). Volume 1 describes textures, structures and methods of classification, while the other three volumes describe rock types in detail, with references to the original literature. One interesting feature is the inclusion of many photographs of the petrologists and geologists of the past.

Igneous and Metamorphic Petrology, by F. J. Turner and J. Verhoogen (2nd edn., McGraw-Hill, 1960), treats rocks as chemical systems and is concerned with all those rocks believed to have been formed or modified at high temperatures and pressures. Problems of petrogenesis are treated from a thermodynamic standpoint and in the light of recent knowledge of the complex behaviour of multi-component systems under a wide range of physical conditions. Part of the book dealing with metamorphic paragenesis is revised and amplified in *Metamorphic Petrology: Mineralogical and Field Aspects*, by F. J. Turner (McGraw-Hill, 1968). Both books require an understanding of thermodynamics. *Solutions, Minerals and Equilibria*, by R. M. Garrels and C. L. Christ (Harper and Row, 1965), also illustrates the application of thermodynamic theory to geological relationships.

The Granite Controversy, by H. H. Read (Murby, 1957), is strongly in favour of the field investigation of igneous rocks. It suggests that experimental studies are not yet advanced enough for the results to be usefully applied to the complex processes taking place at depth in the earth's crust. Each granite is a unit to be discussed by itself and related to its setting; granites can be formed in a number of different ways. *The Geology of Granite*, by E. Raguin (Interscience, 1965), also emphasises the complex origin of this rock and the importance of field relations in deciding how a portion of the earth's crust became mobilised into a granite mass. *Selected Works: Granites and Migmatites*, by J. J. Sederholm (Oliver and Boyd, 1967), contains seven classic papers by this eminent geologist plus a short biography and full bibliography. His work on the origin of the granite and gneiss of Finland and on the Precambrian cycles of sedimentation and orogeny was a major advance in geological understanding. The interrelation of igneous and metamorphic rocks is also discussed in *Migmatites and the Origin of Granite Rocks*, by K. R. Mehnert (Elsevier, 1968). Migmatites are considered to be formed by remelting and mobilisation deep in the earth's crust; their unique nature has only recently been understood. A simple classification scheme for them is proposed in place of the present complex nomenclature.

Metamorphic Textures, by A. Spry (Pergamon, 1969), provides definitions, descriptions and illustrations of the various textures that can occur and treats metamorphism as a series of structural transformations rather than primarily as chemical reactions. There is an extensive bibliography. *The Origin of Metamorphic and Metasomatic Rocks*, by H. Ramberg (University of Chicago Press, 1952), gives a general review of processes of recrystallisation and replacement thought to occur in the earth's crust, while *Metamorphic Reactions and Metamorphic Facies*, by W. S. Fyfe, F. J. Turner and J. Verhoogen (Geological Society of America, 1958), is a detailed account of certain selected topics, mainly concerned with the thermodynamic and kinetic aspects of metamorphism. Results of recent experimental work on the metamorphism and anatexis of various rock types appear in the *Petrogenesis of Metamorphic Rocks*, by H. G. F. Winkler (translated from the German, Springer, 1965).

Layered Igneous Rocks, by L. R. Wager and G. M. Brown (Oliver and Boyd, 1968), gives a full account of research at the classic area of Skaergaard, Greenland, together with shorter descriptions of other layered intrusive complexes and detailed discussion of the concept of fractional crystallisation under gravity.

Recent work on the genesis and geochemistry of the most widespread of all igneous rocks is reviewed in *Basalts: The Poldervaart Treatise on Rocks of Basaltic Composition*, edited by H. H. Hess and A. Poldervaart (2 vols., Wiley/Interscience, 1967).

Ultramafic and Related Rocks, edited by P. J. Wyllie (Wiley, 1967), is a collection of 41 articles dealing with this interesting and enigmatic group; kimberlites and carbonatites are included in the survey. The origin of carbonatites has been described as the most exciting problem in petrogenesis since World War II. Two recent books on the subject are *The Geology of Carbonatites*, by E. W. Heinrich (Rand McNally, 1966), and *Carbonatites*, edited by O. F. Tuttle and J. Gittings (Wiley, 1966). The latter book describes in detail nine carbonatite complexes from different parts of the world. *The Lovozero Alkali Massif*, by K. A. Vlasov, M. Z. Kuz'menko and E. M. Eskova (translated from the Russian, Oliver and Boyd, 1966), provides a comprehensive account of this remarkable layered complex (over 1000 m thick), which has yielded more mineral species than any other complex.

'Studies in Volcanology', by R. R. Coats, R. L. Hay and C. A. Anderson (*Memoir of the Geological Society of America*, **116**, 1968) is a collection of papers dealing with volcanic areas and volcanic phenomena, past and present, in the USA.

Sedimentary petrology

Some aspects of sedimentary petrology are more usefully considered under stratigraphy (Chapter 9). Here we are concerned with sedimentation processes and structures, and with sedimentary petrography.

Sedimentary Rocks, by F. J. Pettijohn (2nd edn., Harper and Row, 1957), covers all main subjects in this field. The emphasis is on rocks rather than sedimentation processes because these are the end-products with which the geologist is primarily concerned. There is a bibliography of 700 carefully selected papers. Methods used in the study of sedimentary rocks are fully discussed in *Sedimentary Petrography*, by H. B. Milner *et al.* (2 vols., 4th edn., Allen and Unwin, 1962). Special reference is made to methods of correlating strata, to petroleum technology and to other economic applications of geology, and there is an extensive bibliography covering work up to the end of 1958. Another comprehensive treatise is N. M. Strakov's *Principles of Lithogenesis* (translated from the Russian, 3 vols., Oliver and Boyd, 1967–70). Here the author is mainly concerned with the processes by which sedimentary rocks are formed—weathering, transport, sedimentation and diagenesis—and how these can be related to past and present climatic conditions.

In *Microscopic Sedimentary Petrography*, by A. V. Carozzi (Wiley, 1960), the author has compiled pictures of 'ideal' thin sections to represent the most frequently encountered sedimentary rock types. A useful supplementary source of information about the origin and properties of minerals which may be formed in sediments after deposition is *Authigenic Minerals in Sedimentary Rocks*, by G. I. Teodorovich (translated from the Russian, Consultants Bureau, 1961). Two recent books on sedimentary techniques are *Methods for the Study of Sedimentary Structures*, by A. H. Bouma (Wiley, 1969), and *Methods in Sedimentary Petrology*, by G. Mueller (translated from the German, Hafner, 1967). Research and methods used in Europe are summarised in the report of a seminar, *Recent Developments in Carbonate Sedimentology*, edited by G. Mueller and G. Friedman (Springer, 1968).

Sedimentary structures are illustrated in *Atlas und Worterbuch der Primären Sedimentstruckturen*, by F. J. Pettijohn and P. E. Potter (Springer, 1964); the text and glossary are in English, French, Spanish and German.

Developments in Sedimentology (Elsevier, 1964–) is a series of books planned to include important developments in sedimentology and works of a review type. Thirteen volumes have already been published (late 1971)—among them *Sedimentary Features of Flysch and Greywackes*, by S. Dzulynski and E. K. Walton (1965); *Sedimentology and Ore Genesis*, a symposium edited by G. C. Amstutz (1964); and *Cyclic Sedimentation*, by P. M. D. Duff, A. Hallam and E. K. Walton (1967).

Clay minerals are difficult to investigate and interpretation of the results obtained requires special reference books. A valued general textbook covering the whole subject is *Clay Mineralogy*, by R. E. Grim (2nd edn., McGraw-Hill, 1968). X-ray diffraction studies are much assisted by *The X-ray Identification and Crystal Structures of Clay Minerals*, edited by G. Brown (2nd edn., Mineralogical Society, 1961), while *Electron-diffraction Analysis of Clay Mineral Structures*, by B. B. Zvyagin (translated from the Russian, Plenum, 1967), describes a recently developed method for investigating the crystal structure of finely powdered material. *Atlas of*

Electron Microscopy of Clay Minerals and their Admixtures, by H. Beutelspacher and H. W. Van Der Marel (American Elsevier, 1967), reviews in English and German the current information available in this field, with photographs of some 260 samples of clays and clay minerals. Later volumes are planned to give infra-red analyses, thermal analyses and X-ray diffraction analyses of the same samples of materials.

The study of soils and soil formation is not within the scope of this survey, but *Fabric and Mineral Analysis of Soils*, by R. Brewer (Wiley, 1964), is recommended for sedimentary petrographers; it includes excellent photomicrographs of soils.

Geochemistry

Geochemistry is a rapidly developing branch of geology in which much of the published work takes the form of reports of conferences and symposia. Thus *Origin and Distribution of the Elements*, edited by L. H. Ahrens (Pergamon, 1969), reports a symposium held in 1967 to review progress in this field. *Researches in Geochemistry*, edited by P. H. Abelson (2 vols., Wiley, 1954, 1967), aims at providing a broad view of advances in this subject for workers specialising in particular aspects. Two books by K. Rankama, *Isotope Geology* (Pergamon, 1954) and *Progress in Isotope Geology* (Interscience, 1963), are an exhaustive and valuable progress report of work in this field and include extensive bibliographies. Unhappily, no third volume is planned, as the time and effort involved in the compilation of the data would be too great. Work on isotopes is also surveyed in *Isotopic and Cosmic Chemistry*, edited by H. Craig, S. L. Miller and G. J. Wasserburg (North-Holland, 1964), a collection of articles made to honour the 70th birthday of H. C. Urey. The ideas of one of the important earlier workers are given in *Geochemistry*, by V. I. Goldschmidt (Oxford University Press, 1954).

The more technical applications of geochemistry are covered in *Geochemistry in Mineral Exploration*, by H. E. Hawkes and J. S. Webb (Harper and Row, 1962), and a valuable guide to techniques is provided in *Methods in Geochemistry*, edited by A. A. Smales and L. R. Wager (Interscience, 1960). *Geochemical Prospecting in Fennoscandia* is a series of progress reports edited by A. Kvalheim (Interscience, 1967), while *Geochemistry of Gallium, Indium, and Thallium*, by D. M. Shaw (Pergamon, 1957), covers the literature to the end of 1955. *Studies in Analytical Geochemistry*, by D. M. Shaw (Royal Society of Canada, 1963), records a symposium at which topics dealt with included trace elements, isotopes of oxygen and sulphur, and statistical methods. Finally, the *Handbook of Geochemistry*, edited by K. H. Wedepohl (Springer, 1958–), being issued in parts in loose-leaf form, will, when complete, provide a comprehensive survey of the distribution of the elements and their isotopes in the earth and in the cosmos.

The Russians, in their search for new mineral deposits, have been most active in promoting geochemical prospecting methods and translations of their books are increasingly available. *Principles of Geochemical Prospecting*, by I. I. Ginzburg (translated from the Russian, Pergamon, 1960), gives a useful general survey. The international conference held in 1963 to commemorate the centenary of V. I. Vernadskii's birth is reported in *Geochemistry of the Earth's Crust*, edited by A. P. Vinogradov (2 vols., Israel Program for Scientific Translations, 1966, 1967). About two-thirds of the papers are concerned with Russian work. Books dealing with particular elements or compounds include *Mineralogy and Types of Deposits of Selenium and Tellurium*, by N. D. Sindeeva (translated from the Russian, Interscience, 1964); *Geochemistry of Beryllium and Genetic Types of Beryllium Deposits*, by A. A. Beus (translated from the Russian, Freeman, 1966); and three volumes of comprehensive reviews by specialist contributors on the rare elements, edited by K. A. Vlasov (Israel Program for Scientific Translations), *Geochemistry of Rare Elements* (1966), *Mineralogy of Rare Elements* (1966) and *Genetic Types of Rare Element Deposits* (1968). The migration and concentration of chemical elements by organic substances is of great interest in relation to other elements such as uranium; this is discussed in *Geochemistry of Organic Substances*, by S. M. Manskaya and T. V. Drozdova (translated from the Russian, Pergamon, 1958).

The need to analyse rock and mineral samples for trace elements has required the development of new methods. *Element Analysis in Geochemistry*, Vol. 1, *Major Elements*, by A. Volborth (Elsevier, 1969), describes the use of classical chemical and gravimetric, and modern instrumental methods, while *Rock and Mineral Analysis*, by J. A. Maxwell (Wiley, 1969), includes both chemical analysis and X-ray emission spectroscopy. Other useful specialist books are *X-ray Emission Spectrography in Geology*, by I. Adler (Elsevier, 1966), which includes both X-ray fluorescence analysis and electron-microprobe analysis; *Atomic Absorption Spectrometry in Geology*, by E. E. Angino and G. K. Billings (Elsevier, 1968); and *Analytical Geochemistry*, by W. D. Evans and L. Brealey (Elsevier, 1969).

Crystallography

It is beyond the scope of this book to provide a comprehensive survey of crystallographic literature, but some of the more important publications with special relevance to mineralogy are described below.

Atomic Structure of Minerals, by W. L. Bragg (Oxford University Press, 1937), was for long the classic reference book, since it gives an account of nearly all the main types of mineral structures then known. It is no longer possible to encompass so much in a single book, and the successor volume, *Crystal Structures of Minerals*, by W. L. Bragg and G. F. Claringbull (Bell, 1965), includes descriptions of most known structures and gives references to others not included in the text. Work up to the end of 1963 is included, but methods of X-ray analysis are omitted since they are dealt with adequately by other books. *Structure Reports*, reporting progress in the determination of structures, are dealt with on p. 313.

Essential for all X-ray structure work are the *International Tables for X-ray Crystallography*, prepared under the guidance of the International Union of Crystallography (3 vols., Kynoch Press, 1952, 1959, 1962). The volume titles are 1. *Symmetry Groups*; 2. *Mathematical Tables*; and 3. *Physical and Chemical Tables*. The three provide, in addition to tabulated data, excellent accounts of techniques in the study of crystals and of the basic theory employed. *Data for X-ray Analysis*, by W. Parrish and M. Mack (3 vols., Philips, 1963), includes charts for the solution of Bragg's equation. *Crystal Data (Determinative Tables)*, by J. D. H. Donnay *et al.* (2nd edn., American Crystallographic Association, 1963), includes almost all results published up to the end of 1960. Another essential publication for anyone concerned with structural studies is *Crystal Structures*, by R. G. Wyckoff (6 vols., 2nd edn., Wiley, 1963–71). It gives space group cell dimensions, atomic positions and interbond angles for each structure; because so much information has become available, only those determinations have been included which define the position of most of the atoms in a crystal.

A useful introductory book for the mineralogist intent on working in this field is *The Determination of Crystal Structure*, by H. Lipson and W. Cochran (Bell, 1966); it assumes a background knowledge of physics and mathematics—more particularly, X-ray optics, structural crystallography and higher algebra.

For specialist use, *X-ray Powder Data for Ore Minerals: The Peacock Atlas* has been compiled by L. G. Berry and R. M. Thomson (Geological Society of America, 1962); it includes authenticated data for nearly 300 opaque minerals, with powder photographs of every mineral listed. *The Handbook of X-ray Analysis of Polycrystalline Material*, by L. I. Mirkin (translated from the Russian, Consultants Bureau, 1964), provides a comprehensive collection of data for use in the interpretation of X-ray powder patterns and a description of Russian methods and materials. New crystallography symmetry theory is described in *Coloured Symmetry*, by A. V. Shubnikov, N. V. Belov *et al.* (Pergamon, 1964). Again, this is a useful introduction to Russian work.

Practical Optical Crystallography, by N. N. Hartshorne and A. Stuart (2nd edn., Edward Arnold, 1969), provides a good general survey, with descriptions of apparatus and techniques. For anyone concerned with the preparation of crystalline material of high purity, *The Art and Science of Growing Crystals*, edited by J. J. Gilman (Wiley, 1963), should prove useful. *The Barker Index of Crystals*, by M. W. Porter and L. W. Codd (3 vols., Cambridge University Press, 1951–64), describes a method of identifying and classifying crystals by measurement of crystal angles.

PERIODICALS

General publications

Short accounts of new or significant developments are found in the general scientific journals such as *Nature, Science, New Scientist, Naturwissenschaften, Comptes Rendus Hebdomadaire des Séances de l'Académie des Sciences* and *Doklady Akademiya Nauk SSSR*.

Other sources of papers on mineralogy, petrology and geochemistry are the general geological journals. Most national journals are concerned mainly with the geology of their own particular country and thus the amount of igneous, metamorphic or sedimentary petrology included depends on which rocks are found there. Official and semi-official publications issued by government departments and the like, particularly the reports of geological surveys, contain much information of interest to the mineralogist and petrologist. It can be difficult to locate this information, since the report is often a description of a particular area, and it is difficult to judge from the title what mineralogical or petrological information is contained within. Nor is it much good relying on abstracting journals, for no long report can be abstracted or indexed to the depth required. If it is known in which areas particular rocks and minerals are found, then it is possible to search through the geological literature for reports of those districts, but this is an uncertain method of progress, particularly for the beginner.

General geological journals dealing with the British Isles include *Journal of the Geological Society, The Geological Magazine, Proceedings of the Geologists' Association* and the official publications of the Institute of Geological Sciences. The universities publish mainly reports of symposia and conferences, but there are a number of local geological societies publishing useful papers. The content of these varies with the location; thus the *Scottish Journal of Geology* contains much igneous and metamorphic petrology, while the *Proceedings of the Yorkshire Geological Society* has papers about sedimentary petrology and evaporites.

There are very many European geological journals and those listed here are the ones that are most likely to carry material on mineralogy and petrology. Details of additional titles appear in Chapter 9.

Scandinavia is an area of marked geological interest, particularly for igneous and metamorphic petrology, and much of the work is described in English. Useful periodicals include *Geologiska Föreningens i Stockholm Förhandlingar*, which includes reviews on topics such as carbonatites and age determinations; *Sveriges Geologiska Undersöking Arsbok*; *Norsk Geologisk Tidsskrift*; *Norges Geologiske Undersøgelse*; and the Finnish *Bulletin de la Commission Géologique de Finlande*. Denmark is represented by *Meddelelser fra Dansk Geologisk Forening*—the field here is more in sedimentary petrology—and, indirectly, by *Meddelelser om Grønland*.

Central Europe is covered by a multiplicity of German journals, published by both East and West, and by the reports of the various German geological societies. Selection among these depends on the locality of interest. *Beihefte zum Geologischen Jahrbuch* deals in detail with the geology of selected regions in all parts of the world. *Geologische Rundschau* has the text in English, French and German, while *Geologische Jahrbuch* and *Geologische Mitteilungen* both include summaries in English and German. Two French journals with special relevance are *Sciences de la Terre*, which describes technical methods for studying minerals and mineral deposits, and *Mémoires du Bureau de Recherches Géologiques et Minières*, which includes lengthy reports on areas outside France. Belgium and Holland are represented by *Bulletin de la Societé Belge de Géologie, de Paléontologie et d'Hydrologie* and by *Leidsche Geologische Mededelingen*. Both are mainly concerned with stratigraphy and sedimentary petrology, while *Geologie et Mijnbouw* also covers mining and economic aspects. Central European areas are covered by, for Austria, *Mitteilungen der Geologische Gesellschaft in Wien* and, for Switzerland, *Eclogae Geologicae Helvetiae.*

In Western Europe *Geologický Sbornik* includes descriptions of the mineralogy and petrology of Czechoslovakia; *Geološki Glasnik* does the same for Yugoslavia; and *Kwartalnik Geologiczny* for Poland. *Przeglad Geologiczny*, also from Poland, deals more with economic minerals and sedimentary deposits. Of the Russian journals, *Izvestiya Akademiya Nauk SSSR—Seriya Geologicheskaya* contains quite a lot of material on geochemistry and igneous petrology and *Geologiya Rudnykh Mestorozhdenii* is concerned with the geochemistry and mineralogy of ore deposits.

American periodicals dealing with general geological and petrological topics include the *American Journal of Science* and the *Journal of Geology.* The latter contains much material on mineralogy and experimental petrology. The US Geological Survey pubishes much important work in many publications such as the *Bulletin of the United States Geological Survey* and the *Professional Papers of the United States Geological Survey.* Most of the states themselves publish reports on the geology and mineral resources of their own regions, in, for example, the *Circular of the Illinois State Geological Survey* and the *Bulletin of the Geological Survey of Virginia.*

Canadian journals are mostly official publications. The *Bulletin of the Geological Society of Canada* is supported by provincial periodicals such as the *Bulletin of the Newfoundland Geological Survey.*

In the same way the *Journal of the Geological Society of Australia* is supplemented by the publications of the states, as in the *Publications of the Geological Society of Queensland* and the *Bulletin of the Geological Survey of South Australia.*

Japanese journals in English include the *Japanese Journal of Geology and Geography* and the *Journal of the Japanese Association of Mineralogists, Petrologists and Economic Geologists.* Many Japanese universities publish their own journals of scientific research, including geology, and these are sometimes available in English (see Chapter 6).

Indian journals include the *Quarterly Journal of the Geological, Mining and Metallurgical Society of India.*

Specialist publications

Essential journals for the mineralogist are *The Mineralogical*

Magazine and *The American Mineralogist*. Similar mineralogical journals are produced in most European countries. These include *Bulletin de la Societé Française de Minéralogie et de Cristallographie*; *Neues Jahrbuch für Mineralogie—Abhandlungen* and *—Monatshefte*; and *Fortschritte der Mineralogie*. Austria has *Tschermaks Mineralogische und Petrographische Mitteilungen*; Switzerland the *Schweizerische Mineralogische und Petrographische Mitteilungen*; and Italy *Rendiconti della Società Mineralogica Italiana* and *Periodica di Mineralogia*.

Outside Europe, two of the most important journals are the Russian *Zapiski Vsesoyuznoge Mineralogischeskoge Obshchestva* and the Ukrainian *Mineralogicheskii Sbornik*. Unfortunately, only the second of these has English summaries of the papers. Another valuable source of mineralogical data is *Trudy Mineralogicheskogo Muzeya, Akademiya Nauk SSSR*.

The new and interesting minerals now being found in the Canadian search for new deposits are described in *The Canadian Mineralogist*.

The rich and varied mineralogical deposits of Japan are reported in English in the *Mineralogical Journal* and in Japanese in the *Journal of the Mineralogical Society of Japan*, and in both languages in the *Journal of the Japanese Association of Mineralogists, Petrologists and Economic Geologists*.

Mineralium Deposita is a new international journal dealing with the geology, mineralogy and geochemistry of mineral deposits. Contributions are published in English, French and German.

Work on petrology is mainly published by the general geological journals. An exception to this is the *Journal of Petrology*, which discusses the theoretical aspects of petrology in more detail than is generally possible in such journals. *Beiträge zur Mineralogie und Petrologie* deals with the petrology and genesis of igneous, metamorphic and sedimentary rocks. Contributions appear in English, French or German.

Sedimentary petrology is such a specialised discipline that it has needed to sponsor special publications. The earliest of these was the *American Journal of Sedimentary Geology*. This has been more recently joined by two new international publications, *Sedimentology* and *Sedimentary Geology*, with texts in English, French and German. Clay minerals likewise are a very specialised field and are covered by periodicals such as *Clay Minerals*, which replaced *Clay Minerals Bulletin* in 1965; *Clays and Clay Minerals* —the report of an annual conference; and *Clay Science*.

Geochemical research is published in the international journals *Geochimica et Cosmochimica Acta* and *Chemical Geology*. The abundant Russian work appears in *Geokhimiya* and from 1956 to 1967 in English translation in *Geochemistry*. The latter journal under its new name *Geochemistry International* also publishes English translations of articles from other periodicals, particularly Russian and Japanese. *Geochemical Journal* is a new English-language quarterly published in Japan. Work in this field also appears in periodicals dealing with geophysical topics, such as the *Journal of Geophysical Research* and *Earth and Planetary Science Letters*.

For work in crystallography the reader is referred to the mineralogical journals described above, and to the appropriate specialist periodicals—in particular, *Acta Crystallographica*; *Zeitschrift für Kristallographie, Kristallgeometrie, Kristallphysik, Kristallchemie*; and *Soviet Physics: Crystallography*.

ABSTRACTING SERVICES

The use of abstracts was discussed in Chapter 5 and the main abstracting journals for earth sciences literature were described. This section considers those abstracting services, both general and specialised, useful for information on mineralogical, petrological, and geochemical topics.

Much of the literature of interest to mineralogists and geochemists is covered by *Chemical Abstracts*. This is without doubt the most comprehensive specialist abstracting service in the world. It started in 1907 and now contains about 250 000 abstracts per year of 12 000 journals, books, conference proceedings and reports. A list of journals covered is published at approximately 5-year intervals. Material is presented in 80 subject sections of which two sections, 53 (Mineralogical and Geological Chemistry) and 70 (Crystallization and Crystal Growth) are of interest to earth scientists. Up to 1967 *Chemical Abstracts* was issued twice a month, but it now appears in two parts issued on alternate Mondays. Sections 1–34 appear one week and Sections 35–80 the next. For those who do not wish to subscribe to the whole journal, five Section Groupings are available separately. Group 4, entitled 'Applied Chemistry and Chemical Engineering Sections', contains Section 53. Each issue of *Chemical Abstracts* possesses a keyword, patent and author index, and the semi-annual volumes also contain a formula index. To help with retrospective searches, several collective indexes have been published. Five of these are decennial indexes covering the years from 1907 to 1956; the sixth covers 1957–61; and the latest 1962–66.

Like all major abstracting services, *Chemical Abstracts* can never hope to be completely up to date and an attempt to solve this problem has been made by publishing *Chemical Titles*. Produced since 1961 and issued twice a month, the journal consists of a keyword-in-context (KWIC) index, an author index and a bibliographic section in which the contents of each journal scanned are entered under the appropriate journal title. Some 700 periodicals form the basis of this service and of these 28 are earth sciences titles. The interval between publication of the original and its coverage in *Chemical Titles* is between 2 and 9 weeks, depending on the place of origin of the journal and whether the publishers are able to obtain proof copies of the periodicals covered.

Referativnyi Zhurnal, Geologiya (1956–   ) has at present some 8000 abstracts a year in the section 'Geokhimiya, Mineralogiya, Petrografiya'. The abstracts are particularly useful to the mineralogist and geochemist, since they include mineral and X-ray data and chemical analyses. Some of the journals abstracted are not readily available in Western countries, and may not be included in other abstracting services. Western literature is also abstracted, usually with about 1 year's delay. Minerals and mineralogical methods are fully indexed, and there is a separate index for localities.

Mineralogical Abstracts (from 1920 to 1958 part of *The Mineralogical Magazine*) contains about 3500 abstracts and book notices a year. These cover all aspects of mineralogy from analytical methods to meteorites and gemstones, with the main emphasis on the occurrence and properties of rock-forming minerals. Mineralogical and chemical data are included in the abstracts when considered to be new or important. World coverage is extensive but not comprehensive, since the journal relies on volunteer abstractors to cover the mineralogical literature of their own countries. The abstracts are classified broadly according to subject field and each issue has an author index. There are annual author and subject indexes and the latter is very good for minerals and mineralogical techniques but does not include much general geology—for example, stratigraphical terms. Mineral localities are grouped under the appropriate country and region, but are also cross-referenced and can therefore be easily traced.

Another specialised abstracting journal of long standing is *Zentralblatt für Mineralogie*. It has two sections: *Teil 1. Kristallographie und Mineralogie* and *Teil 2. Petrographie, Technische Mineralogie, Geochemie und Lagerstättenkunde*. The abstracts are detailed, include much useful data, and are well indexed. The fields of interest are covered selectively rather than comprehensively. Some papers written in English have English abstracts.

The latest of the comprehensive abstracting journals, *Bibliography and Index of Geology*, has one section dealing with 'Geochemistry' and another on 'Mineralogy and Crystallography'. Other relevant fields of interest are 'Igneous and Metamorphic Petrology' and 'Sedimentary Petrology'. The coverage of world literature is good but the very brief annotations do not include any mineralogical or chemical data. There is a full subject index, but it must be used with care until the arrangement becomes familiar. Papers about minerals may be indexed under a number of headings— crystal chemistry, crystal structure, mineral data, mineral deposits and so on, and these references are not necessarily repeated under the index entry for the mineral itself. Countries and regions are fully indexed, but there are no cross-references for minor localities.

The French monthly publication *Bibliographie des Sciences de la Terre* has a section Cahier A. 'Minéralogie, Géochimie et Géologie Extra-Terrestre', which covers, in addition to mineralogy, geochemistry and extra-terrestrial geology, geochronology, and isotope geochemistry. Other work of interest to petrologists is to be found in Cahier C. 'Roches Cristallines' (this includes volcanology); Cahier D. 'Roches Sédimentaires et Géologie Marine'; and Cahier B. 'Gitologie et Economie Minière'. There are no abstracts of the papers and, hence, no record of mineralogical or geochemical data, but the keyword subject index (in each monthly issue) is very detailed, in both general and specific references. This would be a good place to search for the geochemical role of a specific element. Sections are indexed separately, and it is instructive to compare these separate indexes with the main subject index of other abstracting journals. For earlier literature, *Bulletin Signalétique, Sciences de la Terre* is a very useful source of information; the abstracts are only short, but their coverage is world-wide.

IMM Abstracts, published by the Institute of Mining and Metallurgy since 1950, surveys world literature on economic geology, mining and mineral dressing, and may provide additional useful mineralogical information since the service covers technical journals which are not all covered by the more geological abstracting services. There is no index, but the abstracts are arranged according to the UDC classification.

A number of services are of relatively specialised interest. These include *British Ceramic Abstracts* (1924–) and *Journal of the American Ceramic Society* (1918–), both of which cover the literature on clay mineralogy. The former, produced under a variety of other titles from 1905 to 1941, is published monthly by the British Ceramic Research Association. The latter, also published monthly, has contained a section entitled 'Ceramic Abstracts' since 1922.

Papers of interest to mineralogists are also to be found in the bi-monthly *Asbestos Bulletin* (1960–); the fortnightly *Aluminium Abstracts* (1963–), issued by CIDA; and the monthly *Bibliography of Geological Literature on Atomic Energy Raw Material* (1964–), produced by the Institute of Geological Sciences in London.

Structure Reports (1940–), prepared under the supervision of the Commission of the International Union of Crystallography, are annual volumes of critical reports rather than abstracts. They include all works of crystallographic structural interest published during that year. The data may be derived from X-ray, electron diffraction or neutron diffraction, or from other sources. It was at first hoped that it would be possible to present the data in such detail that it would not be necessary to consult the original literature, but the increasing volume of published material has meant that this is no longer practicable. Since Vol. 21 (1957) atomic parameters are not given for structures containing more than about 30 independent atoms. In most instances the information given is quite adequate, and care is taken to see that the text is as far as possible readily understood by non-crystallographers. The standard of editing, production and indexing is uniformly high, but, partly because standards are so high, there is a delay of several years before the annual volumes appear.

Particularly in view of these delays but also because it supplements the information in *Structure Reports*, a recently published bibliography is of great value. This is *Molecular Structures and Dimensions*, edited by O. Kennard and D. G. Watson (2 vols., International Union of Crystallography, 1970). The bibliography deals with organic and organometallic crystal structures and covers the period 1935–69. It provides references to over 4000 compounds whose structures were analysed by X-ray or neutron-diffraction methods. Volume 1 deals with general organic structures and Vol. 2 with complexes, organometals and organometalloids. Author, transition metal and formula indexes are provided.

12

Structural geology and tectonics

P. W. G. Tanner

This chapter is a review of the literature relating to the study of geological structures on all scales from the microscopic to the continental. It covers such diverse aspects as the objective description of these structures, their mechanics of formation and their evolution in time.

Structural geology is concerned with the geometrical form and mode of development of structures in rocks, whereas tectonics is the study of the same phenomena on a regional scale, including fold belts and large portions of the earth's crust. Thus structural geology is concerned more with methods of study and mechanisms of rock deformation, while tectonics is largely the field of synthesis and utilises the data of structural geology, stratigraphy and geophysics to probe into the causes of deformation of the earth's crust.

The two disciplines are complementary and have a considerable area of overlap. Together they constitute 'geotectonics', the application of structural analysis in the broadest sense to geological structures, a usage which is in conformity with that of 'geochemistry' and 'geophysics'. Unfortunately, the terms 'tectonics' and 'geotectonics' are used as synonyms by many authorities and because of possible ambiguity the latter term has been restricted to the text and not used in the title.

In this chapter the literature describing the historical development of structural analysis is briefly reviewed. A section on nomenclature is then followed by a general introduction to the whole field of geotectonics and this in its turn is followed by separate sections on structural geology and tectonics in which both the advanced texts and the primary literature are described. For ease of reference each section is subdivided into separate fields such as petrofabrics, tectonophysics, etc., and selected references to closely related disciplines such as rock mechanics, engineering and economic geology are given.

A separate section on 'Collection and analysis of data' is included which, although largely concerned with field techniques and methods of analysing structural maps and data, also gives sources of regional tectonic maps and map reports. Literature on continental drift and plate tectonics is compiled in the section on tectonics.

Finally, there is a section on sources of abstracts, bibliographies and review articles.

HISTORICAL BACKGROUND

Geotectonics began with attempts to explain the external shape and origin of mountains. This work was largely speculative and was not placed upon a scientific basis until the first detailed geological maps were made in the nineteenth century.

The early history of geotectonics is given in *The Birth and Development of the Geological Sciences*, by F. D. Adams (see p. 401). From around 1830 there followed a period of increasing activity, with large areas of the earth's crust being geologically mapped for the first time. Major structural syntheses resulted from this work and some appreciation of the stimulating atmosphere of that period can be gained from *Chapters on the Geology of Scotland*, by B. N. Peach and J. Horne (Oxford University Press, 1930), and *Tectonic Essays, mainly Alpine*, by E. B. Bailey (see p. 429). Also of value here is *Pre-cambrian Geology of North America*, by C. R. van Hise and C. K. Leith (United States Geological Survey, 1909), and classical texts such as those by Heim and Collet which are referred to in the section on 'Tectonics'.

Meanwhile inquiries into the nature of schistosity and cleavage and the manner in which rocks deform were being carried out but were handicapped by a lack of knowledge of the physics of rock deformation. The development of petrofabric methods which followed, together with the more recent history, is reviewed in Chapter 1 of *Structural Analysis of Metamorphic Tectonics*, by F. J. Turner and L. E. Weiss (McGraw-Hill, 1963).

As knowledge of the major orogenic belts was gradually compiled in such works as *The Face of the Earth*, by E. Suess (see p. 418), the basic problems of orogenesis became apparent, but, as Adams stated in 1930: 'We cannot indeed but recognise that not only is the problem of the origin of mountain ranges still unsolved, but that toward the final elucidation of this subject geological science has made a less satisfactory advance than in many, if not in most, other directions.' Until recently this situation remained unchanged but the advent of the plate tectonics hypothesis now promises to provide a mechanism for orogenesis, and fundamental advances in experimental structural geology and in detailed methods of field analysis have greatly enhanced our understanding of rock deformation.

A selection of the definitive geological literature up to 1950, and including some 24 articles largely concerned with geotectonics, is given in *Source Book in Geology*, by K. F. Mather and S. L. Mason (reprinted 1964, Stechert-Hafner), and *Source Book in Geology 1900–1950*, by K. F. Mather (Harvard University Press, 1967).

TERMINOLOGY

Inconsistencies and deficiencies in structural and tectonic nomenclature have led to considerable confusion and have hampered the development of structural geology. They are still a major obstacle in international communication. The fundamental principle by which the terminology should be judged is that terms used to describe tectonic structures should not carry any genetic connotation whatsoever. Rules of nomenclature and classification, with examples of their use in the classification of faults, are proposed in 'Role of Classification in Geology', by M. L. Hill, in *The Fabric of Geology*, edited by C. C. Albritton, Jr. (Addison-Wesley, 1963).

A major advance achieved corporately by the International Geological Congress, the National Science Foundation and the American Association of Petroleum Geologists is the publication of the 'International Tectonic Dictionary', compiled and edited by J. G. Dennis (*Memoirs of the American Association of Petroleum Geologists*, **7**, 1967). This dictionary only covers English terminology at present but companion volumes of French, German and Russian tectonic terms are being prepared. The history, and, where possible, the first use of each term, is given together with a discussion of its general usage. The dictionary represents an extremely valuable contribution to structural geology and tectonics, and includes a bibliography. A less detailed, but fairly comprehensive, work is the *Glossary of Geology and Related Sciences with Supplement*, by J. V. Howell (2nd edn., American Geological Institute, 1957).

Structural terms are also included in *A Dictionary of Geology*, by J. Challinor (3rd edn., University of Wales Press, 1967); but although the historical comments are often of interest, this work suffers from a less rigorous definition of terms than is given in the works already quoted.

More detailed definitions of terms describing the morphology and geometry of minor structures are given in *Folding and Fracturing of Rocks*, by J. G. Ramsay (McGraw-Hill, 1967), and 'The Description of Folds', by M. J. Fleuty (*Proceedings of the Geologists' Association*, **75**, 461–492, 1964). Microfabric terms are defined in *Metamorphic Textures*, by A. Spry (Pergamon, 1969), and tectonic nomenclature is critically discussed by N. Rast in 'Orogenic Belts and their Parts', in 'Time and Place in Orogeny', edited by P. E. Kent, G. E. Satterthwaite and A. M. Spencer (*Special Publication, Geological Society of London*, **3**, 1969).

GENERAL TEXTBOOKS ON GEOTECTONICS

An excellent introduction to most aspects of geotectonics is given in Chapters 8 and 12 of *Introduction to Geology*, Vol. 1. *Principles*, by H. H. Read and J. Watson (2nd edn., Macmillan, 1968). Much of the same ground is covered in *Principles of Physical Geology*, by A. Holmes (revised edn., Nelson, 1965).

The standard textbooks on geotectonics are largely descriptive and combine an introductory section on structural elements, such as folds, faults and linear and planar fabrics, with sections on regional tectonics and orogenesis. Many do not provide an adequate background in the mechanisms of rock deformation and require revision of the sections on regional synthesis and causes of orogenesis. Textbooks suitable for undergraduate students and those with a basic general knowledge of geology include *Structural and Tectonic Principles*, by P. C. Badgley (Harper and Row, 1965); *Structural Geology*, by L. U. de Sitter (2nd edn., McGraw-Hill, 1964); and *Tectonics*, by J. Goguel (translation of the French edn., 1952, by H. E. Thalmann, Freeman, 1962).

A modern concise treatment of geological structures, rock deformation and plate tectonics can be found in Chapters 3, 9 and 13 of *The Earth, an Introduction to Physical Geology*, by J. Verhoogen *et al.* (Holt, Rinehart and Winston, 1970); while accounts of orogenesis and plate tectonics are given by J. Sutton and E. R. Oxburgh, respectively, in *Understanding the Earth, a Reader in the Earth Sciences*, edited by I. G. Gass, P. J. Smith and R. C. L. Wilson (Artemis Press, 1971).

COLLECTION AND ANALYSIS OF DATA

The geological map provides a basis for most structural interpretations and structural mapping in particular has evolved over the past 20 years or so into a precise scientific exercise. The philosophy which underlies the making of such maps is described in 'Nature and Significance of Geological Maps', by J. M. Harrison, in *The Fabric of Geology* (see p. 317).

An elementary introduction to field mapping techniques is provided in *Manual of Field Geology*, by R. R. Compton (Wiley, 1962), and much useful information, especially for techniques of outcrop mapping, is contained in *Methods in Geological Surveying*, by E. Greenly and H. Williams (Murby, 1930), which was written in honour of that greatest of field geologists, C. T. Clough, 'to publish a few plain instructions for drawing geological boundary lines, a practical matter which seemed to have been somewhat neglected'. Some details of field procedure are also given in *Structural Methods for the Exploration Geologist*, by P. C. Badgley (Harper and Row, 1959), and in *Structural Geology of Folded Rocks*, by E. H. T. Whitten (Rand McNally, 1966), with emphasis upon statistical sampling methods in the latter book. No detailed account has been published of the more recently developed methods based upon the geometric analysis of minor tectonic structures in the field. This information has to be sought in such papers as: 'Superimposed Folding at Loch Morar, Inverness-shire and Ross-shire', by J. G. Ramsay (*Quarterly Journal of the Geological Society of London*, 113, 271–307, 1958); 'Structural Analysis of the Basement System at Turoka, Kenya', by L. E. Weiss (*Overseas Geology and Mineral Resources*, 7, 3–35 and 123–153, 1959); 'The Three Fold-Systems in Metamorphic Rocks of Upper Glen Orrin, Ross-shire and Inverness-shire', by M. J. Fleuty (*Quarterly Journal of the Geological Society of London*, 117, 447–479, 1961); and 'The Tectonic Significance of Small Scale Structures and their Importance to the Geologist in the Field', by G. Wilson (*Annales de la Société Géologique de Belgique*, 84, 423–548, 1961).

The accurate description of tectonic elements is especially dependent upon the correct use of structural terms and the reader is referred to the section on 'Terminology'.

An introduction to map interpretation is given in *Analysis of Geologic Structures*, by J. M. Dennison (Norton, 1968), and in the appendix on 'Representation of Structure' in *The Earth, an Introduction to Physical Geology*. A comprehensive and well-presented account of both map interpretation and elementary data analysis is given in *Structural Geology, an Introduction to Geometrical Techniques*, by D. M. Ragan (Wiley, 1968).

Aerial photographs are often used to select an area of study, make a preliminary structural interpretation and also facilitate both the mapping and the recording of data in the field. For details of available texts reference should be made to Chapter 7.

Structural analysis is concerned largely with the spatial relationship of linear and planar elements and *Angular Relations of Lines and Planes*, by D. V. Higgs and G. Tunell (2nd edn., Freeman, 1966), provides an introduction to both stereographic analysis and the use of spherical trigonometry. Methods of stereographic analysis are described in detail in *The Use of Stereographic Projection in Structural Geology*, by F. C. Phillips (Edward Arnold, 1954), and 'Manual of the Stereographic Projection for a Geometric and Kinematic Analysis of Folds and Faults', by P. J. Haman (*West Canadian Research Publications of Geology and Related Sciences, Series* 1, **1**, 1961); and their use in areas of polyphase deformation is described in advanced textbooks on structural geology such as those by Turner and Weiss and by Ramsay (see pp. 315 and 317). The limitations of these methods are analysed by probability theory in *Structural Diagrams*, by A. B. Vistellius, a translation from the Russian edited by N. L. Johnson and F. C. Phillips (Pergamon Press, 1966).

Structural data are especially amenable to analysis by the techniques of orientation statistics, and computer methods may be used for the storage, retrieval and analysis of such data. One of the earliest publications in this field is *Computer Analysis of Orientation Data in Structural Geology*, by T. V. Loudon (Technical Report No. 13, Office of Naval Research Task No. 389–135, Northwestern University, 1964). A useful source of computer programmes including vector trend analyses of directional data, analysis of subsurface fold geometry and construction of π-diagrams is the series *Computer Contributions, State Geological Survey, University of Kansas* (1966–), which by mid-1971 included around 50 contributions. Recent work in this field is discussed in *Structural Geology of Folded Rocks*, by E. H. T. Whitten (see p. 318), and an introduction to the literature on scale-model simulation studies and finite-element analysis as used in the numerical simulation of folding in rocks is given briefly in *Computer Simulation in Geology*, by J. W. Harbaugh and G. Bonham-Carter (Wiley/Interscience, 1970).

Presentation of the results derived by data processing by means of structural stereograms, block diagrams and profiles is vital to communication with other earth scientists, and some of these methods are described in Turner and Weiss, and Ragan (see pp. 315 and 319), and in *Block Diagrams and Other Graphic Methods Used in Geology and Geography*, by A. K. Lobeck (2nd edn., Emmerson-Trussell, Amherst, Mass., 1958). A classical example of presentation of structural data is 'Geology of the Tovqussap Nuna', by A. Berthelsen (*Meddelelser om Grønland*, **123** (1), 1960).

Geological and tectonic maps, and accompanying bulletins, memoirs and sheet reports, are published by regional and national geological surveys. Details of many of these publications are given in Chapters 7 and 9.

Tectonic maps, in some cases with explanatory notes or treatises, are now available for both individual countries and continents, and a provisional list of these may be obtained from: Le Secrétaire Général, Commission de la Carte Géologique du Monde, 12 rue de Bourgogne, Paris-7. They include:

> *Tectonic Map of Great Britain* (1:1 584 000) (Institute Geological Sciences, 1966)
> *Tectonic Map of North America* (1:5M) (United States Geological Survey, Washington, 1969)
> *Tectonic Map of Africa* (1:5M) (African Association of Geological Surveys and UNESCO, 1968)

STRUCTURAL GEOLOGY

Structural geology was until fairly recently based largely upon observation and description. However, aided by both theoretical and experimental advances in tectonophysics, the study of the mechanics of tectonic deformation, and by detailed field studies of naturally occurring structures, it has evolved over the past 15 years into a true scientific discipline. Some of the fundamental dynamic problems are outlined by F. A. Donath in 'Fundamental Problems in Dynamic Structural Geology', in *The Earth Sciences, Problems and Progress in Current Research*, edited by T. W. Donnelly (University of Chicago Press, 1963); and progress in the various fields is reviewed by N. Rast in 'Recent Advances in Geotectonics' (*Earth-Science Reviews*, **2**, 1–46, 1967). The philosophical approach is reviewed by C. A. Anderson in 'Simplicity in Structural Geology', in *The Fabric of Geology* (see p. 317).

At present, structural geology is concerned with the mechanics of deformation and with the measurement of total strain in rocks. It is also of fundamental value in providing data for tectonic syntheses and in providing datum planes, both geometrical and in the time sense, for investigating other processes such as metamorphism, igneous intrusion and the development of major dislocation zones. General introductory textbooks which deal largely with descriptive aspects of structural geology include *Structural Geology*, by M. P. Billings (2nd edn., Prentice-Hall, 1954), and *Elements of Structural Geology*, by E. S. Hills (2nd edn., Wiley, 1972).

A more detailed treatment of geometry, and the terminology of folds, including a comprehensive bibliography and index, is found in *Structural Geology of Folded Rocks*, by E. H. T. Whitten (see p. 318). Emphasis on the microscopic fabrics of natural and experimentally deformed rocks is given in *Structural Analysis of Metamorphic Tectonites* (see p. 315), while *Folding and Fracturing of Rocks* (see p. 317) provides an excellent advanced treatment of the theories of stress and strain and their geological application, and a quantitative, well-illustrated introduction to structural analysis, excluding microfabrics. A good account of the mechanics of rock deformation is given in *Physical Processes in Geology*, by A. M. Johnson (Freeman, 1970).

The mechanics of rock fracture are examined in *Fault and Joint Development in Brittle and Semi-brittle Rocks*, by N. J. Price (Pergamon, 1966).

Classical texts covering specific aspects of structural geology include:

The Dynamics of Faulting and Dyke Formation with Applications to Britain, by E. M. Anderson (2nd edn., Oliver and Boyd, 1951)
Earth Flexures, their Geometry and their Representation and Analysis in Geological Section with Special Reference to the Problem of Oil Finding, by H. G. Busk (Cambridge University Press, 1929; republished unabridged by William Trussel, New York, 1957)
'Lineation, a Critical Review and Annotated Bibliography', by E. Cloos (*Memoirs of the Geological Society of America*, **18**, 1946)
On the Mechanism of the Geological Undulation Phenomena in General and of Folding in Particular and their Application to the 'Roots of Mountains' Theory, by S. W. Tromp (Sijthoff, Leiden, 1937)

The general features and development of folds are summarised in 'Folds and Folding', by M. J. Fleuty, in *Encyclopedia of Earth Sciences Series*, **5**, edited by R. W. Fairbridge (in press), and a recent attempt to establish strain facies based upon the morphology of folds is 'Strain Facies', by E. Hansen, No. 2 in the monograph series *Minerals, Rocks and Inorganic Materials* (Springer, 1971).

Some of the earliest attempts to understand the mechanisms of folding and faulting were made with models using materials such as sand, Plasticine and putty to represent rock layers. Recent advances are described in *Gravity, Deformation and the Earth's*

Crust as Studied by Centrifugal Models, by H. Ramberg (Academic Press, 1967), and a review and bibliography of work up to 1965 is provided by 'Experimental Structural Geology', by J. B. Currie (*Earth-Science Reviews*, 1, 51–67, 1966).

For a subject to which no single journal is devoted, the published proceedings of conferences and symposia provide useful compilations of recent research and review papers. These have been listed quarterly since 1967 in *Geological Newsletter* (International Union of Geological Sciences) and include:

'Age Relations in High-grade Metamorphic Terrains', edited by H. R. Wynne-Edwards (*Special Papers, Geological Association of Canada*, 5, 1969)

'Proceedings, Conference on Research in Tectonics (Kink Bands and Brittle Deformation)', edited by A. J. Baer and D. K. Norris (*Papers, Geological Survey of Canada*, **68–52**, 1969)

Structural geology encompasses not only phenomena such as folding and faulting but also the mechanisms of igneous intrusion and the movement of large bodies of material such as salt diapirs and glaciers.

Structural techniques as applied to petroleum geology are described in *Structural Geology for Petroleum Geologists*, by W. L. Russell (McGraw-Hill, 1955), and in *Structural Methods for the Exploration Geologist*, by P. C. Badgley (Harper and Row, 1959). Two symposium volumes published by the American Association of Petroleum Geologists which provide useful syntheses of oil-field information are *Possible Future Petroleum Provinces of North America*, edited by M. W. Bell *et al.* (1951), and *The Habitat of Oil*, edited by L. G. Weeks (1958). The 'General Review' by the editor of the latter volume provides a good introduction to the subject.

Salt Deposits, the Origin, Metamorphism and Deformation of Evaporites, by H. Borchert and R. O. Muir (Van Nostrand, 1964), and *The Physics of Glaciers*, by W. S. B. Paterson (Pergamon, 1969), deal with most of the structural aspects of salt and ice, respectively, and both provide extensive bibliographies. An additional more recent work on salt deposits is 'Diapirism and Diapirs, a Symposium', edited by J. and G. D. O'Brien (*Memoirs of the American Association of Petroleum Geologists*, 8, 1968).

Structural aspects of igneous intrusions are discussed in *Structural Behaviour of Igneous Rocks*, by R. Balk (J. W. Edwards, Ann Arbor, originally published as *Memoirs of the Geological Society of America*, 5, 1948), and in *Mechanism of Igneous Intrusion*, edited by G. Newall and N. Rast (Gallery Press, Liverpool, 1970).

Structural mapping and strain analysis are of value in a number of other fields such as civil engineering, mining and mineral exploration. However, the structural control of the ore deposits, for example, has received little detailed attention recently and the contribution which structural geology is able to make in engineering practice is only beginning to be recognised. Such texts as are available include *Structural Geology with Special Reference to Economic Deposits*, by B. Stŏces and C. H. White (Macmillan, 1935), which is superbly illustrated and makes an ideal companion volume to advanced structural textbooks such as that by Ramsay (see p. 317). Also of value is *Ore Deposits, as Related to Structural Features*, by W. H. Newhouse (Princeton University Press, 1942), and symposium volumes such as:

Structural Geology of Canadian Ore Deposits. Vol. II. *Congress Volume* (Canadian Institute of Mining and Metallurgy, 1959)
Stratiform Copper Deposits in Africa. Part II: Tectonics, edited by J. Lombard and P. Nicolini (Association of African Geological Surveys, 1963)
Remobilization of Ores and Minerals (Associazione Mineraria Sarda, Cagliari, Italy, 1969)

Applications of structural geology to engineering are mentioned in Chapter 1 of *Rock Mechanics in Engineering Practice*, edited by K. G. Stagg and O. C. Zienkiewicz (Wiley, 1968). Other useful texts include *Elements of Engineering Geology*, by J. E. Richey (Pitman, 1964), and *Design of Structures in Rock*, by L. Obert and W. I. Duvell (Wiley, 1967).

Petrofabrics

Petrofabric analysis, the study of the optical and dimensional orientation of minerals in a rock, long discredited as a primary tool for understanding the evolution of tectonic structures, is able to make a positive contribution now that more is known about the mechanisms by which rocks and individual minerals deform under stress.

Pioneer work in this field was carried out in Austria in the 1920s and published in *Gefügekunde der Gesteine*, by B. Sander (Springer, 1930), and a later book which has recently been republished in translation as *An Introduction to the Study of Fabrics of Geological Bodies*, by B. Sander (translation by F. C. Phillips and G. Windsor, Pergamon, 1970). Also worthy of mention here is an American work 'Structural Petrology' (*Memoirs of the Geological*

Society of America, **6**, 1938). More recent work is reported in *Structural Petrology of Deformed Rocks*, by H. W. Fairburn (2nd edn., Addison-Wesley, 1949) and by Turner and Weiss (see p. 315).

A related study which has come into prominence during the past 15 years in conjunction with the discovery of the polyphase nature of deformation in orogenic belts is the study of the internal fabrics in minerals and the relationships of mineral growth to successive foldings. *Metamorphism*, by A. Harker (3rd edn., Methuen, 1950), provides well-illustrated descriptions and definitions of mineral textures. The first five sections of *Controls of Metamorphism*, edited by W. S. Pitcher and G. W. Flinn (Oliver and Boyd, 1965), summarise recent work in this field but *Metamorphic Textures*, by A. Spry (Pergamon, 1969), is unique in providing the only comprehensive account of metamorphic textures available at present. A recommended specialist text is 'Rotated Garnets in Metamorphic Rocks', by J. L. Rosenfeld (*Special Papers, Geological Society of America*, **129**, 1970).

Tectonophysics

The study of the mechanisms of deformation of rocks has been approached both experimentally and by the mathematical analysis of the shapes of deformed objects and planar surfaces occurring in rocks. The physical basis for this work was provided by the theories of stress and strain developed in physics and engineering, and an introductory account by J. Handin is to be found as Chapter 11 of 'Handbook of Physical Constants' revised edition, edited by S. D. Clark (*Memoirs of the Geological Society of America*, **97**, 1966). General textbooks which provide a background to the theory of stress and strain include:

Theory of Flow and Fracture of Solids, by A. Nadai (2 vols., McGraw-Hill, 1962)
Studies in Large Plastic Flow and Fracture, by P. W. Bridgeman (Harvard University Press, 1964)
Fundamentals of Rock Mechanics, by J. C. Jaeger and N. G. W. Cook (Methuen, 1969)

The basic theory of the mechanical properties of rocks and examples of its application to geological structures are given in the textbook by Ramsay (see p. 317) and in:

Elasticity, Fracture and Flow, with Engineering and Geological Applications, by J. C. Jaeger (corrected 2nd edn., Methuen, 1964)
Mechanics of Incremental Deformation: Theory of Elasticity and

Viscoelasticity of Initially Stressed Solids and Fluids Including Thermo-dynamic Foundations and Application to Finite Strain, by M. A. Biot (Wiley, 1965)

Tensor analysis is finding increasing application in this field and basic introductions to the theory are given by Ramsay; in *Methods of Mathematical Physics*, by R. Courant and D. Hilbert (Wiley, 1953); and in *Physical Properties of Crystals, their Representation by Tensors and Matrices*, by J. F. Nye (Oxford University Press, 1967).

Results of fundamental importance have been obtained by the experimental deformation of rocks, notably the importance of strain rate and pore fluid pressure in deformation processes occurring over a geological time span. Much of this work is contained in the published proceedings of conferences and symposia—for example :

'Rock Deformation', edited by D. Griggs and J. Handin (*Memoirs of the Geological Society of America*, **79**, 1960)
State of Stress in the Earth's Crust. Proceedings of the International Conference, Santa Monica, California (Elsevier, 1964)
Failure and Breakage of Rock, edited by C. Fairhurst (8th Symposium on Rock Mechanics, University of Minnesota, 1966)

Recent publications also include :

'Experimental Deformation of Crystalline Rocks', by I. Borg and J. Handin (*Tectonophysics*, **3** (4), 1966)
Mechanical Properties of Rocks at High Temperatures and Pressures, by B. V. Baidyuk (translated from the Russian by J. P. Fitzsimmons, Consultants Bureau, 1967)

Primary literature

No single journal is largely or wholly devoted to structural geology. In fact at present papers on structural geology are occasionally found in a large range of journals, of which the *Bulletin of the Geological Society of America*, issued monthly, probably contains the greatest number and diversity. The *Bulletin of the American Association of Petroleum Geologists* carries articles on tectonic structures, largely in sedimentary rocks, and on faulting and diapiric structures, while the *Scottish Journal of Geology* includes an above-average proportion of papers on the structural geology of metamorphic rocks. Russian literature is translated in *Geotectonics*.

Other journals which regularly include notable papers on structural geology are *American Journal of Science, Canadian Journal of Earth Sciences, Geologische Rundschau* and *Journal of Geology*.

The field of rock deformation is covered by *Tectonophysics* and by occasional articles of geological importance in the *Journal of Strain Analysis* and on brittle deformation in the *International Journal of Rock Mechanics and Mining Sciences*. Sections on deformation, faults, fractures and tectonophysics are included in the *Journal of Geophysical Research*.

Papers on structural geology are also published in the four-yearly proceedings of the International Geological Congress. The last, 23rd session, was held in Prague in 1968 and includes a section on 'orogenic belts'. Other sources include *Professional Papers, Geological Society of America*.

A specialist group, the Tectonic Studies Group of the Geological Society of London, has been formed recently to provide a joint forum for structural geologists and workers in related fields such as rock mechanics, engineering and many branches of geophysics. It is hoped to publish abstracts of papers in the *Journal of the Geological Society*.

TECTONICS

The detailed results of structural mapping are compounded into regional descriptions or syntheses which deal with the tectonic development of large portions of the earth's crust. Mechanisms of large-scale folding, in particular the part played by gravity tectonics within the tectonic belts, can be studied by analogy with theoretical and experimental structural geology and simulated by model studies such as those described in *Gravity, Deformation and the Earth's Crust, as Studied by Centrifugal Models*, by H. Ramberg (Academic Press, 1967).

For most of the first half of this century Alpine tectonics had an immense influence on tectonic hypotheses, largely through such masterly works as:

Geologie der Schweiz, by A. Heim (C. H. Tauchnitz, Leipzig, 1919)
The Structure of the Alps, by L. W. Collet (Edward Arnold, 1927)
Bau und Entstehung der Alpen, by L. Kober (Franz Deuticke, Vienna, 1955)

More recent advances in this field are summarised in *Geology of Western Europe*, by M. G. Rutten (Elsevier, 1969), which provides an extensive bibliography of the geology of the continental portion of Western Europe.

Two older orogenic zones—the Caledonian and Appalachian fold belts—have been studied in comparable detail to the Alps. Summaries and bibliographies of work in the Caledonian fold belt in the British Isles are contained in *The British Caledonides*, edited by M. R. W. Johnson and F. H. Stewart (Oliver and Boyd, 1963), and *The Geology of Scotland*, by G. Y. Craig (Oliver and Boyd, 1965). This work is placed in a more general context in *The Geological History of the British Isles*, by G. M. Bennison and A. E. Wright (Edward Arnold, 1969). Structural syntheses of North America, in particular of the Appalachian fold belt, include:

Structural Geology of North America, by A. J. Eardley (Harper and Row, 1962)
'Appalachian Tectonics', edited by T. H. Clark (*Royal Society of Canada Special Publications*, **10**, 1967)
Studies in Appalachian Geology, Northern and Maritime, by E-an Zen, W. S. White and J. B. Thompson, Jr. (Wiley/Interscience, 1968)
The Tectonics of the Appalachians, by J. Rodgers (Wiley/Interscience, 1970)

More recent work on both the Caledonian and Appalachian fold belts is summarised in a number of contributions to *North Atlantic —Geology and Continental Drift*, a symposium volume edited by M. Kay (American Association of Petroleum Geologists, 1969).

Syntheses on a larger scale include 'Age and Nature of the Circum-Pacific Orogenesis', edited by T. Matsumoto (*Tectonophysics*, **4** (4–6), 1967).

In the early part of this century, as the morphology and distribution of the major orogenic belts began to be understood more clearly, a number of writers started to speculate on the causes and controls of orogenesis. These writings include largely discursive works such as:

Earthquakes and Mountains, by H. Jeffreys (2nd edn., Methuen, 1929)
The Deformation of the Earth's Crust, by W. H. Bucher (Princeton University Press, 1933; reprinted 1941)
The Pulse of the Earth, by J. H. F. Umbgrove (2nd edn., Martinus Nijhoff, The Hague, 1950)
Symphony of the Earth, by J. H. F. Umbgrove (Martinus Nijhoff, The Hague, 1954)
Mountain Building; A Study Primarily Based on Indonesia, Region of the World's Most Active Crustal Deformation, by R. W. van Bemmelen (Martinus Nijhoff, The Hague, 1954)
Fundamentals of Geology, by S. von Bubnoff (translated and edited by W. T. Harry, Oliver and Boyd, 1963)

A different viewpoint, emphasising the importance of vertical tectonic forces, has been developed in Russia:

Basic Problems in Geotectonics, by V. V. Belousov (McGraw-Hill, 1962)

Structural Geology, by V. V. Belousov (English translation of 1961 Russian edn., Mir. Publ. Moscow, 1968)

Folded Deformations in the Earth's Crust, their Types and Origin, by V. V. Belousov and A. A. Sorkii (Israel Program for Scientific Translations, Jerusalem, 1965)

The physical problems of orogenesis are discussed in *Principles of Geodynamics,* by A. E. Scheidegger (2nd edn., Springer, 1963), and the relationship of orogenic events to foci of sedimentary deposition in *Geosynclines,* by J. Aubouin (Elsevier, 1965).

With the help of geochronology, the sequence of fold belts from the earliest pre-2500 m.y. greenstone belts to the later linear, mobile belts can be examined. The African continent provides a classic area for such work and much of the early work in Central Africa is reviewed in *The Geochronology of Equatorial Africa,* by L. Cahen and N. J. Snelling (North-Holland, 1966). In the absence of an up to date structural synthesis, *African Magmatism and Tectonics,* edited by T. N. Clifford and I. G. Gass (Oliver and Boyd, 1970), provides a useful bibliography.

Folded portions of the crust can be studied both individually and collectively and ideas of the amount of tectonic shortening across these zones assessed by means of structural analysis or, in some cases, palaeomagnetism. Previous attempts to understand the causes and mechanisms of orogenesis have so far been severely limited by a lack of quantitative data and the compilation of available data into a suitable form. The Council of the Geological Society of London has therefore inaugurated a project whose aim is the systematic compilation of data from orogenic belts on a world-wide scale. The first stage, a review of the problems involved in measuring displacement in orogenic belts, has been published in 'Time and Place in Orogeny', edited by P. E. Kent, G. E. Satterthwaite and A. M. Spencer (*Special Publication, Geological Society of London,* **3**, 1969). Data from Tertiary orogenic belts is at present being compiled and will be subsequently analysed. This will be followed by the consideration of successively older orogenic belts. A synthesis of previous work is given by R. Dearnley in 'Orogenic Fold Belts and a Hypothesis of Earth Evolution', in *Physics and Chemistry of the Earth,* **7** (1966).

The major rift fault systems of the world are comparable in scale and importance with the orogenic belts. They are described and analysed in 'The World Rift System', edited by T. N. Irvine (*Papers, Geological Survey of Canada*, **66**–**14**, 1966), and in *Graben Problems* (Proceedings of the International Rift Symposium, Karlsruhe, 1968), edited by J. H. Illies and St. Mueller (Schweizerbart'sche, Stuttgart, 1970). Data on the East African Rift System are compiled in *Report on the Geology and Geophysics of the East African Rift System* and *Report of the UMC/UNESCO Seminar on the East African System*, both published by UNESCO in 1965. A recent publication, 'A Discussion on the Structure and Evolution of the Red Sea and the Nature of the Red Sea, Gulf of Aden and Ethiopia Rift Junction' organised by N. L. Falcon *et al.* (*Philosophical Transactions of the Royal Society A* (1181), **267**, 1970), provides a bibliography and reference to recent aspects of rift tectonics.

Plate tectonics

The hypothesis that the continents as we now see them were once connected to form a single crustal unit was first expounded by Wegener and du Toit, based upon geological evidence presented in *The Origin of the Continents and Oceans*, by A. Wegener (translated from the 4th revised German edn., 1929 by J. Biram, Dover, 1966), and *Our Wandering Continents*, by A. L. du Toit (Oliver and Boyd, 1937). The continental drift hypothesis has only recently found general acceptance, largely because of the geophysical evidence which has been adduced to support it in, for example, *Continental Drift*, by S. K. Runcorn (Academic Press, 1962). As the geophysical results became available, a number of symposia were held to summarise the changing state of opinion. These provide an interesting commentary and include:

Continental Drift, edited by S. W. Carey (Geology Department, University of Tasmania, 1958)
A Symposium on Continental Drift, edited by P. M. S. Blackett, E. Bullard and S. K. Runcorn (Royal Society, 1965)
'Continental Drift', by G. D. Garland (*Royal Society of Canada Special Publications*, **9**, 1966)
The History of the Earth's Crust, a Symposium, edited by R. A. Phinney (Princeton University Press, 1968)

Geophysical results—in particular, palaeomagnetic data—led to the concept of plate tectonics which provides a mechanism for continental drift. This field is the most rapidly expanding part of the earth sciences, to which a general background is provided by

Continental Drift, by O. H. Tarling and M. P. Tarling (Bell, 1971). Recent accounts of plate tectonics are given in *The Megatectonics of Continents and Oceans*, by H. Johnson and P. L. Smith (Rutgers University Press, 1970), and *The Sea; Ideas and Observations on Progress in the Study of the Seas*, edited by A. E. Maxwell (Wiley/Interscience, 1970).

The value of the plate tectonics hypothesis to tectonics lies in the fact that, in providing a possible mechanism for mountain building, it has given a great impetus to the compilation of tectonic data and an opportunity for a completely fresh look at some of the outstanding problems.

Primary literature

The primary literature on tectonics is largely contained within the same journals as structural geology—in particular, the *Bulletin of the Geological Society of America* and *Tectonophysics*. Most of the recent work on plate tectonics and megatectonics is published in *Nature* but a number of papers are also found in *Science* and in the *Transactions of the American Geophysical Union*. Translations of Russian papers are published bi-monthly in *Geotectonics* (1967–).

ABSTRACTING SERVICES AND REVIEWS

The major current source for abstracts of structural geology and tectonics is the *Bibliography and Index of Geology* (see Chapter 5). Subject index headings include 'tectonics' (which is the most comprehensive), and also 'deformation', 'faults', 'folds', 'fractures', 'petrofabrics' and 'structural geology'. Also of value is Cahier F. 'Tectonique et Géophysique' of *Bibliographie des Sciences de la Terre* (see Chapter 5).

Physical and geophysical aspects of structural geology are in *Geophysical Abstracts* under the sub-heading of 'Geotectonics' and 'Strength and Plasticity'. In the index most of the structural topics are found under 'rock mechanics' and 'tectonics'.

Of more limited use is *Mineralogical Abstracts*, which includes work on petrofabrics, mineral textures and regional studies.

Reviews of structural literature are published in *Earth-Science Reviews* and *Tectonophysics*, as well as in general journals such as *Science*.

13

Applied geology

C. H. James

Although the study of the practical applications of geology was one of the earliest aspects of the earth sciences to be developed, it is also one that has shown an extremely rapid development during the last 20 years. Within that time, not only has there been a spectacular growth in the over-all level of activity but whole new branches of the general field have been developed. As a result, the relevant literature has increased prodigiously. Accordingly, it is impossible within the space of a single chapter to give more than a brief introduction to the scope of the literature that has been written on the subject. Furthermore, it will be appreciated that the references given represent only a selection, and are therefore subjective. Nevertheless, it is hoped that sufficient information will be given to enable a student or non-specialist to make a satisfactory literature search in this field, with a minimum of wasted effort.

In general, only references which refer specifically to the subject in question will be given. A wide variety of general geological publications obviously contain some information pertinent to the study of applied geology, but it is clearly outside the scope of this contribution to refer to these.

Applied geology can be divided into a number of well-defined topics, each of which has its own literature. These are: the geology of metalliferous ore deposits; exploration for metalliferous ore deposits; coal; oil and natural gas; industrial minerals; constructional materials; water supply; and engineering geology. The total volume and rate of growth of the literature of each group is by no means equal but depends on the complexity of the individual topic. For this reason the study of metalliferous deposits and their discovery has expanded particularly rapidly. Indeed, the term 'mining geology' is now widely taken as meaning the geology of metalliferous deposits, notwithstanding the still considerable economic importance of the coal mining industry. Unlike most of the other raw materials, metallic ores are found in a very wide variety of geological environments, and in most cases it is virtually impossible to predict the location of an orebody on purely geological reasoning (as may frequently be done for coal and, to a lesser extent, oil). Accordingly, the search for such deposits has led to the development of highly specialised techniques such as geochemical and geophysical prospecting. For this reason, metalliferous exploration is considered under a separate heading. Techniques of looking for other materials are, where necessary, included under the general heading of the topic in question.

Each topic is considered separately by referring to the textbooks and journals in which most of the information and discussion is to be found. Special mention should be made here of Russian literature. In view of the very small number of geologists who are capable of reading Russian, references to original works in that language have been omitted. However, a number of books have now been translated, while translated versions of a number of Russian journals are regularly obtainable (see Chapter 6).

THE GEOLOGY OF METALLIFEROUS ORE DEPOSITS

Two aspects of the geology of metalliferous ore deposits are of particular importance: namely, descriptions of the geological environments in which individual ore deposits and ore fields are situated, and a more theoretical consideration of both the origins of such deposits and the mechanisms by which they were formed. Frequently, of course, there is a degree of overlap between the two approaches.

A great deal of research work connected with metalliferous mining geology is undertaken by mining companies, who, in the past, have sometimes been reluctant to publish their findings, especially in the case of descriptions of individual orebodies. In recent years, however, there has been a reaction against this somewhat narrow view, and although data on grade or tonnages of orebodies are frequently restricted, most mining companies allow the publication of descriptions of the geology of their mines.

Of outstanding historical interest in the field of ore deposits is Georgius Agricola's *De Re Metallica*, first published in Latin in 1556, but available in an excellent English translation by H. C. and L. H. Hoover (Dover reprint, 1950). This work was one of the first to contain actual descriptions of ore deposits and also provides an interesting insight into early thoughts on the origin of such deposits.

General descriptions of the geology of selected ore deposits throughout the world are to be found in a number of books. In view of the very large numbers of such deposits, it is inevitable that such attempts should be both incomplete and sketchy in nature. Among the more useful works are *Metallic and Industrial Mineral Deposits*, by C. A. Lamey (McGraw-Hill, 1966), and *Die Bergwirtschaft der Erde*, by F. Friedensburg (Ferdinand Enke, 1964), although the latter is written from a viewpoint which tends to be economic as much as geological.

Excellent volumes exist which describe the ore deposits of selected regions of the earth's crust. For instance, accounts of the *Geology of Australian Ore Deposits* have been prepared and published as volumes of the Proceedings of the 5th Commonwealth Mining and Metallurgical Congress, 1953 (edited by A. B. Edwards), and the 8th Congress, 1965 (edited by J. McAndrew). In the case of the 8th Congress, a companion volume was also prepared on the *Economic Geology of New Zealand*.

Three comprehensive accounts of the geology of African ore fields are to be found in *The Mineral Resources of Africa*, by N. de Kun (Elsevier, 1965); *The Mineral Resources of South-Central Africa*, by R. A. Pelletier (Oxford University Press, Capetown, 1964); and *The Geology of Some Ore Deposits in South Africa*, edited by S. H. Haughton (2 vols., Geological Society of South Africa, 1964).

The publication of the 2-volume *Ore Deposits in the United States 1933/67*, edited by J. D. Ridge (American Institute of Mining, Metallurgical and Petroleum Engineers, 1968), represented something of a landmark in the literature of economic geology. Not only does this work provide a comprehensive review of modern thinking on the nature and origin of these deposits but it also contains a most useful historical summary of theories of ore genesis, complete with an extremely full bibliography on this latter subject. Also of note in the literature of the ore deposits of the USA is *Geology of the Porphyry Copper Deposits of South-Western North America*, edited by S. Titley and C. Hicks (University of Arizona Press, 1966).

Unfortunately, no comprehensive account of the geology of Canadian deposits has been published, the nearest thing to this being *Structural Geology of Canadian Ore Deposits*, published by the Canadian Institute of Mining and Metallurgy. Volume 1 of this work was published in 1948 and Vol. 2 in 1957 to coincide with the 6th Commonwealth Mining and Metallurgical Congress. Although this work is not complete in that, as its title suggests, it views the geology from a somewhat specialised aspect, it is nevertheless a most valuable reference book.

Few other countries have detailed accounts in English of the geology of their ore deposits. This is particularly disadvantageous in the case of the USSR and China, since in both countries a considerable number of occurrences of both great economic importance and scientific interest are to be found. In the case of Russia, however, the situation was alleviated somewhat by the publication of English translations of selected parts of *Rudnye Mestorozhdeniya* (*Promyshlennye Tipy Mestorozhdenii Metallicheskikh Poleznykh Iskopaemykh*) [*Ore Deposits* (*Economic Types of Metallic Mineral Deposits*)], by I. G. Magak'yan (Armenian Academy of Sciences, 1961). These appeared as a series in Vol. 10, Parts 7, 8, 9 of *International Geology Review* (1968). Although the sections selected for translation only relate to deposits in the Soviet Union, they represent a most useful summary of the data.

In the case of the UK, perhaps the most comprehensive review of metallic deposits is to be found in the proceedings of a symposium on *The Future of Non-Ferrous Mining in Great Britain and Ireland*, published in 1959 by the Institution of Mining and Metallurgy. In addition, two *Memoirs* of the Geological Survey of Great Britain (now incorporated in the Institute of Geological Sciences) warrant special mention. These are *Geology of the North Pennine Orefield*, by K. C. Dunham (2 vols., 1948), and *The Metalliferous Mining Region of South-West England*, by H. D. Dines (2 vols., 1956), both of which provide great detail of the relevant mining fields, including considerable data on the geology of the two regions.

No comprehensive account of the geology of the recent discoveries in the Republic of Ireland has yet appeared, although descriptions of individual deposits, in varying degrees of detail, have been published in a number of journals. A very old yet valuable account of the occurrences known in Ireland before the present 'boom' is to be found in 'Memoir and Map of Localities of Minerals of Economic Importance and Metalliferous Mines in Ireland', by G. A. J. Coles (*Memoir of the Geological Survey of Ireland*, 1922).

A number of books discuss the processes by which ore deposits are formed, and many of these also provide descriptions of the deposits—often by way of illustration as much as for their own sake. A very useful summary of ideas relating to ore formation, particularly suitable for those unfamiliar with this branch of the science, is to be found in 'The Genesis of Sulphide Ores', which was given as a Presidential Address to the Geologists' Association by David Williams (*Proceedings of the Geologists' Association*, 71 (3), 245–284, 1960). Of an older vintage, but still very widely read are *Mineral Deposits*, by W. Lindgren (4th edn., McGraw-Hill, 1933), and *Economic Mineral Deposits*, by A. E. Bateman (2nd edn., Wiley, 1942), both of which describe the 'classic' theories of ore formation and classification in considerable detail, as well as providing brief descriptions of a wide variety of ore deposits. More recently, C. F. Park and R. A. MacDiarmid have written a very satisfactory student text, *Ore Deposits* (Freeman, 1964), which combines a good theoretical approach with a selection of excellent examples. Continental European writers have also produced a number of very important volumes in this field. *Erzlagerstätten*, by H. Schneiderhöhn (Gustav Fischer, 1962) and *Erzlagerstätten der Erde*. Vol. 1. *Magmatic Deposits*, Vol. 2. *Pegmatites*, both by the same author (Gustav Fischer, 1958, 1961), are fundamental works of great importance, while two French publications of excellence are *Géologie des Gîtes Minéraux*, by E. Raguin (3rd edn., Masson 1961), and *Gisement Métallifères*, by P. Routhier (2 vols., Masson, 1963).

A large number of interesting papers on specific topics in the field of ore deposits are to be found in the *Proceedings of the Symposium on Problems of Post-magmatic Ore Deposition, with Special Reference to the Geochemistry of Ore Veins*, published by the Geological Survey of Czechoslovakia in 1963. Similarly, a collection of papers on sedimentary processes in ore formation is to be found in *Sedimentary Ores: Ancient and Modern (Revised)*, edited by C. H. James (Geology Department, University of Leicester, 1969), which is the proceedings of the 15th Inter-University Geological Congress. A more controversial view of ore genesis is described in *Ore Genesis*, by J. S. Brown (Murby, 1950). Although the opinions expressed in this work have not found general acceptance, the book has particular merit in that it does draw attention to some of the shortcomings of current popular views.

Specific studies of actual processes of ore deposition are naturally more specialised. Perhaps the oldest still widely read book in economic geology after *De Re Metallica* is *The Ore Magmas*, by J. E. Spurr (2 vols., McGraw-Hill, 1923), which was one of the first works systematically to study the relationship between ore sources and ore deposits.

A great deal has been written on the geochemistry of ore formation, but perhaps the best summary of modern thinking on this topic is to be found in *The Geochemistry of Hydrothermal Ore*

Deposits, edited by H. L. Barnes (Holt, Rinehart and Winston, 1968), while a useful discussion of the mechanisms by which ore metals travel in hydrothermal fluids is provided by *Complexing and Hydrothermal Ore Deposition*, by H. C. Helgeson (Pergamon, 1964).

The techniques of reflected light microscopy have long played an important role in the study of ore deposits, to the extent that the methods of ore microscopy now have an extensive literature of their own. Textbooks on the various aspects of this subject include *Ore Microscopy*, by E. N. Cameron (Wiley, 1961), and *Applied Ore Microscopy*, by H. Freund (Macmillan, 1967). It should be mentioned, however, that the latter of these two is a shortened English translation of *Handbuch der Mikroskopie in der Technik*, edited by H. Freund (Umschau, 1960), which was designed for the needs of the mineral dresser rather than the ore microscopist, and as such may be of more restricted use. A further general text on this subject is *Erzmikroskopisches Praktikum*, by H. Schneiderhöhn (Schweizerbart'sche, 1952).

A considerable number of reference books have also been produced. One of the best-known, describing ore texture, is *Textures of the Ore Minerals*, by A. B. Edward (2nd edn., Australasian Institute of Mining and Metallurgy, 1954). However, the most exhaustive treatise on ore minerals under the microscope is the monumental *Die Erzmineralien und ihre Verwachsungen*, by the father figure of ore microscopy, Paul Ramdohr (Akadamie Verlag, 1960). Fortunately for English-speaking ore geologists, an excellent complete translation of this work (edited by C. Amstutz) has been published under the title *The Ore Minerals and their Intergrowths* (Pergamon, 1970).

Tables for the identification of minerals in ore microscopy include *Determination Tables for Ore Microscopy*, by C. Schouten (Elsevier, 1962); *Tables for Microscopic Identification of Ore Minerals*, by W. Uytenbogaarat (Princeton University Press/Hafner, 1951); and 'Microscopic Determination of the Ore Minerals', by M. N. Short (*Bulletin of the United States Geological Survey*, No. 914, 1940).

A comprehensive set of photographic illustrations of ore textures, to which additions are still being made, is *Atlas of the Most Important Ore Mineral Parageneses under the Microscope*, by O. Oelsner (Pergamon, 1966).

Descriptions of the geology of individual ore deposits are most likely to be found in the various scientific and technical journals. Perhaps the most widely read of those devoted to the study of mineral deposits is *Economic Geology* (published eight times a year), in which is included the *Bulletin of the Society of Economic Geologists*. This publication contains a wide variety of papers on all aspects of metalliferous mining geology (including exploration methods) together with some on aspects of the geology of coal. *Mineralium Deposita* is a newer journal, published in Europe, which is also the official organ of the Society for Geology Applied to Mineral Deposits. It is perhaps more orientated towards the study of theoretical aspects of ore genesis than *Economic Geology* but regularly contains a variety of contributions of a high standard. A further publication of this type is *Freiberger Forschungshefte*, published by the Bergakadamie (School of Mines) of Freiberg in East Germany.

Many geological papers of a high calibre are to be found in the publications of the various professional bodies of the metalliferous mining industry. Particular mention must be made here of *Applied Earth Science*, which is published every 3 months and constitutes Section B of *Transactions of the Institution of Mining and Metallurgy*. It represents a regular source of information on all aspects of economic geology related to the metalliferous mining industry. *Mining Engineering*, published by the American Institute of Mining, Metallurgical and Petroleum Engineers, consists of a wide variety of material, including papers on mining engineering, mineral dressing and related topics in addition to those on mining geology and mineral exploration. A similar range of subjects will be found in the *Bulletins* of the Australian, Canadian, Indian, New Zealand and South African Institutes of Mining and Metallurgy.

The main French outlet for information on ore deposits is *Bulletin du Bureau de Recherches Géologiques et Minières*, which almost invariably contains a number of most interesting contributions.

Occasional papers on the geology of individual ore deposits or fields are also to be found in commercial publications relating to the metalliferous mining industry, such as *The Mining Magazine*, *Mining and Minerals Engineering* and *World Mining*. Both *The Mining Magazine* and *World Mining* publish particularly useful annuals, both of which contain reviews of the more important advances in mining geology during the preceding year together with summaries of the main mineral discoveries made during the same period. Two valuable sources of information on the geology of ore deposits of specific regions are the *Proceedings* of both the International Geological Congresses and the Commonwealth Mining and Metallurgical Congresses. The former frequently contain a wider spectrum of papers in economic geology, although, since the topics for discussion vary from meeting to meeting, no general observation can be made. Reference to certain volumes published by the second body has already been made. In addition to these publications, irregular publications containing descriptions of the geology of mines and mining fields are to be found among the bulletins and other publications of geological surveys throughout the world (for details of many of these see Chapter 9). Naturally, this is particularly the case in countries where mining plays an important role in the economy of the country, e.g. Australia, Canada, India, South Africa, (Southern) Rhodesia, the USA and Zambia. In some of these countries and particularly in Canada and the USA, provincial or state surveys also publish considerable quantities of most valuable information.

A great deal of information can also be presented in the form of maps (see Chapter 7) and the number of maps illustrating aspects of economic geology grows annually. Metallogenic maps, showing the distribution and type of ore deposits throughout a region, are now being produced by various authorities. Geophysical and, more recently, geochemical maps are also available, but usually these only cover relatively small areas.

EXPLORATION FOR METALLIFEROUS ORE DEPOSITS

In view of the very great importance of this topic, there are surprisingly few texts which cover it, and most of those which do are far from complete for one reason or another. Perhaps the most widely read is *Mining Geology*, by H. McKinstry (Prentice-Hall, 1948), which, as its title suggests, covers a somewhat wider spectrum than just exploration. This book is now out of date in a number of ways; indeed some of the most important of modern exploration techniques (such as geochemical prospecting) receive hardly a mention. Notwithstanding this fact, however, there is still a great deal of value in it, and some of the fundamental points that it puts forward are still better covered in this than in any other book. Another old but still useful book is *Principles of Field and Mining Geology*, by J. D. Forrester (Wiley, 1946). This is essentially a practical handbook of field techniques, some of which have now been superseded by newer methods. A small volume of interest because it is published in the USSR is *Geological Prospecting and Exploration*, by V. M. Kreiter (MIR Publishing House, Moscow, 1968). Although this work is not of a particularly high level, it could be read with advantage by many geologists in the mining industry, particularly for the chapters on the appraisal of ore deposits, and the quantitative nature of sampling. Although not strictly speaking a textbook, Vol. II of the Proceedings of the 8th Commonwealth Mining and Metallurgical Congress—*Exploration and Mining Geology*, edited by L. J. Lawrence (Australasian Institution of Mining and Metallurgy, 1965)—contains a range of papers that covers the subject so well that it is probably the best single volume obtainable. A similar volume of papers on all aspects of metalliferous ore geology (including exploration) submitted to the 9th Congress, held in London in 1969, was published in 1970 by the Institution of Mining and Metallurgy.

In addition to the more general texts listed above, there is a considerable number of books on specific topics of mineral exploration. Undoubtedly the standard work on geochemical prospecting is *Geochemistry in Mineral Exploration*, by H. E. Hawkes and J. S. Webb (Harper and Row, 1962). Although the subject has advanced considerably since the publication of this book, especially in the interpretation of data, it presents by far the best currently available review of the concepts upon which the technique is based. A slightly older title, *Principles of Geochemical Prospecting*, by I. I. Ginzburg, translated from the original Russian by V. P. Sokoloff (Pergamon, 1960), also has value, although unfortunately it is presented in a less logical form which to some extent detracts from its value.

Three volumes of papers presented at international symposia on geochemical prospecting contain a great deal of information which is both up to date and original. The first was published as *Paper of the Geological Survey of Canada*, **66/54** (1967), while the second appeared as *Quarterly of the Colorado School of Mines*, **64** (1) (1969). The third volume recording a meeting held in Toronto in 1970 has been published as *Special Volume* No. 11 by the Canadian Institute of Mining and Metallurgy.

Details of the analytical techniques employed in geochemical prospecting are found in a wide variety of publications. Methods of instrumental analysis (such as atomic absorption spectrophotometry) are often to be found in manufacturers' publications as well as in journals. Among the many books on atomic absorption spectrophotometry, *Atomic Absorption Spectrophotometry in Geology*, by E. E. Angino and G. K. Billings (Elsevier, 1969), and 'Atomic Absorption Methods of Analysis Useful in Geochemical Exploration', by F. N. Ward *et al.* (*Bulletin of the United States Geological Survey*, No. 1289, 1969), will be found to provide a useful introduction to the subject.

Although the use of colorimetric methods of analysis in geochemical analysis has declined, it is still a very useful approach to a number of geochemical problems. Two very useful and inexpensive books on this subject are 'Analytical Methods Used in Geochemical Exploration by the USGS', by F. N. Ward *et al.* (*Bulletin of the United States Geological Survey*, No. 1152, 1963), and *Rapid Method of Trace Analysis for Geochemical Applications*, by R. E. Stanton (Arnold, 1966).

The application of geophysical methods to the problems of exploration for mineral deposits is a very wide topic. Nevertheless, a number of excellent texts exist. Among the better-known titles of this sort are *Mining Geophysics*, by D. S. Parasnis (Elsevier, 1966); Jakosky's *Exploration Geophysics* (Trija Publishing Co., 1950); *Applied Geophysics in the Search for Minerals*, by A. S. Eve and D. A. Keys (4th edn., Cambridge University Press, 1956); and *Geophysical Exploration*, by C. A. Heiland (Hafner, 1968). A French book of considerable merit is *Prospection Géophysique*, by E. Rothé (2 vols., Gauthier-Villars, 1950, 1952), while a volume entitled *Applied Geophysics USSR*, edited by N. Rast (Pergamon, 1962), contains a collection of contributions by eminent Soviet geophysicists.

As a more specialist evaluation of the interpretation of geophysical data, *Interpretation Theory in Applied Geophysics*, by F. S. Grant and G. F. West (McGraw-Hill, 1965), is a particularly useful text.

Special mention should be made of a 2-volume monograph entitled *Mining Geophysics* prepared by the Society of Exploration Geophysicists. Volume 1 (1966) is devoted to case histories, while Vol. 2 (1967) concerns itself with the theory of geophysical prospecting techniques.

Methods of structural geology as applied to the problems of mineral exploration receive relatively scant attention in the literature, although many texts on structural geology contain some reference to such applications. A volume to mention in this connection, however, is *Structural Methods for the Exploration Geologist*, by P. C. Badgley (Harper, 1959), which consists essentially of a collection of problems illustrating cases in which structural methods can be used in exploration programmes.

The standard work on exploratory diamond drilling is undoubtedly the *Diamond Drill Handbook*, by J. D. Cumming (2nd edn., J. K. Smit and Sons, 1956). Other forms of drilling are poorly represented in the literature, although some information on these and other matters of mineral exploration will be found in *Mining Engineers' Handbook*, by R. Peele (2 vols., 3rd edn., Wiley, 1941), a work which has had many reprints.

The calculation of ore reserves, a problem that is increasingly being studied by economic geologists, is considered in *Mine*

Economics, by S. J. Truscott (3rd edn., Mining Publications, 1962); however, this topic is in the throes of a revolution of thought, and new volumes will almost certainly be appearing in the near future. For those interested in current thinking on this subject, a useful review of the techniques being developed will be found in 'Some Statistical Techniques for Analyzing Mine and Mineral-Deposit Sample and Assay Data', by S. W. Hazen (*Bulletin of the United States Bureau of Mines*, No. 621, 1967).

Two publications on photogeology, which plays an ever-increasing part in mineral exploration are *Photogeology and Regional Mapping*, by J. A. E. Allum (Pergamon, 1966), and *Photogeology*, by V. C. Miller (McGraw-Hill, 1961). The former is particularly useful as a relatively inexpensive introduction to the subject, while the second is notable for the number and quality of stereo-photographic illustrations that it contains.

At the time of writing there is no general textbook available on the methods and techniques of remote sensing, although, in view of the very great interest being shown in the subject, there can be little doubt that one will soon appear.

There are no journals devoted to mineral exploration as such, although a number of specialist publications are available, and virtually all of the journals listed in the previous section contain papers on the subject. Among the specialist publications, some deserve special mention. A French translation of the Russian journal *Razvedka i Okhrana Nedr* was published between 1958 and 1962 under the title *Prospection et Protection du Sous-Sol* and contains many papers on geochemical prospecting. Contributions on this topic are also to be found in *Geochemistry International*, which is an English translation of selected articles from the Russian journal *Geokhimiya* together with translated papers from other sources.

Geophysical prospecting has a number of journals devoted to it, of which *Geophysical Prospecting*, the official journal of the European Association of Exploration Geophysicists, is one of the most interesting. *Geophysics*, published by the Society of Exploration Geophysicists, contains contributions on a relatively wide range of geophysical topics, while *Geoexploration* is devoted to mining geophysics and similar applications.

Special mention must also be made of *Photogrammetric Engineering*, published by the American Society of Photogrammetry, which usually contains very valuable papers on both photogeology and remote sensing. Some papers on photogeology are also to be found in *Photogrammetria*, which is the official journal of the International Society for Photogrammetry, although this latter journal is more specifically concerned with photosurveying and aerial surveys.

COAL

In comparison with the complex nature of the geology of metal-liferous mineral deposits, that of coal deposits is straightforward. Nevertheless, it may be divided into a number of different aspects, and these are broadly reflected in the literature of the topic.

A number of introductory texts on the general geology of coal are available. At a somewhat elementary level *Mining Geology*, by A. Nelson (Thos. Wall and Sons, 1962), provides a useful introduction. *Coal Mining Geology*, by I. A. Williamson (Oxford University Press, 1967), also constitutes a satisfactory general text, although, like the volume by Nelson, it contains a considerable quantity of general geology in addition to its more specific subject matter. At a more advanced level *Coal*, by W. F. Francis (2nd edn., Arnold, 1961), is a particularly useful volume which covers the whole field of coal science, including coal chemistry.

The origin of coal is very much less controversial than is the case for metallic deposits. An excellent, if old, account of the widely accepted processes that lead to coal deposition will be found in *The Nature and Origin of Coal and Coal Seams*, by A. Raistrick and C. E. Marshall (English Universities Press, 1939). A work of particular importance to those wishing to study the nature of coal in some detail is the *International Handbook of Coal Petrography*, published in 1963 by the Centre National de la Recherche Scientifique (Paris) for the International Committee for Coal Petrology. It consists of a glossary of the terms used in the field, each being fully described and in many cases illustrated, and forms a most useful work of reference.

There is no up to date account of the geology of the world's coal-fields, although a number of texts, such as Williamson (see above), do give brief accounts of the major producing areas. This work also gives a reading list which enables references to descriptions of individual fields to be located. For a description of British coal-fields, *The Coalfields of Great Britain*, edited by Sir Arthur Trueman (Arnold, 1954), is still perhaps the most comprehensive source, while *Concealed Coalfields*, by L. J. Wills (Blackie, 1956), is concerned with the deeper occurrences, and particularly those of the English Midlands.

An outstanding publication which contains a great deal of up to date information on all aspects of the geology of coal is *Coal and Coal-bearing Strata*, edited by D. G. Murchison and T. S.

Westoll (Oliver and Boyd, 1968), which derived its material from the 13th International Geological Congress. On the wider aspects of coal and its uses, *Coal Science*, published in 1966 in the 'Modern Chemistry Series' of the American Chemical Society, is the outcome of a meeting held by the American Conference on Coal Science, organised by a number of scientific bodies in the USA.

Because of the very close association of the geology of coal with stratigraphy and sedimentology, many of the papers published on the subject are included in general geological journals. In addition, contributions on the subject are to be found in the *Transactions of the Institute of Mining Engineers* and in the German journal *Fortschritte in der Geologie von Rheinland und Westfalen*.

OIL AND NATURAL GAS

The geology of oil and that of natural gas can usually be regarded as being intimately related; indeed it is only with the increasing exploitation of natural gas as a domestic fuel that any separate texts on its geology have appeared, and most books on the geology of petroleum can be expected to contain information on natural gas. Although the origin of oil is now widely accepted as being essentially biogenic and, hence, related to the sediments in which it was deposited, the mobility of both oil and gas once formed introduces a major complication into the study of their geology. This factor, i.e. the behaviour of fluids within rocks, forms a considerable part of oil geology.

A number of excellent general texts on the geology of petroleum have been published, and these provide the reader with good background reading. Perhaps one of the most widely read introductory texts in this field is *Some Fundamentals of Petroleum Geology*, by G. D. Hobson (Oxford University Press, 1954), which gives a most satisfactory account of the underlying principles of the subject. Other general volumes include *Petroleum Geology*, by E. N. Tiratsoo (Methuen, 1951); *Geology of Petroleum*, by A. I. Levorsen (2nd edn., Freeman, 1962); and *Principles of Petroleum Geology*, by W. L. Russell (2nd edn., McGraw-Hill, 1960). A work more specifically devoted to the geology of natural gas is *Natural Gas*, by E. N. Tiratsoo (Scientific Press, 1967), which includes descriptions of the geology of individual natural gas fields.

The genesis of petroleum products is discussed at some length in *The Origin of Oil and Oil Deposits*, by M. E. Al'tovskii, Z. I. Kuznetzova and V. M. Shvets (Translation of Russian original, Consultants Bureau, 1961), while the nature of environments in which oil is commonly found was the subject of a symposium of the American Association of Petroleum Geologists, published under the title *Habitat of Oil* in 1958.

Exploration for oil is largely based on the location of suitable geological structures which might act as 'traps' for migrating petroleum. These are usually identified by means of geophysical methods which are described in many of the volumes on geophysical prospecting mentioned in the earlier section on exploration for metalliferous mineral deposits. In addition to these, *Introduction to Geophysical Prospecting*, by M. B. Dobrin (2nd edn., McGraw-Hill, 1960), is of particular interest since it is essentially concerned with those geophysical techniques that find application in the search for oil.

In recent years growing attention has been paid to the geochemistry of oil. Much of this work has been described in two volumes: *The Geochemistry of Oil and Oil Deposits*, edited by L. A. Gulyaeva (Translation of Russian original, Israel Program for Scientific Translations, 1964), and *Fundamental Aspects of Petroleum Geochemistry*, edited by B. Nagy and U. Colombo (Elsevier, 1967). The application of geochemical methods to exploration for oil is discussed in *Geochemical Methods of Prospecting and Exploration for Petroleum and Natural Gas*, by A. A. Kartsev *et al.* (University of California Press, 1959).

Investigations into the geological structures found at depths are of particular importance in petroleum exploration. *Subsurface Geology in Petroleum Exploration*, edited by J. D. Haun and L. W. LeRoy, records the proceedings of a symposium held on the topic and was published by the Colorado School of Mines in 1958. A useful contribution to this subject is also to be found in *Handbook of Subsurface Geology*, by C. A. Moore (Harper and Row, 1963).

Other practical considerations relating to oil exploration are to be found in *Formation Evaluation*, by E. J. Lynch (Harper and Row, 1962), which is concerned with the evaluation and identification of strata encountered during oil drilling, and *Oil Property Valuation*, by R. V. Hughes (Wiley, 1967).

Papers describing activity in the general field of oil geology are mostly published in the journals of professional bodies devoted to the subject. Foremost among these perhaps is the *Bulletin of the American Association of Petroleum Geologists*, although other important publications include the *Journal of the Institute of Petroleum* and the *Bulletin of the Alberta Society of Petroleum Geologists*. Worthwhile original articles are also to be found in *The Journal of Petroleum Technology, Petroleum* and *World Petroleum*.

INDUSTRIAL MINERALS

Notwithstanding the very considerable economic importance of the industrial minerals (such as asbestos, mica and talc), the fact that many of these are found in relatively 'normal' rocks leads to a paucity of literature on the subject. Some volumes on the geology of mineral deposits, such as *Metallic and Industrial Mineral Deposits*, by C. A. Lamey (McGraw-Hill, 1966), do contain some information on these minerals, but specific texts are rare. There are, however, two reference works which are of particular value. One of these consists of a series of contributions by eminent authors, in which each particular mineral has its own section. This book, *Industrial Minerals and Rocks* (published in the Seely W. Mudd Series by the American Institute of Mining, Metallurgical and Petroleum Engineers, 1960), forms a particularly useful reference tool, and is of a scope that is somewhat wider than just the minerals described in this section, as it also discusses topics such as the economics and strategic character of each group of minerals. More specifically concerned with the geology of these minerals is *Geology of Industrial Minerals and Rocks*, by R. L. Bates (Harper, 1960).

Gemstones might also be included in the category of industrial minerals since their greatest use is found in industry. Books on gemmology, however, have been mentioned on p. 298.

Relatively few papers are published on industrial minerals, and of those that are, a large proportion appear in journals such as *Mine and Quarry Engineering* and *The Mining Magazine*. Geological surveys throughout the world also publish memoirs devoted to specific industrial minerals of interest in their own country.

CONSTRUCTIONAL MATERIALS

Under this heading are included the geology of building stones, brick and ceramic clays, cement, gypsum (for plaster) and so forth. As with industrial minerals, the literature on the subject is relatively scant. However, information may sometimes be supplemented by contacting such bodies as the British Ceramic Research Association at Stoke-on-Trent, which not only act as research centres but in many cases also publish their findings. It must be appreciated, however, that much of the work of such bodies is only available to contributing organisations.

The UK is notable for the wide variety of building materials used throughout it, and particularly local building stones. An interesting and largely pictorial review of this topic is to be found in *The Stones of Britain*, by B. C. G. Shore (Leonard Hill, 1957). Although this volume is not mainly concerned with geology as such, it contains a great deal of most interesting information.

Both of the books referred to in the previous section contain some information on building materials, particularly those more akin to industrial minerals, such as gypsum. For a general review of the applications and properties of clays, *Applied Clay Mineralogy*, by R. E. Grim (McGraw-Hill, 1962), is of great value, as is *The Geology and Mineralogy of Brick Clays*, by P. S. Keeling (Brick Development Association, 1963), which is a small but useful text. A similarly useful small book is *Limestone as a Raw Material in Industry*, by F. P. Stowell (Oxford University Press, 1963).

Very few papers on constructional materials are published outside the trade journals, and these are seldom concerned with the geology of the deposits in question.

WATER SUPPLY

The geology of water supply is one of the fastest-growing aspects of applied geology, and the view is held by many that it will soon become one of the most important. Both the need to produce more food on a world-wide scale and the development of underdeveloped countries in many cases depends upon meeting a demand for water, while even the USA is finding that a shortage of water is already becoming a source of apprehension in some regions. Numerous texts on the subject have appeared over the last decade. However, an old but still useful introduction will be found in Dixey's *Practical Handbook of Water Supply* (Murby, 1931). Somewhat more recently *The Geology of Water Supply*, by C. S. Fox (Technical Press, 1949), and *Water Resources Engineering*, by R. K. Linsley and J. B. Franzini (2nd edn., McGraw-Hill, 1964), have provided comprehensive reviews of the techniques used in the discovery and supply of water. A volume of use for its quantitative approach to the problem of water supply is *Groundwater Resource Evaluation*, by W. C. Walton (McGraw-Hill, 1970). More theoretical approaches are taken in books on the flow of water in the ground, although these provide essential reading for those interested in water supply. Details of some of these together with a discussion of the periodical literature of the field are given in Chapter 15.

ENGINEERING GEOLOGY

Geology plays a number of parts in engineering. Perhaps the most obvious is in the geological study of engineering sites to provide information on the suitability of a location with regard to foundations, freedom from landslips, drainage, fault movement and so forth. In addition, however, geological information can be of great use in determining the suitability of materials for construction purposes (partly covered in an earlier section), while the application of geohydrology to dam construction is obvious.

A number of texts have been produced which are concerned with general geology but are written especially for the engineer. Among the best-known of these are: *Geology for Engineers*, by F. G. H. Blyth (2nd edn., Arnold, 1945); *Geology for Engineers*, by J. F. Trefethen (2nd edn., Van Nostrand, 1959); and *Elements of Engineering Geology*, by J. E. Richey (Pitman, 1964). General accounts of the application of geology to the problems of engineering are also available. Among these are: *Geology and Engineering*, by R. F. Leggett (2nd edn., McGraw-Hill, 1962); *Application of Geology to Engineering Practice*, edited by S. Paige (Geological Society of America, 1950); *Geology in Engineering*, by J. R. Schultz and A. B. Cleaves (Wiley, 1955); and *Basic Geology for Engineers*, by J. Bundred (Butterworths, 1969). A volume which makes interesting reading is *L'Application de la Géologie aux Travaux de l'Ingénieur*, by J. Goguel (Masson, 1959), while *Principles of Engineering Geology and Geotechnics*, by D. P. Krynine and W. R. Judd (McGraw-Hill, 1957) is a particularly informative volume.

Although soil mechanics is not strictly part of engineering geology, some books on engineering geology do contain limited information on the topic. However, it has a large literature of its own, and although this lies outside the scope of this chapter, two references of interest can be quoted. One of the most widely read books on soil mechanics, which would form an excellent introduction to the subject, is *Fundamentals of Soil Mechanics*, by D. W. Taylor (Wiley, 1948), while *Soil Mechanics: Selected Topics*, by I. K. Lee (Butterworths, 1968), contains a review of some modern concepts in this field.

Titles of a more specific nature include *Rock Mechanics*, edited by C. Fairhurst (Pergamon, 1963), which contains papers presented at the 5th Symposium on Rock Mechanics, and *Landslides and their Treatment*, by G. Zaruba and U. Mencl (Elsevier, 1969). Mention must also be made of the excellent *Applied Geophysics for*

Engineers and Geologists, by D. H. Griffiths and R. F. King (Pergamon, 1965), which discusses the role that geophysics can play in solving some of the problems encountered during the application of geology to engineering.

The number of journals devoted to the subject of engineering geology grows annually. Among those most widely read are *Engineering Geology* and the relatively new *Quarterly Journal of Engineering Geology*. *Géotechnique* (Institute of Civil Engineers, London) is devoted mainly to soil mechanics, while *International Journal of Rock Mechanics and Mining Sciences* is a self-explanatory title. In addition to these, many other journals carry articles on engineering geology. Among these are virtually all of the general geological journals, together with a number of publications devoted to engineering. Such titles include *Engineering Journal, Proceedings of the Institution of Civil Engineers* and *Civil Engineering.*

ABSTRACTING AND INDEXING SERVICES

It has already been pointed out that in a chapter of this length it is obviously impossible to mention all those titles which carry potentially useful information in the field of applied geology. Details of additional books can be located in the various guides referred to in Chapter 4, and both books and periodical articles are abstracted in journals such as *Chemical Abstracts, Bibliography and Index of Geology, Geophysical Abstracts* and *Bibliographie des Sciences de la Terre*. Of particular importance in the applied field are *Annotated Bibliography of Economic Geology* (1928–66) and *IMM Abstracts* (1950–   ), published by Economic Geology Publishing Company and the Institution of Mining and Metallurgy, respectively. The first appeared twice a year (some 3 years after the literature to which it refers) and contained approximately 2000 abstracts per year from some 300 English and foreign-language periodicals. A cumulative index covers the first 25 volumes. *IMM Abstracts* appears bi-monthly and abstracts over 2000 items per year from the world literature (600 journals + conference papers) on economic geology, mining and processing of minerals (except coal); non-ferrous extractive metallurgy, including refining; and allied subjects. Entries are roughly 50 words long and are arranged according to the Universal Decimal Classification.

In 1970 a new abstracting publication, *Geotechnical Abstracts*, was launched by the German National Society of Soil Mechanics and Foundation Engineering. It appears monthly and will announce almost 2000 items per year from over 500 journals and other publications. Abstracts are available in an ordinary paper edition or on cards. For those who wish to use the latter as the basis of an information retrieval system, Geodex International offers a set of appropriately coded peek-a-boo index cards.

Also started in 1970 was *Rock Mechanics Quarterly Abstract Bulletin*, covering about 2500 references per year and published by the American Institute of Mining, Metallurgical and Petroleum Engineers and Imperial College of Science and Technology, University of London. This abstract bulletin is Part 2 of a three-part information service. Part 1 is a *KWIC Index of Rock Mechanics Literature* for the period 1870–1969 and Part 3 is a computerised Rock Mechanics Literature Search Service which will provide a complete list of references on a given subject.

A number of other abstracting and indexing services, although prepared primarily for practitioners in other fields, are of partial interest to applied geologists.

The most general of these is *The Engineering Index Monthly*, which contains abstracts of articles in over 2000 journals and details of selected monographs, books and conference proceedings. Its value to the geologist can be gauged from the fact that the current list of journals scanned contains 316 earth sciences publications, including 54 in the mining field. The monthly issues contain an average of 5500 abstracts arranged under alphabetical subject headings and subheadings selected from an authority list entitled *Subject Headings for Engineering* (SHE). This list contains 16 000 terms and familiarity with the vocabulary significantly aids searching efficiency. An annual cumulation of the abstracts in the monthly issues is available under the title *The Engineering Index Annual* (1885–   ). This contains a complete list of the journals covered as well as an author index. *The Engineering Index Monthly* data base is on magnetic tape (COMPENDEX) and copies of these tapes are available to organisations who might wish to use them as the basis of a mechanised computer retrieval service.

Articles in Civil Engineering (ACE), published monthly since January 1969 by the University of Bradford Library, also covers some applied aspects of geology. It contains about 250 references a year on engineering geology, foundation engineering and mining and quarrying. Entries (bibliographic details only) are arranged in broad subject groups and the purpose of the publication is to provide a current awareness service. The average time delay between publication of the original article and its appearance in ACE is 2–3 months.

Other services in the civil engineering and related fields which contain material on soil and rock mechanics, engineering geology, building materials, etc., are *Highway Research Abstracts* (1931–), a monthly produced by the US Highway Research Board; *HRIS (Highway Research Information Service) Abstracts* (1968–), a quarterly issued by the same body; *Highways—Current Literature* (1921–), a weekly index containing over 7000 references a year and published by the US Department of Transportation; and *Building Science Abstracts* (1928–), a monthly guide to the world's journal, report and conference literature, produced by the UK Building Research Station and containing over 2000 abstracts per year.

There are several guides concerned with fuels which include information on the natural occurrence and winning of petroleum, natural gas and coal. Among these is *Fuel Abstracts and Current Titles* (1960–), produced by the UK Institute of Fuel and issued monthly. Abstracts (over 8000 per year) are approximately 90 words long and the publication covers the contents of 800 journals as well as conference proceedings and reports. *Gas Abstracts* (1945–), produced by the American Institute of Gas Technology, is also issued monthly and covers some 2000 items per year from over 150 journals. Monthly and annual subject and author indexes are available. Articles of interest to applied geologists are to be found mainly in the section entitled 'Production, Supply, Processing'. *Petroleum Abstracts* is published by the Information Services Department of The University of Tulsa and covers over 600 journals, patents, specifications and conference papers. Approximately 10 000 informative abstracts per year appear in weekly issues, which are designed to be cut up to produce index cards. Monthly subject and author indexes with semi-annual cumulations are available. A magnetic tape search service is also offered by the publishers. *Petroleum Abstracts* is somewhat unusual in that subscribers are charged a price which reflects assets. A company with over 1.5 thousand million dollars assets, for instance, is required to pay over $10 000 per annum. A relatively small but useful contribution to the bibliographic control of literature on coal is the UK National Coal Board publication entitled *NCB Abstracts C. Coal and Mining Geology* (1952–). Covering 170 journals and issued gratis every 2 months, the publication contains about 300 abstracts per year. From 1952 to 1958 it was entitled *Coal Measures—Stratigraphy and Microscopy* and from 1958 to 1964 *Carboniferous—Stratigraphy, Palaeontology and Related Literature*.

14

Geophysics

J. W. Clarke

Geophysical information is disseminated within the profession by means of printed texts and illustrations in journal articles and books, by maps that present data through contours, by computer tapes, and by diagrams such as well logs. Each of these communication media fulfils its particular purpose. Most of the geophysical literature is available in any large university library, and the rest can generally be obtained through interlibrary loan.

Although most of the geophysical literature has been published during the last 50 years, many records from the historical past have provided reliable data; this is particularly true for information on such phenomena as earthquakes of China and volcanic activity in the Mediterranean region.

The application of geophysics to exploration for petroleum and mineral resources has justified the commitment of significant expenditure on basic research, and the world-wide nature of many geophysical phenomena has led to remarkable international research programmes. The most notable of the latter have been the International Geophysical Year, the International Year of the Quiet Sun and the Upper Mantle Project.

Strong, well-managed professional organisations have provided the control and inspiration for work of the highest level. The most active groups are the International Union of Geodesy and Geophysics, the American Geophysical Union, the Society of Exploration Geophysicists and the European Association of Exploration Geophysicists. These organisations hold technical meetings, organise symposia, sponsor research and publish on an extensive scale. Their memberships account for most of the recipients of the journals published by each group and this is a crucial factor in making publication of these journals feasible.

BASIC WORKS IN GEOPHYSICS

The following listing was compiled in collaboration with senior research specialists in each field covered. Selection was guided by the question: 'Which books or papers should a practising geologist or geophysicist or advanced student read for basic knowledge in a field that is not his own speciality?' These works in turn carry extensive references that are especially reliable for purposes of acquiring a thoroughly grounded knowledge. In addition, the articles in good encyclopaedias should certainly be utilised. Although these short works are not directed towards the research scientists, they do provide a clear, basic framework for a knowledgeable approach to more technical presentations. The technical articles in the outstanding encyclopaedias have generally been written by scientists with long-established reputations for excellence in their fields.

Ages of Rocks, Planets and Stars, by H. Faul (McGraw-Hill, 1966)
An Introduction to Planetary Physics, by W. M. Kaula (Wiley, 1968)
An Introduction to the Theory of Seismology, by K. E. Bullen (Cambridge University Press, 1963)
Earthquakes and Earth Structure, by J. H. Hodgson (Prentice-Hall, 1964)
Electrical Methods in Geophysical Prospecting, by G. V. Keller and F. C. Frischknecht (Pergamon, 1966)
Interpretation Theory in Applied Geophysics, by F. S. Grant and G. F. West (McGraw-Hill, 1965)
Introduction to Geophysical Prospecting, by M. B. Dobrin (McGraw-Hill, 1960)
Mathematical Aspects of Seismology, by M. Bath (Elsevier, 1968)
Mining Geophysics (2 vols., Society of Exploration Geophysicists, 1966, 1967)
Mining Geophysics, by D. S. Parasnis (Elsevier, 1966)
Paleomagnetism and its Application to Geological and Geophysical Problems, by E. Irving (Wiley, 1964)
Physics and Chemistry of the Earth (a continuing series of volumes), edited by L. H. Ahrens, F. Press, S. K. Runcorn and H. C. Urey (Pergamon, 1959–)
Physics of Geomagnetic Phenomena, edited by S. Matsushita and W. H. Campbell (2 vols., Academic Press, 1968)
Physics of the Earth, by F. D. Stacey (Wiley, 1969)
Radiometric Dating for Geologists, edited by E. I. Hamilton and R. M. Farquhar (Interscience, 1968)
Rock Magnetism, by T. Nagata (2nd edn., Maruzen, 1961)
'Studies in Volcanology', edited by R. R. Coats, R. L. Hay and C. A. Henderson (*Memoir of the Geological Society of America*, **116**, 1968)
The Earth, by H. Jeffreys (4th edn., Cambridge University Press, 1959)
The Earth's Core and Geomagnetism, by J. A. Jacobs (Macmillan, 1963)

The Earth's Mantle, edited by T. F. Gaskell (Academic Press, 1967)
The Fundamentals of Well Log Interpretation, by M. R. J. Wyllie
(Academic Press, 1963)
The Physics of Glaciers, by W. S. B. Paterson (Pergamon, 1969)
Volcanoes and their Activity, by A. Rittman (Wiley/Interscience, 1962)

PERIODICALS

Although geophysical information appears in a wide range of earth
sciences journals, seven primary periodicals account for about 20
per cent of all articles. Details of these publications are given in
Table 14.1.

Two series are of use in connection with the international geophysical programmes mentioned earlier. These are the *Annals of
the International Geophysical Year*, published by Pergamon Press,
and the *Annals of the IQSY*, published by the Massachusetts Institute of Technology. The former comprises 48 volumes, the last of
which is a bibliography of publications produced in connection
with the programme. The *Annals of the IQSY* is a more modest
work consisting of seven volumes. Volume 6 contains a useful
bibliography.

The Upper Mantle Project (UMP), which was initiated in 1960
at the 12th General Assembly of the International Union of
Geodesy and Geophysics, covered the period 1962–70. During these
years and subsequently, over 40 *Upper Mantle Scientific Reports*
have been issued. The last (*Report* No. 41) has recently been published as a book entitled *The Upper Mantle*, edited by A. R.
Ritsema (Elsevier, 1972). The contents are the proceedings of the
final UMP Review Symposium, held in Moscow in 1971, and include a list of earlier *Reports*.

ABSTRACTING SERVICES

References to the articles in the above and many other journals
which carry geophysical information can be located in several
abstracting publications.

Foremost among these is *Geophysical Abstracts**, which began
publication in 1929 as part of the *US Bureau of Mines Information
Circular* series. Starting with issue *87*, responsibility for its production passed to the Geological Survey, and from 1937 to 1942 it was

---

* *Geophysical Abstracts* was discontinued at the end of 1971 and its
function taken over by *Bibliography and Index of Geology* (see Chapter 5).

Table 14.1. DETAILS OF THE MORE IMPORTANT GEOPHYSICAL PERIODICALS

Periodical	Language	Country and publisher	First issue	Frequency	Special features
Journal of Geophysical Research	English	USA (American Geophysical Union)	1896	3 issues per month	abstracts bibliography index
Geophysics	English	USA (Society of Exploration Geophysicists)	1936	bi-monthly	abstracts book reviews index
Bulletin of the Seismological Society of America	English	USA (The Society)	1911	bi-monthly	bibliography book reviews index cumulative index
Geophysical Journal	English	UK (Scientific Publications for the Royal Astronomical Society)	1958	10 issues per year in 2 volumes	abstracts book reviews index
Geophysical Prospecting	English French German	International (European Association of Exploration Geophysicists)	1953	quarterly	abstracts book reviews index
Izvestiya Akademiya Nauk SSSR—Fizika Zemli (translated into English as Izvestiya—Physics of the Solid Earth)	Russian	USSR (Academy of Sciences)	1965	monthly	book reviews index
Geomagnetizm i Aeronomiya (translated into English as Geomagnetism and Aeronomy)	Russian	USSR (Academy of Sciences)	1961	bi-monthly	book reviews index

published in the *Bulletin* series of the Survey. In 1943 the Bureau of Mines resumed production and numbers *112–127* were issued as *Information Circulars*. Beginning with number *128*, the publication was again issued as a *Geological Survey Bulletin*, four issues— A, B, C and D—to the bulletin. In 1963 the Survey started to publish *Geophysical Abstracts* as a completely separate serial issued 12 times per year. Author and subject indexes are available for all issues back to 1929 and cumulative indexes cover those issues produced under the auspices of the Bureau of Mines. Numbers *1–20* are indexed in their *Information Circular* No. 6438; *21–32* in No. 6589; *112–115* in No. 7273; *116–119* in No. 7310; and *124–217* in No. 7414. The complete series of *Geophysical Abstracts* from 1929 to 61 has been reprinted by the Johnson Reprint Corporation. Current issues of *Geophysical Abstracts* are available from the US Government Printing Office. The publication covers the world literature but includes abstracts only of material that is believed to be generally available, i.e. journal articles, books, conference proceedings, etc. Dissertations, open-file reports and memoranda are not normally covered. About 6000 items a year are abstracted and a list of journals commonly cited can be obtained from the US Geological Survey, Washington D.C. 20242. *Geophysical Abstracts* is currently prepared for publication by automatic data processing. The magnetic tapes used for this compilation also constitute a data bank for automatic search of the bibliographic data, abstracts and index terms. This file contained more than 30 000 references at the end of 1970; however, the search system was still in the experimental stage at that time.

Although only a small proportion of its contents is concerned with geophysics, another useful guide to the literature is *Physics Abstracts*. This is part of *Science Abstracts*—a publication which commenced in 1898 and is now available in three sections: A. *Physics Abstracts,* B. *Electrical and Electronic Abstracts* and C. *Computer and Control Abstracts*. *Physics Abstracts* covers 20 000 periodical articles, books, dissertations, reports and patents per year. Section 20 deals with geophysics and gives details of nearly 4000 items per year. Each 2-week issue is accompanied by an author, book, bibliography, conference, patent and report index, and semi-annual author and subject indexes are available. Cumulative indexes covering longer periods have also been produced. *Physics Abstracts* is published by the Institution of Electrical Engineers, who also produce *Current Papers in Physics*, a twice-monthly list covering the same material as *Physics Abstracts* but without the summaries. References find their way into both journals within two or three months of their publication.

A third guide is the *Bibliography and Index of Geology* (see also Chapter 5), which covers approximately 3000 items per year in the geophysical field. Most of these are indexed in Section 19, 'Solid Earth Geophysics'.

Meteorological and Geoastrophysical Abstracts (see also Chapter 15) also contains many references to geophysical material.

For those with a special interest in Russian geophysical research a very useful guide to the literature is produced by the US Government's Joint Publications Research Service. Its title is *Soviet Bloc Research in Geophysics, Astronomy and Space.* It commenced publication in 1961 and approximately 2000 abstracts of journal articles and books appear in 24 issues per year.

A number of foreign-language abstracting journals also cover the world's geophysical literature. Chief among these is *Referativnyi Zhurnal* (see also Chapter 5). The *Geofizika* section is divided into five subseries available in a single volume or separately. These subseries are: (1) Geologicheskie i geokhimicheskie metody poiskov poleznykh iskopaemykh. Metody razvedki i otsenki mestorozhdenii. Razvedochnaya promyslovaya geofizika (geological, geochemical and geophysical prospecting); (2) Geomagnetizm i vysokie sloi atmosfery (geomagnetism and the upper atmosphere); (3) Meteorologiya i klimatologiya (meteorology and climatology); (4) Okeanologiya, gidrologiya sushi, glyatsialogiya (oceanography, hydrology and glaciology); (5) Fizika zemli (solid earth geophysics). *Geofizika* is issued monthly and author and subject indexes are available.

The French abstracting journal *Bulletin Signalétique* (see also Chapter 5) also covers geophysics. The relevant section is 120— *Astronomie, Physique Spatiale, Géophysique.* Geophysical references are listed in Part D under the following subheadings: (1) General features. Instrumentation and methodology; (2) Geodesy, photogrammetry and cartography; (3) Solid earth geophysics; (4) Glaciology; (5) Oceanology; (6) Atmosphere; (7) Meteorology; (8) Aeronomy and geomagnetic disturbances. About 12 000 abstracts per year appear under these headings, and monthly and annual subject and author indexes are available.

Geophysics literature is also covered by another French journal, *Bibliographie des Sciences de la Terre* (see also Chapter 5). Cahier F of this monthly publication carries the title 'Tectonique et Géophysique' and contains some 2800 abstracts per year in three sections—'Tectonique', 'Physique du Globe' and 'Géophysique Appliquée'. Each section has its own author, geographic and KWIC subject index.

An analysis of the contents of abstracting journals can reveal some interesting facts about the growth and origin of the literature of a subject. For example, a detailed study of *Geophysical Abstracts* reveals a startling growth rate in geophysical literature. The number of items cited at decade intervals has been as follows: 1930, 508; 1940, 578; 1950, 835; 1960, 2059; 1970, 5820. The great expansion between 1960 and 1970 is to some extent a reflection of increased government support for research in such subjects as magnetic field and crustal studies, and of extensive developments in the Soviet geophysical exploration programme.

The origin of the literature is revealed in *Table 14.2*, which is a breakdown by language of the citations in the 1970 issues of *Geophysical Abstracts.*

Table 14.2. LANGUAGE BREAKDOWN OF PAPERS ABSTRACTED IN *Geophysical Abstracts* (%)

English	61.0
Russian	26.0
German	3.4
French	2.8
Japanese	2.0
Ukrainian	2.0
Other	2.8

Although the over-all figures show that English is the most frequently used language in formal geophysical communications, a closer examination of the citations reveals that in certain fields this situation does not exist. For instance, 50 per cent of the articles on electrical exploration are in Russian and more than 60 per cent of the articles on seismic exploration are in Russian.

The volume of the geophysical literature is now such that no individual can reasonably be knowledgeable in all phases through reading the primary literature. Secondary literature such as abstracting journals can in part meet the need for current awareness if the abstracts are sufficiently informative. The abstract journal, however, cannot evaluate or synthesise information. The need thus continues to grow for a review literature. In this respect two review journals—*Advances in Geophysics*, produced annually by Academic Press since 1952, and *Reviews of Geophysics*, a quarterly published since 1963 by the American Geophysical Union—are particularly important contributions to the literature. Also of value in meeting the need for reviews is the increasing number of thematic symposia with accompanying publication of the proceedings. Such review volumes not only apprise the scientific community of the state of knowledge of a subject, but also commonly inject a new enthusiasm into the research effort of those participating in the symposium.

15

Hydrology, glaciology, meteorology, oceanography and geomorphology

K. M. Clayton

The wide definition of the earth sciences adopted for this book includes a number of fields that impinge on other subjects or are normally regarded as independent subject fields. Meteorology and oceanography are easily distinguished by their subject matter, although it is not obvious where to place such matters as the geology or the relief of the sea floor. Hydrology and glaciology are less distinct, as they may include not only the problems directly associated with ice or running water but also the associated erosive processes or the interrelationships with climate. Finally, geomorphology is obviously one of the earth sciences, although it is distinguished by the historical accident that in many countries most geomorphologists call themselves geographers. This pattern of allegiance affects the publication of geomorphological literature, which is very different in its distribution from the geological sciences *sensu stricto*.

Before we consider the literature of each of the fields with which this chapter is concerned a number of general points are worth making.

In the first place, although there is relatively little subject overlap between the various fields, there are relatively few journals so specialised that they include material from only one of the divisions. The overlap in terms of published sources is therefore more considerable. Very many journals include material in two or three of these subjects, while a few general journals regularly include articles from all the fields, often of course from all the other earth sciences too. These general journals include both the *Philosophical Transactions* and the *Proceedings of the Royal Society* (strongest on meteorology), the *Transactions of the American Geophysical Union* (particularly strong on hydrology and geomorphology) and *Journal of Geophysical Research* (which includes little glaciology or geomorphology). Concerning the last title, until 1969 the even-numbered parts generally included the material relating to the earth sciences, but since mid-1969 there has been a new 'Oceanography and Meteorology' division. Also of relevance to all the subjects in this chapter are *Nature* and *Science*. These have a number of papers each year in most of the five fields, although both are strongest on geomorphology and oceanography. Commonly the contributions are short reports of work, and these are often published in full at a later stage in specialist journals. *Science*, however, also has longer articles that survey current progress in particular fields.

Rather surprisingly, three American geological journals frequently contain articles on the subjects (other than meteorology) covered in this chapter. These are *Journal of Geology*, *American Journal of Science* and *Bulletin of the Geological Society of America*. In each case geomorphology is most strongly represented, but hydrology, oceanography and occasionally glaciology are also covered. Two other journals that may conclude this brief list of the more general sources are the *Professional Papers, United States Geological Survey* and the Swedish periodical *Geografiska Annaler, Series A*. The former covers geomorphology, hydrology and in some cases oceanography (usually ocean floors, especially the continental shelf of the United States). The latter covers geomorphology, hydrology and glaciology: nowadays the articles are always in English, although until recently French and German were also used.

The second point relates to the areal element to be found in the literature of the fields under discussion. In common with earth sciences literature as a whole, much of the journal literature discussed in this chapter is tied to geographical regions, although this is less so in the case of oceanography, where it is impossible to identify regional journals that cover particular oceans, and in meteorology. In the latter case, however, there are a few national journals which tend to have descriptive or analytical articles on local topics. The areal element is most highly developed in geomorphology, where perhaps half of the articles in any one journal are tied to the country in which it is published. This local literature forms a convenient quarry for the home area, but at times it is an awkward barrier when the area of interest is in a foreign land.

It should be noted that the local language will generally be used, which reinforces the problems of tackling a regional literature of a foreign area. Virtually all the articles on the geomorphology of the USSR are published in Russian and in Russian journals; and for similar reasons, without a knowledge of the appropriate language, one can undertake little detailed reading on the geomorphology of France or Germany. Paradoxically, the smaller or more remote countries tend to make more effort to gain a readership by publishing some articles in a major language or by publishing extended foreign language summaries. Thus English is adequate for the regional geomorphology of Poland, French for Romania and German for Finland. Apart from the language problem, access to regional information suffers from the poor availability of the local journals. Major reference collections or the resources of the National Lending Library in the UK mean that copies can be located, but they are not widespread. Sometimes particular universities will specialise in an area and so hold a good range of the relevant journals. A large proportion of the regional literature consists of memoirs and other government reports, and these can prove even more difficult to locate than the local journals.

The third and final point that must be stressed is the fact that the literature has a relatively long life. This contrasts with the situation in most other scientific fields, where much of the information contained in journals published 10 years ago has already been superseded. *Table 15.1* presents the 'three-quarter lives' (the period of time during which the most recent three-quarters of the literature in current use was published) of the literature of meteorology, etc. The comparative figure for scientific literature in general is 9 years.

Table 15.1. 'THREE-QUARTER LIVES' OF LITERATURE (IN YEARS)

Meteorology	12
Hydrology	12
Glaciology	13
Oceanography	16
Geomorphology	34

The continued value of the older material creates a further difficulty in the use of the literature, for in practice the only way to locate early sources of information is to work through bibliographies or citations in more recent articles. These are not selective and, as any research worker knows, a vast store of irrelevant and outdated material must be read before the critical factual material that is needed is thrown up. There are a few selective bibliographies, but they are not annotated and provide no more than an introduction to this early literature. No modern bibliographic tools include any of the earlier material.

The more recent literature is fairly well indexed, although general standards are somewhat below those in the main science subjects. Each of the five fields here discussed has something in the way of an indexing service in the English language, although the style, comprehensiveness and cost vary considerably.

Alongside this systematic coverage of the subject fields, there are a number of national or regional bibliographies that can be of great value when information is being sought on a particular area. Many of these are published annually by national geological surveys or other appropriate departments. Some of the most useful and most comprehensive come from such smaller countries as New Zealand, Bulgaria, Romania and Poland.

HYDROLOGY

Hydrology has only emerged as an independent sub-discipline in recent years. Until well into this century, it formed part of geomorphology within the earth sciences, and the older literature is scattered through the same sources as geomorphological material. Some aspects of hydrology, theoretical as well as practical, have always been of concern to engineers, and some of the best older material will be found in civil engineering publications or even aspects of applied physics (fluid mechanics). This split continues today, for a large part of the literature is found in publications that would not be classified as earth sciences.

Books

There are a number of specialised texts in this field and the following is a select list, roughly arranged in a sequence of increasing sophistication. The most advanced works tend to be highly mathematical.

> *Streams: Their Dynamics and Morphology*, by M. Morisawa (McGraw-Hill, 1968)
> *Principles of Hydrology*, by R. C. Ward (McGraw-Hill, 1967)
> *Fluvial Processes in Geomorphology*, by L. B. Leopold, G. M. Wolman and J. P. Miller (Freeman, 1964)

Applied Hydrology, by R. K. Linsley, M. A. Kohler and J. L. H.
Paulus (McGraw-Hill, 1949)
Open Channel Hydraulics, by Ven Te Chow (McGraw-Hill, 1959)

On more specialist aspects of hydrology the following are useful
general sources:

Introduction to Hydrometeorology, by J. P. Bruce and R. H. Clark
(Pergamon, 1966)
Hydrogeology, by S. N. Davis and R. J. M. de Wiest (McGraw-Hill,
1966)

Periodicals

Until comparatively recently, the applied and pure aspects of hydrology each maintained a very distinct literature. Study of publications shows that even the citation of articles from one group by
the other is rare, and there seems to have been little intellectual
contact. The emergence of hydrology as a separate sub-discipline
of the earth sciences has identified a sizable body of workers and
their associated publications from within the field of geomorphology. Several new journals, and an international association, have
emerged and these now attract a wide range of contributors. As a
generalisation, it may be suggested that the truly practical material
remains in the applied journals of civil or agricultural engineering
but the more theoretical material has converged on the new
journals.

The major new journals are *Water Resources Research*
(1965–), a publication of the American Geophysical Union, and
Journal of Hydrology (1963–), published by Elsevier and taking
much English material in the absence of a British journal in this
field. The International Association of Scientific Hydrology concerns itself with all aspects of water, including glaciers, but its
Bulletin contains a fair proportion of hydrological material. The
older, well-established journals are outside the field of earth
science; they include: *Civil Engineering, Proceedings of the Institution of Civil Engineers Journal of Fluid Mechanics, Military
Engineer, Proceedings of the American Society of Civil Engineers
Journal of the Irrigation and Drainage Division* and *Transactions
of the American Society of Mechanical Engineers*. A further element is found in journals concerned with such practical problems
as irrigation and soil erosion. These include *Journal of Farm Economics, Journal of Agricultural Research, Journal of Range
Management, Journal of Soil and Water Conservation, Journal of*

Forestry and *Pakistan Development Review.* There is an element in the core journals of soil science and, to a much smaller extent, of meteorology. The two main American soil journals, *Soil Science* and *Proceedings of the Soil Science Society of America*, publish a significant proportion of relevant material. In contrast, the equivalent English journal, *Journal of Soil Science*, is much more restricted in its range of articles.

Perhaps the most remarkable feature of the literature of hydrology is the dominance of workers and publications from the United States. Well over two-thirds of the English-language publications are from the US, and almost every major journal is American. The assembled publications of the American Geophysical Union (its *Transactions, Water Resources Research* and *Journal of Geophysical Research*), the American Society of Civil Engineers (its *Transactions* and *Proceedings*, and, in particular, *Journal of the Hydraulics Division*) and the US Geological Survey (in this field, its *Water Supply Papers, Professional Papers* and *Bulletin*) could form the basis of a very good collection of hydrological literature.

Groundwater (or geohydrology) remains much more the province of the geologist. Very many of the *Water Supply Papers, United States Geological Survey* are concerned with groundwater, as are some of the papers in most of the journals already mentioned. Many of the American states have groundwater sections with a separate publication series and these are particularly important in the drier states, e.g. Nevada. There is also *Journal of the American Water Works Association.* In the UK, groundwater remains at least in part the responsibility of the Institute of Geological Sciences, although prime responsibility for most aspects of hydrology rests with the Water Resources Board. It was only founded in 1964 and is not yet a major publisher.

Abstracting services and bibliographies

With regard to abstracting services and bibliographies, hydrology would seem to be the least well covered of the five fields under discussion. The coverage in *Meteorological and Geoastrophysical Abstracts* is not particularly good (currently 450 citations a year) and it is covered to only a very limited extent in *Geographical Abstracts B* (about 200 citations a year). However, the rise in interest in this field in recent years has already led to the founding of several new journals and improvements in the abstracting services are likely to follow. One useful new source is the series of national hydrological bibliographies being published under the auspices of the International Association for Scientific Hydrology, while a new major US publication is *Selected Water Resources Abstracts*, published semi-monthly since 1968 by the Water Resources Scientific Information Center of the US Department of the Interior. Coverage should approach 10 000 items a year. Finally there is *Hydata*, a monthly current awareness bulletin covering both periodical and non-periodical literature. It has been published by the American Water Resources Association since 1965 and includes about 9000 citations a year. The annual index is published under the title *Hydor*. The same organisation publishes *Water Resources Abstracts*, which gives abstracts of the articles previously announced in *Hydata*. Abstracts are available in loose leaf or card form and in 49 subject sections.

Hydrological data

One aspect of hydrology which should be mentioned here is the availability of run-off or water-balance data. The direct measurement of discharge in rivers involves the construction of special weirs and in most countries has, until recently, been restricted to sites being investigated or used for power generation or water supply. Consequently data are often only available for short-term periods or for scattered and relatively small catchments. There are the added disadvantages that the information is usually published annually on a national basis and has not been collected systematically by most libraries.

The International Association of Scientific Hydrology publishes sources for national hydrological data, and many countries are now realising the importance of adequate information on water resources and are collecting and publishing improved data. The steadily improving coverage of the *Surface Water Yearbook of Great Britain* is an example of this trend.

The US has the finest records of any country and a vast amount of data is published every year, together with periodic guides to its scope; thus the US Geological Survey, Office of Water Data Co-ordination, published in 1966 an *Index to Catalog of Information on Water Data* in two volumes, *Surface Water Stations* and *Water Quality Stations*.

There is no convenient source of data for larger areas than single countries, even for major rivers, and most compilations of this type are published as maps rather than statistical tables (e.g. the map *Superficial Drainage and Hydrology*, scale 1:10 million, compiled by Dr R. Beno and published in 1965 by Cartographia Vallalat, Budapest).

As part of the International Hydrological Decade programme most countries have appointed stations for the collection of data. These stations are listed in *List of Hydrological Decade Stations of the World* (Studies and Reports in Hydrology, No. 6, UNESCO, 1969).

GLACIOLOGY

Books

Glaciology is the smallest of the five sub-divisions discussed in this chapter, and this is reflected in the restricted scope of the journal literature and the very small number of books devoted wholly to it. Most general texts dealing with the Quaternary period or aspects of glacial geology have sections on glaciology, but they are not usually very extensive. *Glacial and Quaternary Geology*, by R. F. Flint (Wiley, 1971), provides a general introduction to glaciology and, although somewhat elementary, so does *Glaciers*, by R. P. Sharp (Oregon State System of Higher Education, 1960). A fuller treatment is to be found in *Glacial and Periglacial Geomorphology*, by C. Embleton and C. A. M. King (Arnold, 1968). The two-volume work by J. K. Charlesworth, *The Quaternary Era* (Arnold, 1957), contains a long section on glaciology and is useful as a bibliography of the literature prior to 1955 as it is essentially a compilation fully referenced to the original sources. One of the classic works is *Der Schnee und Seine Metamorphose*, by H. Bader *et al.*, published in Switzerland in 1939 and available in a translation published in 1954 and entitled *Snow and its Metamorphism*. This translation was published by SIPRE—the Snow, Ice and Permafrost Research Establishment of the US Army, Corps of Engineers. In 1961 this organisation became known as CRREL, the Cold Regions Research and Engineering Laboratory. Another valuable translation is *Principles of Structural Glaciology*, by the Russian author P. A. Shumskii (Dover, 1964). A recent book which includes a very long section on glaciology is *Traité de Glaciologie*, by L. Lliboutry (2 vols, Masson, 1964, 1965). It is not yet available in English.

Periodicals

The distinctiveness of water in the solid form has perhaps given a special character to the literature of glaciology. Most interesting is the absence of American domination of the English-language literature, although *Journal of Glaciology*, the leading journal, has an international editorial board and carries much material by authors from the USA. If the American concern with water resources is reflected in the balance of the literature, perhaps the American glaciological literature reflects America's few, small and scattered glaciers. In recent years US scientists have carried out a large amount of work in Antarctica and the results of this have been published in scattered existing journals, in special series of publications of the American Geophysical Union, in the US Army's Cold Regions Research and Engineering Laboratory (CRREL) publications, in such series as the publications of the Ohio State University's Polar Studies Institute and (for short notes on current work) in *Antarctic Journal of the United States.*

Other work in the Antarctic has been done by the Russians, the British and, to a lesser extent, scientists from Australia, New Zealand and Japan. Russian work is reported in *Informatsionnyi Byulleten—Sovetskoi Antarkticheskoi Ekspeditsii* and the smaller British research effort in *Polar Record* and *British Antarctic Survey Bulletin.* In the northern hemisphere a considerable amount of glaciological work has been done in Greenland and reports of this, partly in English, are to be found in *Meddelelser om Grønland* and in the American journals mentioned above.

Apart from *Journal of Glaciology*, the only English-language journal of major importance is *Geografiska Annaler A.* This has traditionally included quite a high proportion of glaciological material. Important foreign journals exist, notably the two German periodicals *Eiszeitalter und Gegenwart* and *Zeitschrift für Gletscherkunde und Glazialgeologie.*

The theoretical aspects of glaciology involve the physics of ice. Some fundamental material has been published in *Journal of Glaciology*, but much occurs in *Proceedings of the Royal Society of London, Nature, Journal of Applied Physics, Journal of Applied Mechanics, Journal of Colloid and Interface Science* and *Physical Review Letters.* These theoretical aspects of many of the sub-disciplines within the earth sciences tend to develop a distinct literature of their own and can often prove difficult to track down. Fortunately a convenient source is the *Science Citation Index*, where by use of one of the classic articles by Nye, or more recent articles by Lliboutry or Weertman, recent articles citing these sources can be located quickly. A selected list is also to be found in each issue of *Journal of Glaciology*.

Abstracting services and bibliographies

Glaciology is covered in part by sections in *Meteorological and Geoastrophysical Abstracts* and *Geographical Abstracts A*, but the fullest coverage is in the 'Glaciological Literature' section of *Journal of Glaciology*. This is a selective list of the relevant literature, sub-divided by subject headings. Most items have a one-sentence annotation. The total number of citations each year is about 650. They are organised under the following headings:

> General glaciology
> Glaciological instruments and methods
> Physics of ice
> Land ice. Glaciers. Ice shelves
> Icebergs. Sea, river and lake ice
> Glacial geology
> Frost action on rocks and soil. Frozen ground, permafrost
> Meteorological and climatological glaciology
> Snow

The journal is published three times a year by the Glaciological Society, Cambridge, England. Another useful guide to the literature of glaciology is the annual bibliography produced by CRREL under the current title *Bibliography on Snow, Ice and Frozen Ground*. Between 1951, when it started, and 1969 over 28 000 abstracts were published. The publication is available in hard copy or microfiche format. The interest of the American armed services in the cold regions is further reflected by the *Arctic Bibliography*. First published in 1953 by the Department of Defense, this annual guide to the literature is now prepared by the Arctic Institute of North America. Volume 14 was published by McGill-Queens University Press in 1969. Although it is not exclusively concerned with glaciology, there is much information on this and related subjects. A complementary publication entitled *Antarctic Bibliography* is published by the Library of Congress. Three volumes, published respectively in 1965, 1966 and 1968, have been produced, each containing 2000 references. A compilation covering 1951–61 was published in 1970.

The dividing line between glaciology, in the strict sense of the study of ice, and aspects of glacial (or periglacial) geomorphology is not at all clear. All textbooks on the subject go on to consider these geomorphological aspects of the behaviour of ice, and similarly *Journal of Glaciology* includes articles on glacial erosion, glacial deposits and glacial chronology. While the stratigraphy of areas glacierised in the Quaternary might seem to be simply a part of stratigraphical geology, the rather different nature of the deposits and the special problems of establishing stratigraphical units help to keep the literature distinct. The greatest overlap is to be found in such countries as the USA, Canada and The Netherlands, where the geological survey of the country is actively involved in mapping glacial deposits. Here the normal official memoirs and other series (such as the *Professional Papers, United States Geological Survey*) include much material on Quaternary deposits. Overlap is far less in countries such as the UK where the Quaternary deposits have been neglected by the official survey and rarely made the subject of special study. A number of quasi-regional journals devoted to the deposits and the history of the Quaternary carry much of the research output. Examples are *Quaternaria* (Italy), *Folia Quaternaria* (Poland), *Arctic and Alpine Research* (USA), *Revue de Géographie Alpine* (France) and *Quaternary Research* (Japan).

METEOROLOGY

It could be argued that meteorology is not an earth science, although in the widest sense of that term it must obviously be included. As the science dealing with the atmosphere, meteorology is readily distinguished from most of the other earth sciences, and articles are found in an equally distinct range of publications. Its literature has some distinctly 'scientific' aspects, citations commonly being of works less than 10 years old, with virtually no citations of works published before 1920 and, finally, the predominance of a few particularly important journals.

Books

A number of textbooks survey the field of meteorology and climatology and contain useful bibliographies. A recent British text is *Atmosphere, Weather and Climate*, by R. G. Barry and R. J. Chorley (Methuen, 1968), while *Viewing Weather from Space*, by E. C. Barrett (Longmans, 1967) introduces the new dimension brought about by satellite photographs. A very useful survey of part of the applied field is *Agricultural Meteorology*, by Jen-Yu Wong (Pacemaker Press, 1963). Summaries of progress of research have been published by the American Meteorological Society in their *Meteorological Monographs* series. Volume 3 (numbers 12–20, 1957) covered reviews of progress in nine fields for the period 1951–55.

Climatic diagrams and small-scale maps of climatic regions of continents or the world are to be found in many atlases. Major works of this type include the 'Soviet World Physical Atlas' (*Fiziko-Geograficheskii Atlas Mira*), the German series *World Atlas of Climatic Diagrams* and Elsevier's *Agricultural Climatic Atlas of Western Europe*. A very useful source is a volume in the series of encyclopaedias being produced by Reinhold, New York, *Encyclopedia of Atmospheric Sciences and Astrogeology*, edited by R. W. Fairbridge (1967). This includes about 300 brief (and some much longer) articles on meteorology and also has a 13-page summary of units, numbers, constants and symbols.

There are a number of dictionaries or glossaries in meteorology. They include: T. F. Malone's *Compendium of Meteorology* (American Meteorological Society, 1951); D. Brazol's *Dictionary of Meteorological and Related Terms* (Hachette, 1955); R. E. Huschke's *Glossary of Meteorology* (American Meteorological Society, 1959); and D. H. McIntosh's *Meteorological Glossary* (HMSO, 1963). The *Glossary of Meteorology* attempts to define every important meteorological term found in the literature today, with definitions that are 'understandable to the generalist and yet palatable to the specialist'. The definitions do not include bibliographic references. The *Meteorological Glossary* is a more modest publication but some of the explanations are longer and it will be found adequate for all but the most specialised topics. The World Meteorological Organisation has published an *International Meteorological Vocabulary* (1966) with lists of equivalent terms in English, French, Spanish and Russian, and definitions duplicated in English and French.

Periodicals

A number of meteorological articles, particularly those of a rather theoretical nature, are found in such 'general' journals as *Proceedings of the Royal Society*, *Nature* and *Journal of Geophysical Research*. However, far more articles appear in the recognised meteorological journals, notably *Quarterly Journal of the Royal Meteorological Society of London*, *Journal of Meteorology*, and

Journal of the Atmospheric Sciences. Other journals of major importance are *Tellus, Monthly Weather Review, Journal of Applied Meteorology* and *Meteorological Magazine.* Most countries have at least one meteorological journal in the national language. Examples are *Zeitschrift für Meteorologie* (Berlin), *Journal de Mécanique et de Physique de l'Atmosphère* (Paris) and *Meteorologiya i Gidrologiya* (Moscow). Other foreign journals such as *Tokyo Journal of Climatology* are published in English.

Articles of an 'applied' nature are an important aspect of the meteorological literature. These can range from irrigation and flooding, through agricultural meteorology to air pollution. Typical journals here are *International Journal of Biometeorology, Public Health, Air Pollution Control Association Journal, Atmospheric Environment, Journal of Applied Meteorology, Journal of Agricultural Meteorology* and *Rivista di Meteorologia Aeronautica.*

Abstracting and indexing services

Meteorology has the most ambitious and professional abstracting service of the five fields covered in this chapter—*Meteorological and Geoastrophysical Abstracts.* Dating from 1950 (up to 1960 it was called *Meteorological Abstracts and Bibliography*), it is published every month by the American Meteorological Society with the support of various US Government agencies. Covering reports, conferences and about 800 journals, it contains about 10 000 abstracts a year; and while it is strongest on meteorology (general and applied), there are also sections on oceanology (c.650 items a year), glaciology (c.150) and hydrology (c.450). Subject, author and geographic indexes are available in each issue and annually. G. K. Hall have recently published an 8-volume *Cumulated Bibliography and Index to Meteorological and Geoastrophysical Abstracts, 1950–1969.* Another regularly published guide to the literature of meteorology is the *Meteorological Office Library—Monthly Bibliography* (1919–   ). A duplicated list with entries arranged according to the UDC, it contains bibliographic details (but no abstracts) of about 10 000 items per year received by the Office Library. It is intended primarily as a current awareness tool and no cumulative indexes are issued.

Meteorological data

Another important aspect of the meteorological literature is the recording and publishing of data. This is a most complex procedure, which is exceedingly well organised as a result of the need to compile observations several times a day for forecasting purposes. The central organising body here is the World Meteorological Organisation, and apart from its data-gathering activities it also publishes *Technical Notes*. Inevitably, most of the observations collected are never published but are accumulated in national and local records.

However, the production of daily forecast maps by many national meteorological offices normally involves the compilation of basic data from a number of stations, and these daily maps consequently become an important archival source. The British Meteorological Office publishes two such maps each day, the *Daily Weather Report* (*DWR*) and the *Daily Aerological Record* (*DAR*—Upper Air Data). It also publishes long-term (monthly) forecasts, monthly reports on irrigation needs and so on. The equivalent US publication is the *Daily Weather Map*. Worth noting is that both the US and the British publications include summary charts of the whole of the northern hemisphere. In 1965 the World Meteorological Organisation published a *Catalogue of Meteorological Data for Research*, a loose-leaf work to which supplements will be added from time to time. This lists, principally by countries, the availability of synoptic and climatological data. The same organisation also sponsors (and the US Weather Bureau publishes) *Monthly Climatic Data for the World*. Other valuable collections of meteorological data include *Climatic Summary of United States to 1930* (3 vols., US Weather Bureau) plus supplements covering 1931–52 and 1951–60; 'World Weather Records' (*Smithsonian Miscellaneous Collections*, **79**, **90** and **105**); and the British Meteorological Office's publication *Tables of Temperature, Relative Humidity and Precipitation for the World* (HMSO, 1960).

OCEANOGRAPHY

Oceanography has many of the characteristics summarised in the introductory paragraphs of the section on meteorology, and even shares one or two of the same journals, notably *Tellus*. However, that part of the subject which involves the gradual collection of data on particular areas shows the persistence of early literature which is a feature of the earth sciences proper. Of the five subjects treated in this chapter, only geomorphology makes use of the nineteenth-century literature more frequently.

Books

The field of oceanography is extremely broad and effectively there has only been one attempt to tackle it in a single, comprehensive volume: *The Oceans*, by H. U. Sverdrup, M. W. Johnson and R. H. Fleming (Prentice-Hall, 1942). This is now largely of historical interest, although some sections still represent the best accounts available today. *The Sea*, edited by M. N. Hill (3 vols, Wiley, 1962–63), is a collective work designed to replace *The Oceans*. It is a valuable source, although rather incomplete and lacking the cohesiveness of the earlier work. Two major texts have been translated into English: *Allgemeine Meereskunde*, by G. Dietrich (Borntraeger, 1957), appeared as *General Oceanography* (Wiley/Interscience, 1963) and has a good bibliography of papers up to the mid-1950s; a translation of a work by A. Defand is *Physical Oceanography* (2 vols., Pergamon, 1960). A more modern text covering this aspect is G. Neumann and W. J. Pierson's *Principles of Physical Oceanography* (Prentice-Hall, 1966). A number of monographs cover more restricted topics and G. Neumann's *Ocean Currents* (Elsevier, 1968) is perhaps particularly noteworthy for its extensive bibliography.

For reference purposes one of the most useful books is *Encyclopedia of Oceanography*, edited by R. W. Fairbridge (Reinhold, 1966). It includes about 400 articles written by specialists and each has a brief bibliography. Among several dictionaries B. B. Baker's 'Glossary of Oceanographic Terms' (*Special Publication, US Naval Oceanographic Office*, No. 35, 1966), and *A Glossary of Ocean Science and Undersea Technology Terms*, edited by L. M. Hung and S. G. Groves (Compass Publications, 1965), have been found most useful. The latter includes over 3500 terms in fields including oceanography and marine sciences.

A collection of tables embodying basic relationships for the processing of raw oceanographic data is 'Handbook of Oceanographic Tables', by E. L. Bialek (*Special Publication, US Naval Oceanographic Office*, No. 68, 1966). These tables are grouped into four main divisions: (1) general mensuration information related to the oceans; (2) data on oceans not related to geography; (3) data on oceans related to geography; and (4) tables for computation and conversions.

Periodicals

A peculiar difficulty of oceanography is that reports on a particular area of the oceans can come from very scattered sources. There can be no 'local' literature such as is so characteristic of land areas. Some countries may traditionally carry out their research in particular oceans, but many of the more active visit all parts of the globe. Instead of the characteristic 'memoirs' of geological surveys, we find multivolume reports of particular expeditions, e.g. the *Challenger Reports* and the *Swedish Deep Sea Expedition Report*. Research organisations normally publish their own series of 'reports', such as *Johns Hopkins Oceanographical Studies, Publications of the Institute of Marine Sciences* and *Bulletin de l'Institut Océanographique*. An interesting feature of the literature in this field is the tendency for the major research institutions to publish their collected papers as a series. Examples are the *Collected Reprints* of the *National Institute of Oceanography* (UK), and the *Woods Hole Oceanographic Institution (US)*, and *Contributions, Scripps Institution of Oceanography*.

The major regular journals are *Deep Sea Research and Oceanographic Abstracts, Journal of Marine Research* and *Limnology and Oceanography*. *Deep Sea Research and Oceanographic Abstracts* is published by Pergamon Press with editors from the Woods Hole Oceanographic Institution, Massachusetts, and the National Institute of Oceanography, Surrey, England. This is the most important single source for bibliographical search (in English) in oceanography, and many of the references are cited with abstracts. It is perhaps worth noting that, rather like *Journal of Glaciology*, *Deep Sea Research and Oceanographic Abstracts* functions as an international journal with an editorial board drawn from many countries.

The main 'general' journals that include articles on oceanography are *Journal of Geophysical Research, Tellus, Nature, American Journal of Science, Science, Journal of Geology, Geophysical Journal of the Royal Astronomical Society* and *Bulletin of the Geological Society of America*. Although US geological journals frequently include sea-floor geophysics and other marine geology articles, the British geological journals tend to be more conservative. Recently, with the appearance of journals such as *Marine Geology*, new outlets have grown up for the rapidly expanding interest in the ocean floor. Nevertheless much of this material continues to appear in the more general journals such as *Science* and *Journal of Geophysical Research*. Other journals worth listing here are *Deutsche Hydrographische Zeitschrift, Journal of Fluid Mechanics* and *Progress in Oceanography*. The last is a review journal and its articles can form a convenient introduction to the current literature. Most issues cover a wide range of topics, but it is worth mentioning the review of major deep-sea expeditions, 1873–1960, by G. Wüst, that appeared in Vol. 2, 1964.

Much of the literature that is included under oceanography in the broadest sense cannot really be considered part of the earth sciences. There is, for instance, a large volume of material on fisheries and on other biological aspects of the ocean environment. However, data and other observations in this literature contribute to chemical and physical oceanography, so that journals in this field are often cited as sources. Typical examples from among a large number are *Oceanography and Marine Biology Annual Review*, *Journal of the Marine Biological Association* (UK), *Ecological Monographs*, *Journal de Conseil*, *Australian Journal of Marine and Freshwater Research* and *Bulletin of the Fisheries Research Board of Canada*. There are also a number of journals serving the specific field of chemical oceanography, such as *Analytica Chimica Acta* and *Analytical Chemistry*. One of the major Russian journals is *Okeanologiya*, a cover-to-cover translation of which is published in the USA under the title *Oceanology* (see Chapter 6).

Abstracting services and bibliographies

There is a section on oceanology in *Meteorological and Geoastrophysical Abstracts* with about 650 citations a year. A more complete service is in the journal *Deep Sea Research and Oceanographic Abstracts*. This covers about 2500 abstracts a year, most of them in English, but a few in the French or German of the original article. A third bibliographic source is *Oceanic Index* and *Oceanic Abstracts* (formerly *Oceanic Citation Journal*), published by the Oceanic Research Institute, La Jolla, California. The *Abstracts* lists over 12 500 items a year, usually with key-words, and these are cumulative indexed in *Oceanic Index* by a permuted key-word system. Another guide in this field is the current awareness publication *Marine Science Contents Tables*. Formerly entitled *Current Contents in Marine Sciences*, it has been issued monthly since 1966 as a joint service of FAO Fishery Resources Division and UNESCO Office of Oceanography. It reproduces the contents pages of about 70 leading oceanographic and marine biology journals. In addition to these serial bibliographies two one-off guides to the literature are available. The first is *Oceanography: A Report Bibliography*, by E. E. Thompson (CFSTI, AD 404 750, 1963) and the second is *Bibliography of Oceanographic Publications*, by M. W. Pangborn (Interagency Committee on Oceanography, 1963).

GEOMORPHOLOGY

Books

There are many books covering the field of geomorphology. *The Encyclopedia of Geomorphology* has been mentioned in Chapter 4 and serves as quite a useful bibliographic source. The standard advanced American text is *Principles of Geomorphology*, by D. W. Thornbury (2nd edn., Wiley, 1969), and part of the field is covered in a modern approach in *Fluvial Processes in Geomorphology*, by L. B. Leopold *et al.* (Freeman, 1964). A useful broad-ranging approach is to be found in *Environment and Archaeology*, by K. W. Burter (2nd edn., Methuen, 1972). The best bibliographically orientated guide to the history of the subject is the 3-volume study *The History of the Study of Landforms*, by R. J. Chorley *et al.* (Methuen, 1964–). So far only Vol. 1, *Geomorphology Before Davis*, has appeared. Occasional review articles appear in *Progress in Geography* started in 1969.

Periodicals

Very few journals are devoted entirely to articles in geomorphology, yet the field is a well-developed one with a considerable literature, perhaps 3000–4000 articles each year. The result of an absence of major specialised journals and the considerable volume of literature means that geomorphology is a microcosm of the earth sciences literature with very scattered sources. Add to this the wide range of languages involved in the local area-orientated literature together with the long threequarter-life (see *Table 15.1*) of geomorphological literature, and we have a difficult situation.

Two journals can claim to be international serials in this field, *Geografiska Annaler A*, and *Zeitschrift für Geomorphologie*. The latter, although published in Germany, also uses the French and English translations of the title, *Annales de Géomorphologie* and *Journal of Geomorphology*, and these are occasionally used in references. The *Zeitschrift* carries articles in German, English and French. Recently English has become a little more frequent and about half the current articles are in English. Brief summaries in the other two languages are usually given. *Geografiska Annaler A* (series B covers human geography) is now almost entirely in English, with very occasional articles in French. Like the *Zeitschrift*, it has an international editorial board. The title *Journal of*

Geomorphology was also used between 1938 and 1942 for an international English-language journal edited in the USA by D. W. Johnson. Despite several recent attempts, no American successor has yet emerged.

The major UK and US geological and geographical journals are an important source of English-language literature in geomorphology. The American sources carry more articles, largely because they publish far more articles a year than their English counterparts. Proportionately, however, more geomorphological articles are probably found in the English geographical journals, particularly the *Transactions of the Institute of British Geographers*. However, if the journals are ranked according to the number of geomorphological articles they publish, the order would be:

Bulletin of the Geological Society of America
Journal of Geology
American Journal of Science
Transactions of the Institute of British Geographers
Professional Papers, United States Geological Survey
Journal of the Geological Society
Proceedings of the Geologists' Association
Geographical Journal
Geological Magazine
Geographical Review
Annals of the Association of American Geographers

A number of other British and US sources are frequently referred to, but apart from incidental geological or geomorphological data, the actual articles they print are not really part of the literature of geomorphology. Examples are *Memoirs of the Geological Survey of the United Kingdom*, *Transactions of the American Geophysical Union*, and *Journal of Sedimentary Petrology*.

The distinction between what is of local interest and what is of more general (or 'systematic') interest is often made in geomorphology, not least by authors and the editors of journals. It is an attractive simplification to suggest that there are a number of new ideas of wide application that deserve wide circulation and discussion, while conversely much material is descriptive or analytical geomorphology, concerned with a local area, and of interest only to a local readership. The latter will then appear in a local language in a national (or local) journal, the former in a language such as English and in a journal taken by libraries all over the world. The contrast may be taken further by suggesting that the article of more general interest will concern itself with generalisations and, in so far as it adduces local evidence at all, will keep this to a minimum, or perhaps seek to relate observations from several different areas.

This division of the literature has not developed to any great extent in the UK or the USA, because the major journals already listed include both types of article. Thus some of the 'local' items receive as wide a circulation as those that are more general in their approach. The division has developed in smaller countries, although shorter versions of some of these 'local' papers do appear in places where they will receive a wider readership, e.g. the 'proceedings' of international conferences, journals such as *Zeitschrift für Geomorphologie* or special journals published largely in English and designed for a foreign readership. A good example of the latter is *Geographia Polonica*.

The great disadvantage of this arrangement, which seems so sensible and convenient at first sight, is that geomorphological concepts cannot usually be presented effectively without the detailed field evidence which gave rise to them. The interpretation of this evidence is not straightforward, so that, shorn of their detailed observational support and of the local facts on which the argument is based, the general article seems unconvincing and makes less impact than it should. There are signs that this difficulty is being recognised, and in many cases articles are now longer and include more local detail. This has been particularly true of *Geographia Polonica*, which now includes many full translations of articles already published in Polish. It is also true of countries such as Sweden or Japan, which have traditionally published an appreciable proportion of their literature in English and where articles full of local detail are available. Many of these articles are published in university series such as *Lund Studies in Geography (Series A)*, *Upsala Physical Geography Publications* and *Bulletin of the Geological Institution of the University of Upsala*. Japanese journals with an appreciable geomorphological content in English include *Science Reports, Tohoku University, Series 7, Geography* and *Japanese Journal of Geology and Geography*. Israel publishes the *Israel Journal of Earth Sciences*.

Alongside this ready availability of literature from some countries must be set the lack of material, other than publications in the local language, for many countries. In some cases this is offset by the publication of summaries or national abstracts such as those available for Rumania and Bulgaria, and (until 1966) China. In other cases material is often available in French or German; much Finnish material, for instance, is published in German. But there are some countries that do virtually nothing to aid the dissemination of their geomorphological literature abroad. This is particularly true of the USSR and this is most serious, because a very large proportion (about 30 per cent of the world's literature) is published in Russian. It is true that a few journals are published abroad in cover-to-cover translation, such as the *Proceedings of the Academy of Sciences of the USSR, Earth Science Sections*, but the major geographical journals (*Izvestiya, Akademiya Nauk SSSR, Seriya Geograficheskaya*; *Vestnik, Leningradskogo Universiteta, Seriya Geologii i Geografii*; and *Voprosy Geografii*), which contain most of the geomorphological articles, are not systematically translated. However, a number of the articles in these journals do appear in *Soviet Geography: Review and Translation* and a few more items can be found in *International Geology Review*.

Other countries that have inadequate bibliographies and little available outside the national language are Hungary and Czechoslovakia (however, French, German or English abstracts are usually printed with the articles) and the Latin American countries. So far most of the Latin American countries publish very little in this field, although the Brazilian journal *Revista Geografica* has recently been much improved and should provide a useful foreign outlet for that country.

Two other countries that publish mainly in their national language remain to be discussed, France and Germany. One French journal, *Revue de Géomorphologie Dynamique*, attracts an international range of authors and publishes in English as well as French, although the latter is the usual language. There are some geomorphological articles in *Revue de Géographie Physique et de Géologie Dynamique* and material on France can also be found in *Comptes Rendus de l'Académie des Sciences Paris* (*Série D*) and *Bulletin du Service de la Carte Géologique de la France*. Other important sources are the several regional journals published from various provincial universities. These include *Revue de Géographie Alpine, Méditerrannée, Norois*; *Revue Géographique de L'Est*; and *Revue de Géographie des Pyrénées et du Sud-Ouest*. The German literature, apart from *Zeitschrift für Geomorphologie*, which has already been noted, appears in quite a large number of regular serials and university series. The regular journals include *Erdkunde* and *Die Erde*. *Petermanns Geographische Mitteilungen* was revived in 1948 and is approximately equally divided between physical and economic geography. Specialised journals include *Meyniana, Die Küste, Zeitschrift für Gletscherkunde und Glazialgeologie, Kölner Geographische Arbeiten* and *Münchner Geographische Hefte*. One useful publication for the English reader is *Mundus*, edited by Dr Jürgen Holnholz at Tübingen. This quarterly journal reviews, in English, German books and journal articles in a wide range of subjects, including geomorphology.

For the main English-speaking countries it will probably be most helpful to give a list of the main journals that include geomorphological material. This list is presented as an Appendix to this chapter. It will be noted that outside the USA most of these are primarily geographical.

Some specialised branches of geomorphology have their own journals. The most remarkable example is limestone scenery (karstic landforms), which is covered by a range of professional journals and many rather amateur caving journals. In the UK there are perhaps as many as 30 or 40 of these, and in the USA many individual states have their own publications. Indeed, there are few countries that do not have at least one such journal, such as *Die Hohle* (Austria), *Helictite* (Australia) and *Ceskoslovensky Kras*. A relatively new journal with an international range of authors is *Studies in Speleology*, published in England by the William Pengelly Cave Research Association. The more important British caving journals include *Proceedings of the University of Bristol Speleological Society, Journal of the British Speleological Association* (formerly *Cave Science*) and *Transactions of the Cave Research Group of Great Britain*. The most important American journal is *Caves and Karst*.

Another strongly specialised field is that of periglacial geomorphology. Here the main journal is *Biuletyn Peryglacjalny*, edited by Jan Dylik in Poland. It is mainly in English but includes articles in French and German and normally appears once a year. Topics involving permafrost appear in such journals as *Problemy Severa* (translated in Canada as *Problems of the North*); *Arctic*; the publications of the Division of Building Research (National Research Council of Canada); and *Canadian Geotechnical Journal*. Since they are of particular interest to many continental geomorphologists, the European literature also contains a good deal on periglacial topics. Some relevant material is listed in *Ground Water in Permafrost Regions. An Annotated Bibliography*, by J. R. Williams (US Geological Survey, 1965).

Another field with several specialised journals is coastal geomorphology. *Shore and Beach* is a rather glossy journal aimed at the general public and concerned particularly with coastal defence and coastal conservation. More academic is *Coastal Research Notes*, a duplicated journal published in the USA. Much detailed work is reported in the *Technical Memoranda* and other publications of the US Army Coastal Engineering Research Center. The German journal *Die Küste* is published from Kiel and tends to specialise on the local North Sea coast, while the Danish publication *Folia Geographica Danica* frequently contains articles on areas such as the Wadden Sea. A bibliography in this field is *Selected Bibliography of Coastal Geomorphology of the World*, by J. T. McGill (1960), and the Commission of Coastal Geomorphology of the International Geographical Union covers the field with a publication every 4 years. *Bibliography 1963–1966*, edited by A. Guilcher and A. Morec (Centre National de la Recherche Scientifique, 1968), is the most recent.

The final example of a specialised sub-field within geomorphology is one that overlaps with stratigraphical geology: the study of the Quaternary period (see also Chapter 9). This is a field of study that has expanded rapidly and there are a number of specialised journals. The German journal *Eiszeitalter und Gegenwart* is one of the longest-established. Publication in English has lagged behind, but a new journal, *Quaternary Research*, commenced in 1970. The International Association for Quaternary Research holds congresses every 4 years. The 8th and most recent was in Paris in 1969.

It would be possible to mention many other journals restricted to particular topics, or more commonly primarily concerned with particular areas. But enough has been included to demonstrate the very scattered nature of geomorphological material and the special problems of many 'local literatures'. It was these problems which led to the founding of an abstracting journal in this field and it is to be hoped that *Geographical Abstracts A* (see below) provides an adequate guide for the student of geomorphology.

Abstracting services and bibliographies

Fortunately three abstracting services include geomorphology: *Referativnyi Zhurnal*; *Bulletin Signalétique*; and *Geographical Abstracts A, Geomorphology*. However, in all cases they offer a better guide to the contemporary literature than to the range of literature that is still in use, and the older material has to be located through the citations at the ends of other articles. *Geographical Abstracts A* currently publishes about 2200 abstracts a year. This is one of the four parts of the monthly *Geographical Abstracts*, published by Geo Abstracts, University of East Anglia. The Russian abstracting journal has a special part for geomorphology with nearly 3000 abstracts a year. As this total suggests, the coverage is very good, although a good deal of marginal and ephemeral material is included.

There are three bibliographies without abstracts (or adequate indexing) which cover geography and include quite a lot of geo-morphological material. These are the American Geographical Society's *Current Geographical Publications* (1938–), the Royal Geographical Society's *New Geographical Literature and Maps* (1918–) and *Bibliographie Géographique Internationale* (1891–).

A more specialised publication of use primarily to those with an interest in karst geomorphology is *Speleological Abstracts* (1962–). It is produced annually by the British Speleological Association and contains approximately 800 abstracts of papers from about 150 British journals, and details of books, monographs and newspaper articles.

National geographical or geological bibliographies are also useful sources in geomorphology and a number have been published. The UK literature is covered in *A Bibliography of British Geomorphology*, by K. M. Clayton (George Philip, 1964), and this seems to be the only example of a national bibliography restricted to this field.

APPENDIX: PERIODICALS FROM THE MAIN ENGLISH-SPEAKING COUNTRIES, CONTAINING GEOMORPHOLOGICAL MATERIAL

UNITED KINGDOM AND IRELAND
Amateur Geologist
East Midland Geographer
Essex Naturalist
Geographical Journal
Geography
Geological Magazine
Irish Geography
Journal of the Geological Society
Mercian Geologist
Proceedings of the Cumberland Geological Society
Proceedings of the Geologists' Association
Proceedings of the Prehistoric Society
Proceedings of the Royal Irish Academy
Proceedings of the Ussher Society
Scottish Journal of Geology
Transactions of the Cardiff Naturalists' Society
Transactions of the Institute of British Geographers
Transactions of the Norfolk and Norwich Naturalists' Society
Transactions of the Suffolk Naturalists' Society

UNITED STATES
American Antiquity
American Journal of Science
Annals of the Association of American Geographers
Arctic and Alpine Research
Bulletin of the Geological Society of America

Geographical Review
Journal of Geology
Ohio Journal of Science
Proceedings of the Indiana Academy of Science
The Professional Geographer
Professional Papers, United States Geological Survey
Southeastern Geology
Tebiwa

CANADA

Albertan Geographer
Arctic
Canadian Geographer
Canadian Journal of Earth Sciences
Maritime Sediments
Ontario Geography
Papers of the Geological Survey of Canada
Revue de Géographie de Montreal

AUSTRALIA

Australian Geographer
Australian Geographical Studies
Australian Journal of Science
Journal and Proceedings of the Royal Society of New South Wales
Journal of the Geological Society of Australia
Papers and Proceedings of the Royal Society of Tasmania
Transactions of the Royal Society of South Australia

NEW ZEALAND

Journal of Hydrology, New Zealand
New Zealand Geographer
New Zealand Journal of Geology and Geophysics
New Zealand Journal of Science

16

Soil science

D. A. Jenkins and R. I. J. Tully

Soil science is unified by its subject matter rather than by any particular scientific discipline. In this respect it is a microcosm of the geological scene, in that it involves the application of a range of disciplines to the study of a natural phenomenon; however, like geology, it too has to some extent evolved its own unique philosophy. The literature relating to soils is therefore diverse in character and approach, and although some of it is concentrated in publications specific to soils, a portion of it is dispersed through the general chemical, biological, geological and civil engineering literature. As in most other branches of science, this literature is expanding rapidly. According to Jacks (1966), who presents an interesting analysis of this expansion and the trends embodied in it, the volume has more than doubled over the post-war years.

Historically, early investigations into the nature of soil, such as those summarised by Humphry Davy in 1813 in his book *Elements of Agricultural Chemistry*, were essentially agricultural in nature. They involved a chemical approach to soils and plant nutrition which was carried through to the latter half of the century, receiving impetus from the publication in 1840 of Liebig's *Chemistry in its Application to Agriculture and Physiology*. In studying the nature and variability of soil as a naturally occurring material, great emphasis was placed on the geological setting until the relative importance of climate as a controlling factor was recognised by Russian scientists, such as Dokuchaev (Glinka, 1928) towards the end of the nineteenth century. Brief accounts of these early developments can be found in the introductory chapters of Russell (1961) and Joffe (1949).

In this century, the study of soil has expanded rapidly to incorporate an increasing range of expertise, from mineralogy, through physical, inorganic and organic chemistry, biochemistry and micro-biology, to agronomy, ecology and geomorphology. Although some of the interrelationships between these various aspects are as yet tenuous, there is sufficient cohesion for soil science to constitute a viable unit of study.

Owing to the importance of climate as a factor in soil develop-ment, there is a strong national or regional flavour to soil investiga-tions and this is reflected in the resulting literature. This pattern is also seen in geomorphology but is not evident to the same degree in geology as a whole. In coping with the non-English literature the following publications are of value: *Multilingual Dictionary of Soil Science*, by G. V. Jacks (FAO, Rome, 1960), and *Elsevier's Dictionary of Soil Mechanics*, by A. D. Visser (Elsevier, 1965).

The present position of soil science within the framework of geology is uneasy. Soils, *per se*, receive scant attention in geological circles, even with geology re-styled as 'earth sciences'—this despite the fact that soils are essentially a geological material and the focus of the sedimentary geochemical cycle. The main incentive in understanding the nature and properties of soil remains agricultural, and this reflects the economic significance of soil fertility. This aspect of soils carries through into geography and ecology. It has led to the establishment of an increasing number of national soil surveys, although the latter have subsequently acquired varying degrees of independence from agriculture. There is yet a further aspect, and that is soil considered as a material involved in civil engineering. Here soil is grouped with other superficial deposits of the regolith, the prime considerations being its physical, mech-anical or structural, properties. Soil mechanics and the agricultural aspects comprise two distinct fields of study that are surprisingly independent both in practice and in their respective literatures.

In this chapter, soil will be considered primarily in its own right as a naturally occurring material. This is, therefore, a geological rather than an agricultural or engineering viewpoint and conforms to the definition of 'pedology' given by Robinson (1924). After a consideration of the relevant general texts and primary sources, additional information relating to specific aspects such as mapping and classification, mineralogy, chemistry, physics, mechanics and biology will be dealt with briefly. Finally, the secondary literature —general reviews, abstracts and bibliographies—will be discussed. Since the subject is so diverse and ramifying, this review is neces-sarily selective, rather than comprehensive, in nature. Its aim is simply to provide an initial point of entry for those with geological interests who are not familiar with the literature of soil science. For those wishing to develop a more general philosophical background to soil science a fascinating reading list has been suggested by Kellogg (1940).

GENERAL TEXTBOOKS

Brief accounts of the nature of soils will be found in many of the standard geological textbooks, particularly those on physical and sedimentary geology, but their treatment is generally limited and disappointing. At a basic level there are several introductory books aimed at either the layman or first-year student, for example: *The World of the Soil*, by E. J. Russell (Fontana paperback, 1957); *Soil*, by G. V. Jacks (Nelson, 1954); *An Introduction to Soil Science*, by G. Leeper (4th edn., University Press, Melbourne, 1964); and *An Introduction to the Scientific Study of Soil*, by W. N. Townsend (5th edn., Arnold, 1972).

At a more advanced level, however, there is a lack of modern textbooks dealing with soils as a whole, although a consideration of soils in their agronomic context is well provided for by: *The Nature and Properties of Soils*, by H. O. Buckman and N. C. Brady (7th edn., Macmillan, 1969); *Soils: An Introduction to Soils and Plant Growth*, by K. L. Donaghue (2nd edn., Prentice-Hall, 1965); *Fundamentals of Soil Science*, by H. D. Foth and L. M. Turk (5th edn., Wiley, 1972); and *Soil Conditions and Plant Growth*, by E. W. Russell (9th edn., Longmans, 1961). Condensed surveys of soils in a geographical context are presented in: *The Geography of Soils*, by B. T. Bunting (Hutchinson, 1965), and *World Soils*, by M. Bridges (Cambridge University Press, 1970). However, a recent book dealing with pedology, with special reference to structure, micromorphology and (an individualistic) classification, is: *Pedology: A Systematic Approach to Soil Science*, by E. A. Fitzpatrick (Oliver and Boyd, 1971).

There are also several texts issued some 20–30 years ago which are still valuable for their presentation of the philosophical and historical aspects of soil science, although the technical and conceptual advances made since their publication limit their present usefulness. Included among such books are: *Factors of Soil Formation*, by H. Jenny (McGraw-Hill, 1941); *Pedology*, by J. S. Joffe (2nd edn., Pedology Publications, New Brunswick, 1949); and *Soils, their Origin, Constitution and Classification*, by G. W. Robinson (3rd edn., Murby, 1949).

As has been mentioned earlier, there are a number of recognisable schools of thought in soil science usually with a national basis. Of the foreign textbooks, several important Russian works have been translated into English and published by the Israel Program for Scientific Translations, Jerusalem, including *Soil Science* by D. G. Vilenski (3rd edn., Moscow, 1959; translated 1960); *Soil Science*, by A. A. Rode (Moscow, 1955; translated 1962); and *Treatise on Soil Science*, by K. D. Glinka (Moscow, 1931; translated 1963). Unfortunately, however, several important texts in other languages are as yet unavailable in English translation—for example: *Précis de Pédologie*, by P. Duchaufour (3rd edn., Masson, 1970); *L'Evolution des Sols: Essai sur la Dynamique des Profils*, by P. Duchaufour (Masson, 1968); and *Lehrbuch der Bodenkunde*, by F. Scheffer and P. Schachtschabel (7th edn., Ferdinand Enke, 1970). Finally, specific categories of soils have been dealt with individually in such works as: *Tropical Soils*, by E. C. J. Mohr and I. A. van Baren (Interscience, 1954); *An Introduction to the Study of Soils in Tropical and Subtropical Regions*, by H. Buringh (Centre for Agricultural Publication and Documentation, Wageningen, 1969); *Forest Soils*, by S. A. Wilde (Ronald Press, New York, 1958); and *West African Agriculture*, Vol. 1 of *West African Soils*, by P. M. Ahn (Oxford University Press, 1970). Other selected aspects have also recently received specialised treatment and this, perhaps, is a reflection of the present phase of expansion and diversification in soil science; they will be noted where relevant in the sections that follow.

PRIMARY LITERATURE DEALING WITH GENERAL SOIL SCIENCE

An indication of the range of publications covering soil science, in conjunction with fertilisers and general agronomy, is given by the fact that reference was made to some 700 journals and periodicals in the preparation of abstracts for *Soils and Fertilizers* by the Commonwealth Bureau of Soils in 1962. However, the main bulk of the soil science literature is contained within a limited number of primary journals. Details of some of the more important journals are given in *Table 16.1*

Of the English-language journals, *Proceedings of the Soil Science Society of America* carries the largest portion of the literature. It spans soil science and agronomy, and its articles are classified under nine headings, which include: soil physics; chemistry; mineralogy; genesis; morphology; and classification. Of the other three major English-language journals, the American *Soil Science* also has an agronomic bias, whereas the British *Journal of Soil Science* and the recently introduced international *Geoderma* are devoted specifi-

Table 16.1. DETAILS OF THE MOST IMPORTANT SOIL SCIENCE PERIODICALS

Periodical	Language*	Country	First issue	Issues/vols p.a.	Articles in 1969	Pages in 1969
Proceedings of the Soil Science Society of America	E	USA	1936	6/1	200	1000
Soil Science	E	USA	1916	12/2	125	950
Soviet Soil Science	E	USSR	1958	12/-	200	750
Zeitschrift für Pflanzenernährung. Düngung und Bodenkunde	G(E)e	Germany	1922	3/3	55	800
Anales de Edafologia y Agrobiologia	Se	Spain	1942	6/1	57	900
Journal of the Indian Society of Soil Science	E	India	1953	4/1	80	500
Geoderma	E(G)e	Internat.	1967	6/1	35	500
Journal of Soil Science	E	UK	1949	2/1	40	350
Canadian Journal of Soil Science	E(F)ef	Canada	1957	3/1	45	400
Australian Journal of Soil Research	E	Australia	1963	3/1	30	350
Soil Science and Plant Nutrition	E	Japan	1955	6/1	37	300
Agrochimica	I(EFG)iefg	Italy	1956	6/1	50	550
Annales Agronomiques	F	France	1931	6/1	30	650
Pédologie	F(E)fge	Belgium	1951	3/1	20	400
Pakistan Journal of Soil Science	E	Pakistan	1965	2/1	15	100

* E, English; G, German; F, French; S, Spanish; I, Italian.
Letters in parentheses denote occasional articles only. Lower case denotes
translated summaries.

cally to soil science. In these last three journals the articles are unclassified.

The literature on tropical soil science is concentrated in the Indian and Pakistan journals. There is, in addition, an extensive Japanese literature, of which the main English-language journal is *Soil Science and Plant Nutrition* (formerly *Soil and Plant Food*), and this also incorporates English abstracts of Japanese articles in the current *Journal of the Science of Soil and Manure, Japan*. There is a further large volume of primary literature in languages other than English (for example, from Central Europe), which space precludes dealing with here.

In 1958 a valuable addition was made to the literature available in English by the production of *Soviet Soil Science* under the auspices of the Soil Science Society of America. This was a cover-to-cover translation of the Russian *Pochvovedenie*, with a delay of the order of 6 months. The articles are classified under headings similar to those in the *Proceedings of the Soil Science Society of America*; it should be pointed out that the 12 numbers of this journal issued per year up to 1968 were not designated by any volume number; in that year translations of important articles from other Russian language journals were also published as a supplementary issue to *Soviet Soil Science* under the title of *Doklady Soil Science*. Since 1969, however, only a selection of the principal articles from *Pochvovedenie* and other Russian journals has been published, with the remaining articles from *Pochvovedenie* included in abstract form only.

Further sources of information include the publications of a large number of government and university research stations or departments and of regional soil surveys and discussion groups.

Apart from these specific journals and publications, occasional articles concerning soil science are also contained in such general journals as *Nature*, *Science*, etc.

SOIL MAPPING AND CLASSIFICATION

Soil survey maps and publications constitute the raw data of soil science. Most of this 'literature' is obviously specific to particular areas, but it also embodies much of the philosophy and concepts current in soil science through its involvement with genesis and classification. The situation may be compared with that in geological survey, where three-dimensional structural and stratigraphical complications are often dominant, although the geological units involved are generally simpler, lacking the lateral heterogeneity of soils.

In Britain, mapping is now carried out by the Soil Survey of England and Wales and the Soil Survey of Scotland at a scale of $1:25\,000$ ($2\frac{1}{2}$ inches to the mile) and the results published at a scale of $1:63\,360$ (1 inch to the mile). The maps are accompanied by a *Memoir of the Soil Survey of Great Britain*, in which the characteristics, distribution and utilisation of the 'Soil Series' recognised are presented, together with relevant analytical data. Special maps at other scales are occasionally published. Some 20 *Memoirs* and *Bulletins* have now been issued, the details being available in the *Annual Reports* of the *Rothamsted Experimental Station* (for England and Wales) and the *Macaulay Institute for Soil Research* (for Scotland). A brief description of the methods employed and classification system used (evolved from Avery, 1956) is included in each *Memoir*, and further details will be available from the *Soil Survey Handbook* (HMSO), which is in preparation. More general information can be obtained from such books as *The Study of Soil in the Field*, by G. R. Clarke (5th edn., Oxford University Press, 1971).

Similar surveys are carried out by other nations. For example, those published by the US Department of Agriculture (USDA) are based on individual state counties, but differ in that the area is covered not by a map but by a set of aerial photographs (scale $1:20\,000$) upon which the distribution of 'Soil Series' is superimposed in outline. The relevant methodology is contained in the 'Soil Survey Manual' (*USDA Agricultural Handbook*, No. 18, 1951) and the scheme of classification in *Soil Classification: A Comprehensive System, 7th Approximation* (USDA, 1960) and its supplement (1967). The *7th Approximation* is important, since it introduced a new and systematic terminology which is now passing into general international usage: a definitive version of this system will be available in the near future. A recent publication which presents the current position in the survey and classification of soils in Australia is *A Handbook of Australian Soils*, by H. C. T. Stace, G. D. Hubble *et al*. (Rellim Technical Publication, Adelaide 1968). An equivalent publication for New Zealand is 'Soils of New Zealand', published in 3 volumes as *Soil Bureau Bulletins, New Zealand Department of Scientific and Industrial Research*, **26** (1968).

On the international scale, the production of a *Soil Map of the World* at a scale of $1:5\,000\,000$ is now nearing completion. This is being carried out under the auspices of UNESCO and FAO in conjunction with the International Society of Soil Science, whose publications also include *Soil Map of the World—Catalogue of*

Maps (3rd edn., FAO, 1965). *A Soil Map of Europe* at a scale of 1:2 500 000 has already been published (FAO, 1966) together with an explanatory text. An important contribution to the characterisation of European soils is *Die Wichtigsten Boden der Bundesrepublik Deutschland*, by E. Mückenhausen (Kommentator, Frankfurt, 1959). This carries colour illustrations of type profiles similar to those of its classic predecessor *The Soils of Europe*, by W. L. Kubiena (Murby, 1953). An account of soil distribution on a world-wide basis is also presented in *Atlas zur Bodenkunde*, by R. Ganssen and F. Hädrich (Bibliographisches Institut, Mannheim, 1965).

SOIL MINERALOGY AND MICROMORPHOLOGY

The increasing importance of soil mineralogy can be seen in its separation from soil chemistry as an individual section in the *Proceedings of the Soil Science Society of America* in 1969. Soil mineralogy can be broadly divided into two parts: that of the sand fraction, comprising mainly minerals inherited from parent materials; and that of the clay fraction, the most important components being the clay minerals and amorphous materials. The literature for both overlaps with that of sedimentary mineralogy, but that of the former part is usually restricted to considerations of relative mineral stabilities and determination of soil parentage, while the evolution of clay mineralogy has to a large extent taken place actually within soil science.

Sections on soil mineralogy will therefore be found in geological texts—for example, the chapter by D. Carroll in *Sedimentary Petrography*, edited by H. B. Milner (4th edn., Allen and Unwin, 1962). They are also found in soil science texts—for example, the chapter by M. L. Jackson in *Chemistry of the Soil*, edited by F. E. Bear (2nd edn., Reinhold, 1964), and Vol. 1 of *The Physical Chemistry and Mineralogy of Soils*, by C. E. Marshall (Wiley, 1964). *Mineralogical Investigations in Soil Science,* by E. I. Parfenova and E. A. Yarilova (Israel Program for Scientific Translations, Jerusalem, 1965), in parts deals exclusively with soil mineralogy, as does the first section of *Fabric and Mineral Analysis of Soils*, by R. Brewer (Wiley, 1964). The latter half of this book, however, is perhaps more important, since it is the first comprehensive treatment of micromorphology since the publication of *Micropedology*, by W. L. Kubiena (Collegiate Press, Iowa, 1938), although this has recently been supplemented by *Micromorphological Features of Soil Geography*, by the same author (Rutgers Univ. Press, 1970).

The literature dealing with clay mineralogy (see also Chapter 11) alone is more concise. A standard reference dealing with the nature, properties and occurrence of clay minerals is *Clay Mineralogy*, by R. E. Grim (2nd edn., McGraw-Hill, 1968), and standard texts dealing with the special techniques required in their identification include *The X-ray Identification and Crystal Structures of Clay Minerals*, edited by G. Brown (2nd edn., Mineralogical Society, London, 1961), and *The Differential Thermal Analysis*, Vol. 1, edited by R. C. Mackenzie (Academic Press, 1970). Two English-language journals cater specifically for clay mineralogy, each carrying a large amount of data relevant to soil science. They are the British *Clay Minerals* and the American *Clays and Clay Minerals*. The former is a restyled version of *Clay Minerals Bulletin* (1951–64); and the latter, of *Proceedings. National Conference on Clay and Clay Minerals* (1952–67). Important papers on soil mineralogy are also occasionally published in *Journal of Sedimentary Petrology*, *American Mineralogist* and similar journals.

SOIL CHEMISTRY AND ANALYSIS

Considered as a chemical system, soil is notoriously complex. This has resulted in a concentration of the chemical literature within soil science, with less overlap with that of general chemistry than might be expected. The main theme has been the application of chemistry to the understanding of the behaviour of nutrients in soils, particularly of macro-nutrients such as phosphorus, potassium and nitrogen, and of phenomena such as ion exchange. This theme is in fact central to many of the general texts that have already been referred to above, and it also forms the main contribution to such journals as *Proceedings of the Soil Science Society of America* and *Soil Science*. Additional texts concerned directly with the chemical properties of soils include various chapters in *Chemistry of the Soil*, edited by F. E. Bear (2nd edn., Reinhold, 1964) and *The Physical Chemistry and Mineralogy of Soils*, Vol. 1. *Soil Materials*, by C. E. Marshall (Wiley, 1964). The chemistry of the organic fraction, which to a large extent distinguishes soils from sediments, is well covered by *Soil Organic Matter*, by M. M. Konanova (2nd English edn., Pergamon, 1966).

A large portion of the soil chemical literature is devoted to various types of analysis. Several good manuals are available for routine soil analysis, including: *Soil Chemical Analysis*, by M. L. Jackson (Constable, 1962); *Textbook of Soil Analysis*, by P. R. Hesse (Murray, 1971); *Methods of Chemical Analysis for Soil*

Survey Samples, by A. J. Metson (NZDSIR, Wellington, 1956); and *Methods of Soil Analysis*, edited by C. A. Black (American Society of Agronomists, Wisconsin, 1965). The last-named text is a comprehensive treatment in two volumes comprising over 10 chapters contributed by specialists on physical, mineralogical and microbiological, as well as chemical, analysis.

Certain trace elements are of importance as essential micronutrients and have therefore received special attention. In addition to the standard textbooks on geochemistry, further information specific to soils can be found in *Geochemistry of the Rare and Dispersed Chemical Elements in the Soil*, by A. P. Vinogradov (English translation, Chapman and Hall, 1959), and also in the following *Technical Communications of the Commonwealth Bureau of Soils*: 'The Spectrographic Analysis of Soils, Plants and Related Materials', by R. L. Mitchell (No. 44a, 1964); and 'The Trace Element Content of Soils', by D. Swaine (No. 48, 1955). Articles concerned with soil analysis will be found in most of the primary journals already cited, and also occasionally in such journals as *Journal of the Science of Food and Agriculture* and *Analyst*. Articles on soil geochemistry are also occasionally carried by *Geochimica et Cosmochimica Acta*.

SOIL PHYSICS, MECHANICS AND CONSERVATION

The complexity of soil as a naturally occurring material is reflected in its physical as well as its chemical properties. The structural arrangement of its solid components and the distribution and mobility of the liquid and gaseous phases are important characteristics, as are the thermal properties. General introductions to this aspect of soil science are given in: *Soil Physics*, by L. D. Baver (3rd edn., Wiley, 1956); *Soil Physics*, by H. Kohnke (McGraw-Hill, 1969); *An Introduction to Soil Behaviour*, by R. N. Yong and B. P. Warkentin (Macmillan, 1966); and *Physical Properties of Soils*, by R. E. Means and J. V. Parcher (Constable, 1964). Of these, the latter two have a more mechanical bias, with a greater emphasis on methodology in the last-named text.

Further consideration of the physical properties of soils leads into two fairly well-defined fields. In an agronomic context these properties characterise the physical environment of plant growth, and also determine soil drainage and other practical management characteristics. In this, the aqueous phase is a dominant factor, and its importance is well covered in *An Introduction to the Physical Basis of Soil Water Phenomena*, by E. C. Childs (Wiley/Interscience, 1969). Other specific topics are dealt with in such texts as *An Introduction to Clay Colloid Chemistry*, by H. Van Olphen (Interscience, 1963), and *Drainage of Agricultural Lands*, edited by J. N. Luthin (American Society of Agronomists, 1957).

Consideration of those soil properties relevant to civil engineering is a separate field with its own considerable literature. Introductions to this may be found in *Fundamentals of Soil Mechanics*, by D. W. Taylor (Wiley, 1948), and *Basic Soil Engineering*, by B. K. Hough (Ronald Press, New York, 1957).

The primary literature relating to soil physics (water, atmosphere, drainage, temperature, etc.) is mostly contained in the general journals already referred to. Occasional articles are also carried by *Journal of Hydrology* and *Journal of Agricultural Engineering Research*. In the field of civil engineering, important articles concerning soil physical properties can be found in a number of journals—for example, *Géotechnique* and its American counterpart, *Proceedings of the American Society of Civil Engineers, Journal of Soil Mechanics and Foundations Division*.

Literature on the conservation of soils, particularly of tropical soils, is scattered through the primary journals. However, the topic is specifically covered in *Soil Conservation*, by H. Kohnke and A. R. Bertrand (McGraw-Hill, 1959) and in *Soil Conservation*, by N. Hudson (Batsford, 1971).

SOIL BIOLOGY AND BIOCHEMISTRY

Soil organisms are an integral and essential component of soils. Apart from their population varying with, and being characteristic of, a particular soil environment, soil fauna and flora are an important factor in the actual processes of pedogenesis. This is particularly evident in the case of microbes which control chemical reactions within the soil system: it is probably true for sedimentary geochemistry as a whole, but is more patently so in the case of soils. The literature tends to fall into the fairly well-defined areas of soil biology (macro- and mesofauna), and microbiology and biochemistry, with little overlap with other areas of soil science.

Although some of the wide range of macro- and mesofauna are effectively unique to soil, some also occur in other environments. The literature dealing with individual organisms, therefore, merges with other zoological literature. There are, however, several textbooks dealing specifically with soil organisms, of which *Ecology of Soil Organisms*, by J. A. Wallwork (McGraw-Hill, 1970), is a good introduction and guide to the recent literature, and *Soil Biology*, by W. Kühnel (English edn., Faber, 1961), a useful reference work. Articles on soil organisms occur sporadically through the primary zoological journals, but there are also two journals devoted specifically to soil biology. They are *Pedobiologia*, a German quarterly with articles in German, French and English; and *Revue d'Ecologie et de Biologie du Sol*, a French quarterly containing about 40 articles per year written in French or English with English, French and German summaries.

A general outline of soil microbiology is given in: *Micro-organisms in the Soil*, by A. Burges (Hutchinson, 1958); *Introduction to Soil Microbiology*, by M. Alexander (Wiley, 1961); and *Soil Microorganisms*, by T. R. G. Gray and S. T. Williams (Oliver and Boyd, 1971). A classic text, though unfortunately unavailable in English translation, is *Microbiologie du Sol*, by S. Winogradsky (French translation; Masson, 1949). Further specific aspects have been covered in the proceedings of international symposia published by Liverpool University Press, namely *The Ecology of Soil Fungi*, edited by D. Parkinson and J. S. Waid (1960), and *The Ecology of Soil Bacteria*, edited by T. R. G. Gray and D. Parkinson (1968).

The biochemistry of soils, however, has only recently evolved into a distinct subject, and although no basic textbook dealing directly with this section of soil science is yet available, a useful review of the subject is presented in *Soil Biochemistry*, Vol. 1, edited by A. D. McClaren and G. H. Peterson (1967), and Vol. 2 (1972), edited by A. D. McClaren and J. Skujins (Arnold). The primary literature in this area is dispersed through such journals as *Journal of Bacteriology*, *New Phytologist* and *Biochemical Journal*, although a channel has recently been provided by the introduction of the new quarterly journal *Soil Biology and Biochemistry*, produced by Pergamon and containing French and German summaries of the English articles.

No attempt will be made here to do justice to the extensive literature on the important subject of soil-plant relationships, where soil science links through to ecology and agriculture. As has already been pointed out, the repercussions are in any case seen in the emphasis of many of the introductory texts in soil science and in the literature on soil chemistry in particular. However, further detailed information and references can be found in such books as *Soil-Plant Relationships*, by C. A. Black (2nd edn., Wiley, 1968), and in such journals as *Plant and Soil*, *Journal of Ecology*, and *Agronomy Journal*.

REVIEWS AND SYMPOSIA

At the present time general review articles are regrettably sparse and sporadic, and there are unfortunately no review journals actually centred on soil science. However, journals such as *Soil Science* and *Proceedings of the Soil Science Society of America* occasionally carry invited review articles and there are often such articles preceding the abstract section in individual numbers of *Soils and Fertilizers*. The only review journal to incorporate regular contributions on soil science is *Advances in Agronomy*, published annually since 1949 by Academic Press. This generally contains from five to ten articles, each of some 25–100 pages in length, and although the over-all emphasis is agronomic, one or two articles per issue are usually on some aspect of pedology, soil chemistry or soil mineralogy. *Earth Science Reviews* also occasionally carries articles relating to soil science.

The major international meeting, whose contributed papers are subsequently published, is the International Congress of Soil Science. A full congress is held every 4 years, the ninth being that in Adelaide, Australia, 1968. The papers presented totalled over 80 (760 pages) and are grouped under seven commissions, namely: soil physics; soil chemistry; soil biology; fertility and plant nutrition; genesis, classification and cartography; technology; and mineralogy. The *Transactions* are usually published in five volumes and are obtainable via the International Society of Soil Science (c/o Royal Tropical Institute, 63 Mauritskade, Amsterdam, Netherlands). Conferences on selected themes are also held by individual commissions in the intervening years.

Other symposia are held by various universities or institutions on a wide range of topics, some regularly, some sporadically, the proceedings being published as monographs or reports. Examples of regular meetings on varying topics are the Easter Schools in Agricultural Science held by Nottingham University. *Soil Zoology*, edited by D. K. McKevan (Butterworths, 1955), and *Experimental Pedology*, by E. G. Hallsworth and D. V. Crawford (Butterworths, 1965), are the proceedings of the first and eleventh Easter Schools, respectively.

A number of individual topics have been covered at regular intervals by international conferences and symposia. The resultant publications are typically of the order of 300–500 pages, containing 40–50 articles. Some papers delivered at European meetings are in French or German, but there are usually summaries in English. These books form valuable sources of data, particularly in those branches of soil science which are developing rapidly. Some examples are: *Soil Organisms*, edited by J. Doeksen and J. van der Drift (North-Holland, 1965); *Soil Micromorphology*, edited by A. Jongerius (Elsevier, 1964); and *Soil Clay Mineralogy*, edited by C. I. Rich and G. W. Kunze (University of North Carolina Press, 1964).

ABSTRACTING SERVICES AND BIBLIOGRAPHIES

Without doubt, the most valuable and comprehensive system of abstracting in soil science is that operated by the Commonwealth Bureau of Soils. Their publications include the abstract journal *Soils and Fertilizers*; *Bibliography of Soil Science, Fertilizers and General Agronomy*; and annotated bibliographies on selected topics, in addition to the *Technical Communications* to which reference has already been made.

Soils and Fertilizers was initiated in 1938 replacing various earlier publications covering this field, and it is issued every two months. Each issue now contains some 800 abstracts which are designed to be informative rather than indicative, and which, according to Jacks (1966), represent approximately 60 per cent of the total number of articles actually reviewed by the Bureau. The abstracts are arranged in groups by subject matter, according to a simplified version of the Universal Decimal Classification system, and fall into one of five main classes, of which the first, 'Soils', is relevant in this context. This class comprises pedology, soil chemistry, analysis, physics, classification and biology, and it accounts for over half of the abstracts in each issue (i.e. approximately 500), the remainder covering the cultivation, fertilisation and management of soils, plant diseases and pesticides. The time lag between publication in a primary journal and appearance of the abstract varies from under a year for articles in the major English language journals, to two or three years in the case of the more obscure journals in foreign languages. In addition to the abstracts, each issue also contains an author index.

All these data are periodically rationalised and published as *Bibliography of Soil Science, Fertilizers and General Agronomy*. This was a triennial publication, starting with the period 1931–34, the last volume being that for 1959–62. Information is presented in four sections. The first is a bibliography arranged in numerical order according to the UDC, and amounting to well over 10 000 references. The second section is a list of abbreviations of the 700 periodicals consulted, and sections three and four are author and alphabetical subject indexes, respectively. These bibliographies are invaluable in following the literature on a particular topic, and it is unfortunate that they only cover the period up to 1962.

Another valuable form of publication produced by the Commonwealth Bureau of Soils is the *Annotated Bibliography*, which has now replaced the earlier triennial *Bibliographies*. This is a duplicated typed series of abstracts of the literature on selected topics. The bibliographies cover a specified period which ranges from 4 to 30 years, and on important topics supplements are issued to maintain the abstracts up to date. A bibliography may contain from 10 to 100 abstracts and range from 20p to £2 in price: the first requested, however, is available free of charge. Well over 1000 such bibliographies have now been issued, those currently available being listed in the front of *Soils and Fertilizers*. They are obtainable direct from the Commonwealth Bureau of Soils (Rothamsted Experimental Station, Harpenden, Hertfordshire, England).

Soil science literature is also covered by several other abstracting journals, although none is as comprehensive or convenient as *Soils and Fertilizers*, nor is the rationale behind their selection systems so evident. *Biological Abstracts* (see also Chapter 10) contains a total of over 100 000 abstracts per year, of which 1000 come under the section headed 'Soils'—that is, around 40 abstracts per issue. The section, however, is subdivided into three parts only —general, inorganic and organic: abstracts on soil microbiology are included in a separate section. *Chemical Abstracts* (see also Chapter 11) operates on a similar scale and contains some 20 abstracts per issue relating to soils in the section headed 'Mineralogical and Geological Chemistry'. In this case there is no subdivision of the soil abstracts. *Mineralogical Abstracts* (see also Chapter 11) operates on a smaller scale but is particularly valuable for the section on 'Clay Minerals' introduced in 1960. This section is in turn subdivided into two parts—'Techniques; Structures; Properties' and 'Petrology; Weathering; Soils'. Both of these parts average some 20 abstracts each per issue, and other relevant abstracts may also be found in the sections on 'Apparatus and Techniques' and 'Sedimentary Geochemistry'. Similarly, *Geographical Abstracts B*, which bears the title 'Biogeography and Climatology', contains a section headed 'Pedology'. Each issue carries about 60 abstracts relating to soils. Abstracts of the papers on soils up to 1964 have been collated by B. T. Bunting and published as *An Annotated Bibliography of Memoirs and Papers on the Soils of the British Isles* (Geographical Abstracts, 1964).

REFERENCES

Avery, B. W. (1956). 'A Classification of British Soils'. *Report of the 6th International Congress of Soil Science,* **E279**

Glinka, K. D. (1928). 'Dokuchaev's Ideas on the Development of Pedology and the Cognate Sciences'. *Proceedings and Papers of the 1st International Congress of Soil Science,* **1,** 116

Jacks, G. V. (1966). 'The Literature of Soil Science'. *Soils and Fertilizers,* **29** (3), 227–230

Joffe, J. S. (1949). *Pedology,* 2nd edn. Pedology Publications, New Brunswick

Kellog, C. E. (1940). 'Reading for Soil Scientists, together with a Library'. *Journal of the American Society of Agronomy,* **32,** 867–876

Robinson, G. W. (1924). 'Pedology as a Branch of Geology'. *Geological Magazine,* **61** (9), 444–455

Russell, E. W. (1961). *Soil Conditions and Plant Growth,* 9th edn. Longmans

17

History of geology

D. A. Bassett

Ideally a bibliographical essay on the history of geology and its allied sciences should contain references to the original publications and to the commentaries of contemporary and later authorities—both geologists and historians. In the space available, however, it is impossible to do justice to both primary and secondary sources. Since the former are adequately listed in most of the major textbooks on the history of the subject and there is no comprehensive guide to the ever-increasing volume of commentaries, attention is concentrated on the latter and mention of the former is confined largely to works which are now readily available as facsimile reprints.

The two most widely known English-language textbooks on the history of the subject, both written in the last decade of the nineteenth century, are: (a) *Geschichte der Geologie und Paläontologie bis Ende des 19. Jahrhunderts*, by Karl Alfred von Zittel (Oldenbourg, 1899; Johnson reprint, 1965). A slightly abridged translation, excluding almost all the bibliographical references and biographical footnotes, by Marie M. Ogilvie-Gordon, a former pupil of Zittel, was published under the title *History of Geology and Palaeontology to the End of the Nineteenth Century* (Scott, 1901; facsimile reprint, Wheldon and Wesley, 1962). (b) *The Founders of Geology*, by Sir Archibald Geikie (Macmillan, 1897) —being the Williams Memorial Lecture delivered at the Johns Hopkins University in 1896. A second edition, revised and enlarged, was published in London in 1905 and was reprinted by Dover Publications in paperback form in 1962. The volumes complement each other in that the former is systematic and encyclopaedic; the latter, biographical and selective. The greater portion of each volume is, however, concerned with the period from the second half of the eighteenth century to the middle of the nineteenth century, and the emphasis is on stratigraphy, palaeontology and dynamic geology.

Immediately before the Second World War a new history was published giving greater emphasis to the earlier periods in the development of the science and to mineralogy and petrography. This was: *The Birth and Development of the Geological Sciences*, by Frank Dawson Adams (Baillière, Tindall and Cox, 1938; Dover reprint, 1954). Written by one of the few geologists sufficiently familiar with classical and mediaeval languages to study the manuscripts and records, it contains generous quotations and translations, many illustrations and extensive references, but, unfortunately, many minor errors.

All three of the above volumes are still in use. Zittel's work is in many ways still the most satisfactory one-volume history of the subject, though much less readable than Geikie's charming *Founders*; while Adams' book is still the best single source for the early history of the subject.

A less well-known, but nevertheless useful, work is *History of Geology*, by Horace B. Woodward (Watts, 1911). A much earlier work, which is cited regularly in the literature, is the historical survey which occupies the first chapter in *Principles of Geology*, by Charles Lyell (3 vols., Murray, 1830–33; reprint, 3 vols., Wheldon and Wesley, 1970) and which is reprinted without alteration in all the editions of the work.

The following are the better-known foreign-language textbooks. *La Science Géologique, ses Méthodes, ses Resultats, ses Problèmes, son Histoire*, by Louis de Launay (Colin, 1905), which contains an excellent summary of the history of geology in the first 92 pages. *Histoire de la Géologie*, by André Cailleux (2nd edn., Presses Universitaire Français, 1968)—an outline of the beginnings and later history of the science of geology and geological concepts relative to those of other scientific disciplines and to the environment of the different civilisations and centuries. *Geschichte der Geologie und des Geologischen Weltbides*, by Carl Christoph Beringer (Enke, 1954), in which the author's principal concern is to outline the growth of the historical feeling for the earth in the world of ideas and its influence and interactions with science and philosophy, with the emphasis on continental philosophy. Much the largest and best portion of this book is devoted to the early nineteenth century. *Geologie und Paläontologie in Texten und ihrer Geschichte*, by Helmut Holder (Alber, 1960), in which the development of the geological sciences is presented as a series of thorough discussions of the major problems with their successive interpretations in the light of the observations and concepts prevailing at a given time. The second part of the book, for example, is devoted to a very original synthesis of mountain building. There are a number of long quotations and elaborate bibliographies.

Geology is poorly represented in most of the general histories of science. Two important exceptions, however, are: *History of the Inductive Sciences, from the Earliest to the Present Time*, by William Whewell (3 vols., Parker, 1837; facsimile reprint of 3rd edn., Cass, 1967), which is generally recognised as the first major history of science, in which over 100 pages are devoted to geology and another 60 pages or so to mineralogy; and the four volumes, *La Science Antique et Médiévale* (1957), *La Science Moderne* (1958), *La Science Contemporaine* (1961), *La Science Contemporaine II* (1964), under the editorship of René Taton of the Centre National de la Recherche Scientifique. An English translation of the latter, by A. J. Pomerans, is issued under the over-all title *A General History of the Sciences* (Thames and Hudson, 1963–66). It contains chapters on twentieth-century geophysics and mineralogical sciences, geology and comparative anatomy and palaeontology, by Pierre Tardi, Raymond Furon and Jean Piveteau, respectively; on nineteenth-century mineralogy, geology, invertebrate and vertebrate palaeontology by Jean Orcel, Raymond Furon, Mlle. Andrée Tétry and Jean Piveteau, respectively; on eighteenth-century geology by Raymond Furon; and on Renaissance geology by Paul Delaunay.

It has been claimed, by a number of contemporary historians of science, that Adams, Geikie, Lyell and Zittel all shared a tendency to view the history of geology as a succession of important discoveries which led inevitably to the abandonment of 'wrong' ideas and to the formulation and acceptance of 'correct' theories. This kind of thinking, variously described as 'Whig' or 'inductivist' —see, for example, *The Whig Interpretation of History*, by Herbert Butterfield (Bell, 1931)—leads to the selection and discussion of those figures and ideas deemed good, while it precludes any reconstruction of the climate of opinion characteristic of an historical era. Such criticisms of the historical writings of geologists by professional historians of science are countered by practising geologists who point out that the historians of science go too far in the other direction, disregarding modern findings altogether, and that historians are commonly not well enough versed in the science itself. In one sense these conflicting criticisms reflect the considerable growth of interest in the history of the subject since the Second World War among both historians and geologists.

Three representative major works by eminent historians of

science published since the Second World War are:

(a) *Genesis and Geology: A Study in the Relations of Scientific Thought, Natural Theology, and Social Opinion in Great Britain, 1790–1850*, by Charles C. Gillispie (Harvard University Press, 1951; Harper Torchbook, 1959). The book traces the changes in religious interpretation of nature made necessary by the growth of geological knowledge from the days of Abraham Werner and James Hutton. In a paraphrase of Gillispie's words, the scientific literature of the years before 1850 makes it apparent that neither the conflict between religion and science, as maintained in the book *History of the Conflict between Religion and Science*, by John W. Draper (King, 1875), nor even that between theology and science, as set out in the classic *A History of the Warfare of Science with Theology in Christendom*, by Andrew D. White (Appleton, 1896; Dover reprint, 1960), was the simple, universal, black-and-white affair that it seemed in the optimistic perspective of the late nineteenth-century positivist rationalism. Indeed, from the birth of modern geology to the publication of Charles Darwin's *Origin of Species*, the difficulty as reflected in scientific literature appears to be one of religion (in a crude sense) *in* science rather than one of religion *versus* science. A synopsis of Gillispie's thesis is given in his later book *The Edge of Objectivity: An Essay in the History of Scientific Ideas* (Princeton University Press and Princeton Paperback, 1960; Oxford University Press, 1960).

(b) *The Death of Adam: Evolution and its Impact on Western Thought*, by John C. Greene (Iowa State University Press, 1959; Mentor Book, 1961). As the title suggests, the book is concerned mainly with speculation on the theory of evolution, including a discussion of such problems as the early debates about the extinction of species. Much space is, however, given to the evolution of the earth as such, including an analysis of the ideas of many eighteenth-century geologists.

(c) *Natural Law and Divine Miracle. The Principle of Uniformity in Geology, Biology and Theology*, by Reijer Hooykaas (Brill, 1959). Because of criticisms of the rather misleading title, the subtitle or second title *The Principle of Uniformity . . .* is printed in bolder letters in the second impression (1963) and 'may now be considered as the main title'. The book is probably the first major attempt to analyse the philosophical structure of geological science, using the historical evidence of the period during which its methodology was still a matter for open debate. Its analysis of the great geological controversies pays equal attention to the strictly scientific and the metaphysical factors, and shows how they were related.

A very recent book that reflects the views of both geologists and historians is *Toward a History of Geology*, edited by Cecil J. Schneer (MIT Press, 1969), which includes the contributions to the New Hampshire Inter-Disciplinary Conference on the History of Geology held in September, 1967. The work provides the most useful cross-section of current research available in book form—with particular reference to the sixteenth, eighteenth and early nineteenth centuries. In spite of its 450 pages, however, the Symposium Volume is very restricted in its coverage, even within the periods which it purports to cover.

It follows from this very brief survey of secondary literature that no volume or group of volumes provides anything like a comprehensive and up-to-date coverage of the history of geology. The remainder of this chapter is accordingly an attempt to provide a provisional bibliographical essay on the subject with particular reference to publications in the English language. The references are grouped according to the co-ordinates 'time', 'subject' and 'location' suggested by George Sarton in his *The Study of the History of Science* . . . (Harvard University Press, 1936; Dover reprint, 1957). Such an arrangement does involve a certain amount of duplication because the sections are obviously not mutually exclusive, but there are obvious merits in arranging the material in this fashion, especially as geology differs from most sciences in that very many of its investigations are geographically determined.

REFERENCES ARRANGED CHRONOLOGICALLY

Classical times and the dark and middle ages

The most accessible works which deal with these early periods are: *The Birth and Development of the Geological Sciences*, by F. D. Adams in which there is a 42-page chapter on 'Geological Sciences in Classical Times' and a 26-page chapter on 'The Conception of the Universe in the Middle Ages'; Archibald Geikie's *Founders*, in which there is a 40-page chapter on 'Geological Ideas among the Greeks and Romans'; *A Source Book in Greek Science*, compiled by M. R. Cohen and I. E. Drabkin (McGraw-Hill, 1948), which contains 600 pages of extracts on geography, geology, meteorology, physics, etc.; 'The First, First Principles of Geology', by John W. Harrington (*American Journal of Science*, **265**, 449–461, 1967), which discusses Herodotus' understanding of geological principles; 'Pliny's *Historia Naturalis*: The Most Popular Natural History Ever Published', by E. W. Gudger (*Isis*, **6**, 269–281, 1924); 'Ptolemy's Geography, Book VII, Chapters 6 and 7', by O. E. H. Neugebauer (*Isis*, **50**, 22–29, 1959); 'Geology in Tenth Century Arabic Literature', by Rushdi Said (*American Journal of Science*, **248**, 63–66, 1950); 'Geology in Embryo (up to 1600 A.D.)' by C. E. N. Bromehead (*Proceedings of the Geologists' Association*, **56**, 89–134, 1945), which contains a useful bibliography of works by contemporary and modern authors; and the chapter on 'The Development of Geological Thought and the Scientific Study of Scenery', by R. W. Clayton in *Science in its Context: A Symposium with Special Reference to Sixth-form Studies*, edited by John Brierley (Heinemann, 1964), which also contains a list of classical works, classified according to their subject matter, and a very useful chart illustrating the growth of geology and the names of the major authorities.

The three volumes of George Sarton's *Introduction to the History of Science* (Williams and Wilkins, for the Carnegie Institution, 1927–48) provide a very important reference series for the period up to the fourteenth century, being rich in biography and bibliography. The titles of the volumes are: 'From Homer to Omar Khayyam' (1927); 'From Rabbi Ben Ezra to Roger Bacon' (1931: in two parts); and 'Science and Learning in the Fourteenth Century' (1948).

Three earlier works which are of particular importance are: *Della Storia Naturale delle Gemme, delle Pietre è di tutti i Minerali, ovvero della Fisica Sotteranea*, by D. Giacinto Gimma (2 vols., Naples, 1730), which, in F. D. Adams' words, provides one of the best guide-books for the student who wishes to explore the mazes of the ancient literature of the geological sciences; *The Failure of Geological Attempts Made by the Greeks from the Earliest Ages Down to the Epoch of Alexander*, by Julius Schvarcz (1868), which reviews the whole body of ancient Greek literature prior to the time of the expedition of Alexander the Great (334 B.C.); and *History of the Inductive Sciences, from the Earliest to the Present Time*, by Whewell (see above). References to other early commentaries are given by Adams.

The works of Aristotle (384–322 B.C.); *Historia Naturalis*, by Pliny (A.D. 23–79); *De Rerum Natura*, by Lucretius (99–55 B.C.); *On Architecture*, by Vitruvius (?37 B.C.); and the *History* of Herodotus are available in English translation in the Loeb Classical Library and in the 'Great Works of the Western World' series (Encyclopaedia Britannica, 1952).

Secondary sources for geographical ideas are contained in: *The Geographical Lore of the Time of the Crusades. A Study in the History of Medieval Science and Tradition in Western Europe* [1095–1270], by J. K. Wright (American Geographical Society, 1925; Dover reprint, 1965); and *Geography in the Middle Ages*, by G. H. T. Kimble (Methuen, 1938), which contains a chapter on 'Physical Geography in the Middle Ages'. The Dover edition of the former contains a new introduction by C. J. Glacken which attempts to bring the reader up to date.

Fifteenth and sixteenth centuries

A useful introduction to the study of geological phenomena in the fifteenth century is given by Evelyn Stokes in the paper 'Fifteenth Century Earth Science' (*Earth Science Journal*, **1**, 130–148, 1967), in which the parts dealing with earth science phenomena in two late mediaeval encyclopaedias— *Mirrour of the World* and Ranulph Higden's *Polychronicon* (both printed by William Caxton in the 1480s)—are examined in relation to fifteenth-century ideas about the physical nature of the earth and the universe. Topics such as the four elements, the earth and the spheres, gravity, the arrangement of the continents and oceans, the unity of waters, earthquakes and volcanoes, erosion, fossils, mountain building and climatic zones are summarised and reference is made to the Biblical and classical Greek sources of these ideas.

The outstanding figure of the period was the universal genius Leonardo da Vinci (1452–1519). Reference is made to his work in: 'Some Early References to Geology from the Sixteenth Century Onwards', by W. P. D. Stebbing (*Proceedings of the Geologists' Association*, **54**, 49–63, 1943); *Geografia e Geologia Negli Scritti di Leonardo da Vinci*, by Agostino Gianotti (Museo Nazionale Scientia, Milan, 1953?); and 'Leonardo da Vinci's Geologische Studien', by Richard Weyl (*Natur und Volk Senckenbergische Naturforschende Gessellschaft*, **79**, 2–10, 1949). *Leonardo da Vinci's Note Books, Arranged and Rendered into English*, by Edward McCurdy (1906), should also be consulted.

The most contentious figure, on the other hand, was probably Phillipus Aureolus Theophrastus Bombastus von Hohenheim, usually referred to as Paracelsus (1493–1541). The immense literature on his work and influence can be readily traced in general works on the history of science and of medicine. Three books which are of particular interest in the present context are: *Paracelsus. An Introduction to Philosophical Medicine in the Era of the Renaissance*, by Walter Pagel (Karger, 1958), a thoroughly documented study which is a reliable guide to the writings and biographical studies of Paracelsus; *The English Paracelsians*, by Allen G. Debus (Oldbourne Press, 1965); and *The Chemical Dream of the Renaissance*, also by Debus (Heffer, 1968).

The contribution by A. G. Debus in *Toward a History of Geology*—namely, 'Edward Jorden and the Fermentation of Metals: An Iatrochemical Study of Terrestrial Phenomena'—outlines the 'chemical, medical, vitalistic and religious backdrop' against which to view the work of the English physician-chemist Edward Jorden (1569–1632), who sought a new explanation of metals and the source of the internal heat of the earth.

Theories of the earth in the cosmologies of the period approximately delimited by Nicolas of Cusa (1401–64) on the one hand and Nathanael Carpenter (1589–1628) on the other are outlined in 'Theories of the Earth in Renaissance Cosmologies' by Suzanne Kelly in *Toward a History of Geology*. Of the many topics surveyed, three are discussed in some detail—namely, the origin and future destruction of the earth; the structure of the earth, involving the material or materials of which the earth was made and their arrangement; and the changes in the surface of the earth, both minor and major. Other topics, such as the origin of metals, rivers and springs are discussed in Adams' *Birth and Development*.

Three of the main names in sixteenth-century geology are given in the title of Pierre Brunet's article 'Les Premier Linéaments de la Science Géologique: Agricola, Palissy, George Owen' (*Revue d'Histoire des Sciences et de leurs Applications*, 3, 67–79, 1950), which forms a useful starting point for the study of the period.

The best-known reference to the work of Georgius Agricola (1494–1555) is the translation by Herbert Clark Hoover and Lou H. Hoover of the impressive *De Re Metallica* (The Mining Magazine, 1912; Dover reprint, 1950), which includes important biographical and bibliographical material as well as extensive discussions of the early history of geology and the mineral industry. *Agricola on Metals . . .* , by Bern Dibner (Norwalk, Conn.; Burndy Library, 1958), on the other hand, is a slim book which contains a historical introduction, summaries of the contents of each chapter of the original placed in their proper historical setting and explained in language intelligible to the non-technical reader, and an explanation of the relation of the work of Agricola to that of the later geologists Abraham Werner and James Hutton. Another important translation is that from the first Latin edition (1546) of *De Natura Fossilium* by Mark C. and Jean A. Bandy for the Mineralogical Society of America, issued as *Special Paper*, 63, by the Geological Society of America (1955). Two other essential references are: the series of seven books under the collective title *Georgius Agricola—Ausgewählte Werke*, edited by H. Prescher and sponsored by the Staatliches Museum für Mineralogie und Geologie at Dresden; and the commemorative volume *Georgius Agricola, 1494–1555, zu seinem 400. Todestag 21* (1955), issued by the Central Agricola Commission of the German Democratic Republic of East Germany. A comparatively recent paper in English is: 'Georgius Agricola (1494–1555)' by Joan M. Eyles (*Nature*, **176**, 949–950, 1955), which draws attention to Agricola's earlier work *De Ortu et Causis Subterranerrum*, published in 1546.

'Palissy the Potter' (1514 or 1520–90) is known as an important authority on certain aspects of earth history. Notable among the subjects in which he was interested are the hydrological cycle, the use of mineral fertilisers, fossils, pottery clays and alchemy. His discussion of alchemy illustrates particularly what seems to be his most remarkable characteristic—his philosophical iconoclasm. In geology he was a premature Neptunist. Aurèle La Rocque has provided a translation of Palissy's best-known work, *Discours Admirables, de la Nature des Eaux et Fontaines, tant Naturelles qu'Artificielles, des Métaux des Sels et Salines, des Pierres, des Terres*, etc. (Paris, 1580) under the title *The Admirable Discourses* (University of Illinois Press, 1957). La Rocque's accompanying notes are chiefly philological. In his article in *Toward a History of Geology*, entitled simply 'Bernard Palissy', however, he outlines the sources of Palissy's work and the important influences in his life. He also catalogues the comments by subsequent authorities on Palissy's work and thus provides a useful bibliographic key. A supplement to La Rocque's work is the paper on 'The Geographical and Geological Observations of Bernard Palissy the Potter', by H. R. Thompson (*Annals of Science*, **10**, 149–165, 1954). The rare edition of Palissy's *Oeuvres Complètes* (J. J. Dubochet et Cie, 1844) was reprinted in 1961 by the Librairie Scientifique et Technique Albert Blanchard. The original notes and historical notice by Paul-Antoine Cap are reproduced and a new preface is added, by Professor Jean Orcel. The preface is designed as a popular introduction.

The descriptions of Pembrokeshire by George Owen (1552–1613), the 'Patriarch of English geology', remained in manuscript until 1796, when they were printed in Vol. 1 of the *Cambrian Register*, by Richard Fenton, and then reissued in Fenton's *A Historical Tour Through Pembrokeshire*, edited by E. Davies (1903).

The importance of Owen's work was appreciated by W. D. Conybeare and W. Phillips in their *Outlines of the Geology of England*

and Wales . . . (Phillips, 1822), who maintain that it was 'undoubtedly the earliest attempt to establish the important and fundamental geological fact that the same series of rocks succeed each other in a regular order throughout extensive tracts of country, and to elucidate the geological structure thus indicated'. His work has been described as perhaps the earliest attempt to give a full description of boulder clay in Britain. The best general reference to Owen's geological work is the article 'From Giraldus Cambrensis to the Geological Map. The Evolution of a Science', by F. J. North (*Transactions of the Cardiff Naturalists' Society*, **64**, 20–97, 1933).

A name which should have been included in the title of Pierre Brunet's article (see above) is that of Conrad Gesner (1516–65), whose work on minerals is discussed in Adams' *Birth and Development* and on fossils in W. N. Edwards' *The Early History of Palaeontology*. See also *Conrad Gesner, Physician, Scholar, Scientist —A Quatercentenary Exhibit*, by Richard Durling (National Library of Medicine, 1968) and the biography by Hans Fischer cited in the section on 'Biography'.

Secondary sources of more general interest for the fifteenth and sixteenth centuries include: *The Legend of Noah*, by Don Cameron Allen (University of Illinois Press, 1949)—a thoroughly documented history of the Flood controversy; *Tudor Geography 1485–1583*, by E. G. R. Taylor (Methuen, 1930), which contains a 'Catalogue of English Geographical or Kindred Works (Printed Books and Mss.) to 1583'; and *Science and Thought in the Fifteenth Century*, by Lynn Thorndyke (Columbia University Press, 1929).

Seventeenth century

The British philosopher Alfred North Whitehead (1861–1947) has written of the seventeenth century that: '. . . a brief, and sufficiently accurate description of the intellectual life of the European races during the succeeding two centuries and a quarter' is that succeeding generations 'have been living upon the accumulated capital of ideas provided for them by the genius of the seventeenth century'. Applied to geology, the statement seems, on first acquaintance, to be too bold. During the first half of the century the subject was largely ignored, but in the second half discoveries were deliberately undertaken, collections were made, and consequently museums rose to a position of importance. Furthermore, the germs of the basic principles underlying modern stratigraphy and palaeontology can certainly be found in the writings of the period.

A survey of geology around the middle of the century is given in 'La Géologie au Milieu du XVIIe Siècle', by Robert Lenoble (*Les Conférences du Palais de la Découverte, Ser. D*, No. 27, 1954).

The extremely important contributions of Nicolas Steno (Niels Stensen) (1638–86) and Robert Hooke (1653–1703) are the subject of an extensive literature. To quote some of the more recent examples: *Nicolaus Steno and his Indice*, edited by Gustav Scherz (Munksgaard, Copenhagen, 1958), contains numerous articles on Steno as well as the text of his *Indice di Cose Naturali, Forse Dettato da Niccolò Stenone*. Steno's other geological treatises are submitted in their original texts, in translation with annotations and with critical reviews, in 'Steno: Geological Papers', edited by Gustav Scherz and translated by Alex J. Pollock (Odense University Press, 1969), being Vol. 20 of *Acta Historica Scientiarum Naturalium et Medicinalium*. Steno's relations with the Royal Society are outlined in 'Nicolas Stenon et la Royal Society of London', by Remacle Rome (*Osiris*, **12**, 244–268, 1956).

Micrographia: or Some Physiological Descriptions of Minute Bodies made by Magnifying Glasses with Observations and Inquiries Thereupon, by Robert Hooke, was printed for the Royal Society by John Martyn and James Allestry of London in 1665. The work was reproduced in Vol. 13 of R. T. Gunther's series *Early Science in Oxford* (Oxford University Press for the Subscribers, 1938; reprinted 1968). Attention to the enlightened geological views of Robert Hooke was drawn by A. P. Rossiter in 'The First English Geologist' (*Durham University Journal*, **29**, 172–181, 1935). A later paper, 'Robert Hooke and his Conception of Earth History', by Gordon L. Davies (*Proceedings of the Geologists' Association*, **75**, 493–498, 1964), and the biography *Robert Hooke*, by Margaret 'Espinasse (Heinemann, and University of California Press, 1956), together provide much additional material and useful bibliographies of both primary and secondary sources.

Bibliographic data relating to various theories of the earth published in the seventeenth century, with special reference to contemporary discussion and published commentaries, in the light of their significance for the study of the history of geology, are given in 'Bibliography and the History of Science', by V. A. Eyles (*Journal of the Society for the Bibliography of Natural History*, **3**, 63–71, 1955).

The general scientific setting of the century is given in 'The 17th-Century Scientific Revolution', by Robert Lenoble in René Taton's *The Beginnings of Modern Science from 1450 to 1800* (Thames and Hudson, 1964).

Eighteenth century (the first three quarters)

V. A. Eyles' contribution in *Toward a History of Geology*—'The Extent of Geological Knowledge in the Eighteenth Century, and the Methods by which it was Diffused'—provides a comprehensive guide to the primary literature of the period. He maintains that contrary to popular opinion a great deal of geological work was achieved during the century, and his contention is confirmed in the other eight papers in this Symposium Volume which deal with various aspects of eighteenth century geology in America, France, Germany, Russia and Sweden.

Eyles makes little reference to secondary sources, however, and for this reason the following two general papers are very useful as supplements: 'Problems and Sources in the History of Geology, 1749–1810', by Rhoda Rappaport (*History of Science*, 3, 60–78, 1964), and 'The History of Geology: Suggestions for Further Research', by V. A. Eyles (*History of Science*, 5, 77–86, 1966). Furthermore, Eyles makes little direct reference to two important aspects of eighteenth century geology—the growth of geomorphology and the search for order. The former is dealt with in a comprehensive way in *The Earth in Decay, A History of British Geomorphology 1578–1878*, by G. L. Davies (Macdonald, 1969). The latter was a search shared by most of the sciences. One aspect is discussed in Jacob Bronowski's short chapter on 'The Eighteenth Century and the Idea of Order' in his book *The Common Sense of Science* (Heinemann, 1951; Penguin Books, 1960); another in *The Great Chain of Being: A Study of the History of an Idea*, by A. O. Lovejoy (Harvard University Press, 1936; Harper Torchbook, 1960), the classic study of the *Scala Naturae* or 'Ladder of Being'.

The last three decades of the eighteenth and the first two of the nineteenth century are commonly recognised by historians as being of critical importance in the history of the subject of geology and referred to as the Heroic Age. The term was first coined by Zittel in his *History* to denote the period 1790–1820 but was later extended by George Sarton in his 'La Synthèse Géologique de 1775 à 1918' (*Isis*, 11, 357–394, 1919) to cover the period from 1775 to the 1820s.

Before outlining the activities of the Heroic Age it is appropriate to cite the details of recently issued facsimile reprints and/or translations of the famous publications of the century, other than those of James Hutton and Abraham Werner which are mentioned later. These include: *An Introduction to the Natural History of the Terrestrial Sphere*—the first English translation of Rudolph Erich Raspe's *Specimen Historiae Naturalis Globi Terraquei* (1763), by A. N. Iversen and A. V. Carozzi (Hafner, 1970). 'Buffon, Les Epoques de la Nature (Edition critique avec le Manuscrit, une Introduction et des Notes)', edited by Jacques Roger (*Mémoires du Muséum National d'Histoire Naturelle Série C: Sciences de la Terre*, **10**, 1962). *The Lying Stones of Dr. Johann Bartholomew Adam Beringer being his 'Lithographiae Wirceburgensis . . .'*, translated and annotated by Melvin E. Jahn and Daniel J. Woolf (University of California Press, 1963). *Telliamed or Conversations between an Indian Philosopher and a French Missionary on the Diminution of the Sea*, by Benôit de Maillet (University of Illinois Press, 1968), being a translation by A. V. Carozzi from the recently established original text *Telliamed, ou Entretiens d'un Philosophe Indien avec un Missionnaire Français sur la Diminution de la Mer, la Formation de la Terre, l'Origine de l'Homme . . .* (Amsterdam, 1748). *Italian Journey* (1786–88), by Johann Wolfgang von Goethe, translated by W. H. Auden and Elizabeth Mayer (Collins, 1962; Penguin Classics, 1970), which contains many references to geological and mineralogical subjects.

Secondary sources of particular relevance include: *Les Sciences de la Nature en France au XVIII Siècle. Un Chapitre de l'Histoire des Idées*, the classic study by Daniel Mornet (Colin, 1911); *Cosmogonies of our Fathers: Some Theories of the Seventeenth and Eighteenth Centuries*, by K. C. Collier (1934); *The Lisbon Earthquake*, by T. D. Kendrick (1956), which provides interesting background material; *English Naturalists from Neckham to Ray: A Study of the Making of the Modern World*, by C. E. Raven (Cambridge University Press, 1947); *Mountain Gloom and Mountain Glory*, by Marjorie Nicolson (Norton Library, 1963); and *A History of Science, Technology and Philosophy in the Eighteenth Century*, by A. Wolf (Macmillan, 1939), which contains a chapter on geology.

The 'Heroic Age of Geology' (1770s–1820s)

The activities of this relatively short but revolutionary period can be appreciated by referring to some of the very many publications dealing with the lives and works of three of the major formative figures—James Hutton (1726–97), Abraham Gottlob Werner (1749–1817) and William Smith (1769–1839).

Hutton's works are probably more accessible today than at any time during the past 100 years. The two volumes of *Theory of the*

Earth, with Proofs and Illustrations (Cadell and Davies, 1795) are available in facsimile reprint (Wheldon and Wesley, 1959); *James Hutton, System of the Earth, 1785; Theory of the Earth, 1788; Observations on Granite, 1794; together with Playfair's Biography of Hutton*, published as Vol. 5 in Contributions to the History of Geology Series (Hafner, 1970) brings together all of Hutton's geological writings not previously reprinted. Extensive quotations from Hutton's works, with explanatory notes, are given in numerous works, the most recent of which is *James Hutton —The Founder of Modern Geology*, by E. B. Bailey (Elsevier, 1967).

Although a great deal has been written about Hutton, it is surprising how little scholarly attention has been paid to his *Theory*. Apart from the general texts mentioned early in this chapter, the most readily accessible studies are: first, the 1949 volume of the *Proceedings of the Royal Society of Edinburgh*, which commemorates the 150th anniversary of Hutton's death; second, Chapter 6, 'The Huttonian Earth Machine 1785–1802', in G. L. Davies' book *The Earth in Decay* (Macdonald, 1969); third, 'James Hutton and the Concept of a Dynamic Earth', by R. H. Dott in *Toward a History of Geology*; and fourth, 'James Hutton und die Ewigkeit der Welt', by R. Hooykaas (*Gesnerus*, **23**, 55–66, 1966).

A very useful introduction to Werner's work is given in 'Abraham Gottlob Werner (1749–1817) and his Position in the History of the Mineralogical and Geological Sciences', by V. A. Eyles (*History of Science*, **3**, 102–115, 1964). This paper is an essay review of A. V. Carozzi's translation of *Von den Ausserlichen Kennzeichen der Fossilien* (Leipzig, 1774)—Werner's first book. The translation—*On the External Characters of Minerals* (University of Illinois Press, 1962)—is based on Werner's personal and annotated copy of the original edition and is therefore a rendering of what could have been the 'second edition' of the work. Werner's second book was the 28-page pamphlet *Kurze Klassifikation und Beschreibung der Verschiedenen Gebirgsarten* (1785), which has been translated into English as *Brief Classification and Description of the Different Rocks* by A. M. Ospovat (Hafner, 1970). The contents are discussed by Ospovat in *Toward a History of Geology* and in a short note of his in *Isis* **58**, 91–95 (1967).

A wide range of papers was published to celebrate the 150th anniversary of Werner's death. Fifteen of these, including several in English, are contained in 'Abraham Gottlob Werner; Gedenkschrift aus Anlass der Wiederkehr seines Todestages nach 150 Jahren am 30 Juni 1967', edited by H. J. Rösler *et al.* (*Freiberger Forschungshefte, Reihe* C, No. 223, 1967).

As Geikie has remarked, Werner was a great teacher of geology. The list of his students includes Leopold von Buch (1774–1853); Alexander von Humboldt (1769–1859), author of the massive five-volume *Cosmos: A Sketch of a Physical Description of the Universe*, translated by E. C. Otté (London: Bohn, 1849); Johann Karl Wilhelm Voigt (1752–1821); Johann Friedrich Blumenbach (1752–1840); Baron Ernst von Schlotheim (1764–1832), one of the few pupils who devoted himself to the study of fossils and author of *Die Petrefactenkunde* (Becker, 1820); Jean D'Aubuisson de Voisins (1769–1841), author of one of the three most trustworthy reports on Werner's 'geognosy'—*Traité de Géognosie* (Strasburg and Paris, 1819); and Robert Jameson (1774–1854), Werner's British advocate and author of another of the reports on Werner's 'geognosy'—*Elements of Geognosy* (Edinburgh, 1808).

The much-publicised dispute between Hutton the vulcanist (or plutonist) and Werner the neptunist is a very complicated matter and needs much further study. Among the works dealing with the controversy are 'The University of Edinburgh's Natural History Museum and the Huttonian–Wernerian Debate', by Arnand C. Chitnis (*Annals of Science*, **26**, 85–94, 1970), and 'Hutton and Werner Compared: George Greenough's Geological Tour of Scotland in 1805', by Martin J. S. Rudwick (*British Journal for the History of Science*, **1**, 117–135, 1962).

The extensive list of papers issued on the Werner anniversary dwarfs the equivalent one for the 200th anniversary of William Smith's birth in 1769. Compared to the many studying Werner, it appears that one historian only—Joan Eyles—is currently studying the works of the 'Father of English Geology and Stratigraphy'. Mrs. Eyles has almost completed a biography of Smith, a work which will help fill a very big gap in the literature of the history of British geology. Existing works on Smith include 'William Smith: Some Aspects of his Life and Work', in *Toward a History of Geology* and 'William Smith, the Father of English Geology and of Stratigraphy: An Anthology', by D. A. Bassett (*Geology*, **1**, 38–51, 1969).

Nineteenth century (second, third and fourth quarters)

The period 1820 to 1840 has been referred to as the period of the 'Great Masters of Geology', and as the 'Golden Age of Geology'. Others claim that the period should be extended to 1875 so as to include the great geological syntheses of Charles Lyell, Leopold von Buch, Alexander von Humboldt and Elie de Beaumont, as well as the evolutionary synthesis of Charles Darwin. Contemporary reviews of these developments include W. D. Conybeare's 'Report on the Progress, Actual State, and Ulterior Prospects of Geological Science' (*Report of the British Association for the Advancement of Science* for 1832, 365–414, 1833); *Histoire des Progrès de la Géologie de 1834 à 1845*, by Le Vicomte E. J. Adolphe d'Archiac (La Société Géologique de France, 1847–60), a work which changed the final date in its title while it was in progress, extending eventually to 1859; and the series of brilliant articles which appeared anonymously in *The Edinburgh Review* but is now known to have been written by Dr. W. H. Fitton. The first of these, which represents one of the very first histories of British geology, was reissued, under the author's name, expanded to twice its length, but carrying the story no further in time, as 'Notes on the History of English Geology' (*Philosophical Magazine*, (1–2), 1832–33). Assessments or reassessments of the events chronicled in these contemporary reviews are given in a number of later summaries usually written to commemorate some centenary or other. They include 'The Rise of Geology and its Influence on Contemporary Thought', by H. H. Thomas (*Annals of Science*, **5**, 325–341, 1947).

Societies devoted exclusively to geology appeared for the first time in the early decades of the nineteenth century. The work of the very first—The Geological Society of London—founded in 1808, is described in *The History of The Geological Society of London*, by H. B. Woodward (Geological Society, 1907), which is concerned mainly with the first 50 years of the Society's existence. Unlike Woodward, who had to rely almost exclusively on published sources, Martin Rudwick has had access to the private papers and correspondence of George Bellas Greenough (1778–1856), the first President of the Society (from 1807 to 1813), and has consequently been able to reconstruct the early events of the Society in much better detail. His paper 'The Foundation of the Geological Society of London: Its Scheme for Co-operative Research and its Struggle for Independence' (*British Journal for the History of Science*, **1**, 325–355, 1963) provides a very useful introduction to the state of geological study and to the conflict between the practical and the theoretical geologist and consequently to the influence of Francis Bacon.

References to the history of other national societies include: *La Société Géologique de France de 1880–1929. Centenaire de la Société*, by E. de Margerie (Livre Jubilaire, 1930). The history of the first half-century is given by A. de Lapparent in the third series of the Society's bulletin (1880). 'The Geological Society of Dublin and the Royal Geological Society of Ireland 1831–1890', by G. L. Davies (*Hermathena. A Dublin University Review*, No. 100, 66–76, 1965). 'Aus der Geschichte der Deutschen Geologische Gesellschaft', by Karl Andrée (*Deutsche Geologische Gesellschaft, Zeitschrift*, **100**, 1–24, 1950), which outlines the history of the society from its founding in 1848. *The Geological Society of America, 1888–1930. A Chapter in Earth Science History*, by H. L. Fairchild (Geological Society of America, 1932).

The Geological Survey of Great Britain was founded in this period and the story of its work and development is given in two books: the 'official' history, *The First Hundred Years of the Geological Survey of Great Britain*, by Sir John Smith Flett (HMSO, 1937), and the much more personal *Geological Survey of Great Britain*, by Sir Edward Bailey (Murby, 1952).

The success of the British Survey resulted in the initiation of a large number of national surveys, and it is tempting to refer to the period following the 'Golden Age' as the age of national geological surveys. Details of the various European surveys are given in 'Report Upon the National Geological Surveys. Part I. Europe', by W. Topley (*Report of the British Association for the Advancement of Science* for 1884, 221–237, 1885); and the story of some critical decades in the history of the US Survey is given in *Government in Science. The U.S. Geological Survey, 1867–1894*, by T. G. Manning (University of Kentucky Press, 1967).

In spite of the emphasis on empirical work, theory inevitably played a major role in the development of the subject, as demonstrated in studies of the work of Charles Lyell. The publication of Lyell's *Principles of Geology: An Attempt to Explain the Former Changes of the Earth's Sculpture by Reference to Causes now in Operation* (3 vols., Murray, 1830–33) was one of the major events of the century and one which had an effect on geological thought similar to that of Charles Darwin's *Origin of Species* on biological thought some three decades later. Recent important analyses of the work are: 'The Strategy of Lyell's Principles of Geology', by M. J. S. Rudwick (*Isis*, **61**, 5–33, 1969); and 'The Intellectual Background to Charles Lyell's *Principles of Geology* (1830–1833)', by L. G. Wilson in *Toward a History of Geology*.

The majority of references to Lyell's work, in both geological and historical textbooks, imply that it completely or almost completely resolved the catastrophist–uniformitarian controversy. That the controversy has not yet been fully resolved is clear from the remarkable defence of Cuverian catastrophism in the two-volume *Synthetische Artbildung*, by N. Heribert-Nilsson (Lund, Gleerup, 1953). Furthermore, Toulmin demonstrates in his paper 'Historical

Inference in Science: Geology as a Model of Cosmology' (*Monist*, **47**, 142–158, 1962) that the controversy is still very much part of contemporary cosmological thinking. Brief references to the controversy are extremely common but adequate historical studies are few. Important modern introductions are the two papers by W. F. Cannon: 'The Uniformitarian–Catastrophist Debate' (*Isis*, **51**, 38–55, 1960); and 'The Impact of Uniformitarianism: Two Letters from John Herschel to Charles Lyell, 1836–1837', by W. F. Cannon (*Proceedings of the American Philosophical Society*, **105**, 301–314, 1961).

It is generally suggested that the study of Lyell's *Principles* was a seminal factor in Charles Darwin's intellectual development. This link is strongly upheld by Michael T. Ghiselin in his book *The Evolution of Charles Darwin. The Triumph of the Darwinian Method* (University of California Press, 1969). Additional information on Darwin's development is given by H. E. Gruber and V. Gruber in 'The Eye of Reason; Darwin's Development during the *Beagle* Voyage' (*Isis*, **53**, 186–200, 1962), in which they state that, of the 2530 pages of notes written by Darwin during the voyage, 1383 were on geological topics. An earlier but still useful work is *Charles Darwin as Geologist*, by A. Geikie (Cambridge University Press, 1909). The centenary of the publication of *On the Origin of Species by Means of Natural Selection* (John Murray, 1859) led in 1959 to a spate of biographical and other studies on the importance of Darwin's work. Details are given in the chapter 'History and Biography of Biology', by J. L. Thornton and R. I. J. Tully, in the companion volume *The Use of Biological Literature*, edited by R. T. Bottle and H. V. Wyatt (2nd edn., Butterworths, 1971). An inventory of Darwin's publications is available in *The Works of Charles Darwin. An Annotated Bibliographical Handlist*, by R. B. Freeman (Dawsons, 1965).

A work which undoubtedly did great service in familiarising the reading public with the idea of evolution and thus prepared them for the more complete and more efficient *Theory* presented by Darwin was the anonymous *Vestiges of the Natural History of Creation* (Churchill, London, 1844), written by the essayist, publisher and amateur geologist, Robert Chambers. It proved to be 'one of the most roundly hated works of its time' and yet 'alarmingly' popular. Four editions were issued in the first seven months and by 1860 it had reached the 11th edition (which represented the author's final revision of the text) and about 24 000 copies had been sold. The authorship was first acknowledged in the 12th edition in 1884 after Chambers' death. A facsimile of the book was recently issued by Leicester University Press (1969) with an introduction by Sir Gavin de Beer.

As a result of the theory enunciated by Darwin and the formulation of the second law of thermodynamics, the swing of the pendulum from catastrophism to uniformitarianism was very effectively dampened—see, for example, Thomas Huxley's famous presidential address to the Geological Society of London in 1869 'Catastrophism, Uniformitarianism, Evolutionism' and M. King Hubbert's article 'Critique of the Principle of Uniformity' in 'Uniformity and Simplicity, A Symposium on the Principle of the Uniformity of Nature', edited by C. C. Albritton, Jr. (*Geological Society of America Special Paper*, No. 89, 1967).

A very useful inventory of early nineteenth-century literature is given in the 'Bibliographical Essay' in Charles Gillispie's *Genesis and Geology* (see p. 403). It includes sections on works of a theological character, scientific works and popular science, as well as details of scientific societies and their publications, general periodicals and journals of opinion.

Developments during the second half of the nineteenth century are outlined in some of the general papers referred to earlier. Professional historians of science have not, as yet, been attracted to the period. The main exception appears to be George Sarton's article 'La Synthèse Géologique de 1775 à 1918' (*Isis*, **11**, 357–394, 1914–19), which is devoted in the main to the work of Edward Suess (1831–1914), author of the monumental *Das Antlitz der Erde* (Prague, F. Tempsky; Leipzig, G. Freytag), the three volumes of which appeared in 1885, 1888 and 1908, respectively. The English translation by H. B. C. Sollas under the direction of W. J. Sollas was published by Clarendon Press between 1904 and 1924. A measure of the influence of Suess' theories is that they formed the theme of the presidential addresses by Charles Lapworth and James Geikie to the geological and geographical sections, respectively, of the British Association meeting at Edinburgh in 1892.

Two striking developments of the second half of the nineteenth century were the advances in palaeontology and geomorphology which resulted from the opening up of the American Far West by pioneers such as E. D. Cope (1840–97), G. K. Gilbert (1843–1918), C. King (1842–1901), O. C. Marsh (1831–99), J. W. Powell (1834–1902) and others. The book *Great Surveys of the American West*, by R. A. Bartlett (University of Oklahoma Press, 1962), provides a useful introduction to the surveys and includes a very good bibliography of published and unpublished government documents as well as of secondary sources. *Exploration and Empire. The Explorer and the Scientist in the Winning of the American West,* by W. H. Goetzmann (Alfred Knopf, 1966), is equally useful. The third section of *The Earth We Live On: The Story of Geological Discovery*, by Ruth Moore (Cope, 1957), contains an attempt to correlate and assess the achievements of the scientific discoverers of the West. An introduction to the geomorphological story is given in *The History of the Study of Landforms or the Development of Geomorphology*. Vol. 1, *Geomorphology before Davis*, by R. J. Chorley, A. J. Dunn and R. P. Beckinsale (Methuen, 1964); and to the palaeontological story in Jean Piveteau's contribution to *La Science Contemporaine*, edited by René Taton (see above), and in the papers on the history of the study of vertebrate palaeontology cited later.

There were major developments in other branches of geology, particularly in petrography and glacial geology. The general picture was one of unprecedented growth, the diversity of which is well illustrated in the titles of the committees set up by Section C (Geology) of the British Association for the Advancement of Science, an Association in which geologists played a very important part from its inception in 1831 to the end of the century.

During the 1870s geologists began discussing the need for an international gathering to bring order to the language and literature of geology. The Société Géologique de France was approached with a view to organising a Congress in connection with the Paris Exposition of 1878. They accepted and the meeting proved to be the beginning of the International Geological Congress. The other early congresses were at Bologna (1881), Berlin (1885), London (1888), Washington (1891) and Zurich (1894). The activities at the first of these are described briefly by William Thurston in 'The First International Geological Congress' (*Geotimes*, **13**, 16–17, 1968).

Another development was that of geology in education, a subject which has not been studied in detail but some introduction to which can be gleaned from the works of Thomas Huxley as described in *T. H. Huxley: Scientist, Humanist and Educator*, by C. Bibby (Watts, 1959).

Twentieth century

As with the second half of the nineteenth century, professional historians of science have not attempted to study the plethora of developments in the subject during the twentieth century. A major step towards a history of geology has, however, been taken by Emmanuel de Margerie in his monumental *Critique et Géologie:*

Contribution à l'Histoire des Sciences de la Terre (1882–1947) (4 vols., Colin, 1943, 1946, 1946, 1948). The first volume is autobiographical and bibliographical, with historical information regarding the author and his publications. Many personal letters relating to his principal publications are reproduced in facsimile and many portraits of distinguished geologists are interspersed. The second volume deals with the oceans and oceanic terrains, the general bathymetric map of the oceans and polar regions, geodesy, topography and cartography, and the international map of the world; the third, with the regional geology of the Pyrenees, of France and Spain; and the fourth, with the regional geology of the Jura region of France and Switzerland.

Reviews of the development of particular branches of the subject by practising geologists are obviously numerous. The main review journals have been discussed in Chapter 4.

Works containing extracts from some of the important publications of the century include: *Source Book in Geology 1900–1950*, by K. F. Mather (Harvard University Press, 1967); *Study of the Earth. Readings in Geological Science*, edited by J. F. White (Prentice-Hall, 1962); and *Adventures in Earth History. Being a Volume of Significant Writings from Original Sources, on Cosmology, Geology, Climatology, Oceanography, Organic Evolution, and Related Topics of Interest to Students of Earth History, from the Time of Nicolaus Steno to the Present*, selected, edited and with introductions by Preston Cloud (Freeman, 1970).

REFERENCES ARRANGED BY SUBJECT

Palaeontology

There is no single volume devoted to the history of palaeontology in the English language. A useful introduction to the subject is, however, given in the British Museum handbook *The Early History of Palaeontology*, by W. N. Edwards (British Museum (Natural History), 1967). The much earlier work 'History and Methods of Palaeontological Discovery', by O. C. Marsh (*American Journal of Science*, Ser. 3, **18**, 323–59, 1879), which reviews the progress of palaeontology from Zenophanes, 500 B.C. to 1879, is still useful. Raymond Furon devotes a part of his book *La Paléontologie. La Science des Fossiles, son Histoire, ses Enseignements, ses Curiosités* (2nd edn., Payot, Paris, 1951) to a review of the history of palaeontology from biblical times to the present. Otto H. Schindewolf discusses the history of the subject in his *Wesen und Geschichte der Paläontologie* (Wiss. Editionsges., Berlin, 1948). *Osnovy Paleontologii (Fundamentals of Palaeontology)*, edited by Y. A. Orlov (Izdatel'stvo Akademii Nauk SSSR, 1959; translated from Russian and published for the National Science Foundation, Washington, D.C., by the Israel Program for Scientific Translations, 1962), has a 20-page chapter on 'The Development and Present State of Palaeontology' and a 35-page chapter on 'Problems of Darwinism in Palaeontology' which is in part a historical summary.

For reviews of the more recent developments reference should be made to C. E. Weaver's 'Invertebrate Paleontology and Historical Geology from 1850 to 1950' in *A Century of Progress in the Natural Sciences 1853–1953* (California Academy of Sciences, 1955) and, in British journals, to three presidential addresses: 'British Palaeontology; A Retrospect and Survey', by L. R. Cox (*Proceedings of the Geologists' Association*, **67**, 209–220, 1956–57); 'Recent Developments and Trends in Palaeontology', by Oliver M. B. Bulman (*Advancement of Science*, **16**, 33–42, 1959); and 'Relationship of Paleontology to Stratigraphy', by Sir James Stubblefield (*Advancement of Science*, **11**, 149–159, 1954). In American journals a useful collection of papers is available in the 'Symposium on 50 Years of Palaeontology' in *Journal of Paleontology*, **33** (1959). Developments in micropalaeontology are also outlined in 'Zur Geschichte der Angewandten Mikropaläontologie', by Heinrich Hiltermann (*Naturhistorische Gesellschaft zu Hanover, Bericht*, **109**, 23–47, 1964), which traces the history of the development and application of the subject and which contains a bibliography of 100 titles.

Developments in palaeontology are given in E. A. Newell Arber's contribution to *Studies in the History and Method of Science*, edited by C. Singer (Clarendon Press, 1921); in W. T. Gordon's Presidential Address 'Plant Life and the Philosophy of Geology' (*Report of the British Association for the Advancement of Science*, 50–82, 1934); and in 'Half a Century of Modern Palynology', by A. A. Manten (*Earth-Science Reviews*, **2**, 277–316, 1966).

Aspects of the development of vertebrate palaeontology are summarised in 'On the Development of Vertebrate Palaeontology', by R. S. Lull (*American Journal of Science*, Ser. 4, **46**, 193–221, 1918).

The professional German writer Herbert Wendt has written a popular history of palaeontology under the title *Ehe die Sintflut Kam* (Gerhard Stalling, 1965), which has been translated from the German by Richard and Clara Winston as *Before the Deluge: The Story of Palaeontology* (Gollancz, 1968) and which tells the story primarily in terms of the men who carried out the explorations. There is also some useful palaeontological material in *Dawn of Zoology*, by Willy Ley (Prentice-Hall, 1968) and in the other standard histories of botany and zoology. The latter are cited in the companion volume *The Use of Biological Literature*.

Evolution. The history of the theories of organic nature are outlined in A. O. Lovejoy's *The Great Chain of Being. A Study of the History of an Idea* (Harvard University Press, 1936; Harper Torchbook, 1960). The stages leading up to Darwin's concept are discussed briefly in the last chapters of this book and in greater detail in three papers in *Popular Science Monthly:* 'Herder: Progressionism without Transformism' (**66**, 1904); 'Buffon and the Problem of Species' (**79**, 1911); and 'Kant and Evolution' (**79**, 1911). These three articles, in revised form, are included in the volume *Forerunners of Darwin: 1745–1859*, edited by B. Glass, O. Temkin and W. L. Straus, Jr. (Johns Hopkins Press, 1959; Johns Hopkins Paperback edition, 1968). The paperback edition contains a new introduction by Richard Macksey which provides a very useful bibliographical note on some of the more important contributions to the intellectual history of evolution since the appearance of the original volume. Two other relevant references from among the very considerable number of books on evolution are: the one by Greene mentioned on p. 403 and *Darwin's Century: Evolution and the Men who Discovered It*, by L. Eiseley (Gollancz, 1959).

Stratigraphy and historical geology

Stratigraphy and historical geology is the subject that receives greatest prominence in the textbooks by Zittel and Geikie. In the former, for example, the second longest chapter is devoted to an elaborate catalogue of developments up to the end of the nineteenth century. Subsequent publications, which provide additional information and much better historical analyses, are: 'The Rise of Historical Geology in the Seventeenth Century', by C. J. Schneer (*Isis*, **45**, 256–268, 1954), in which the author relates the initial stages of historical geology to the flourishing of historical studies; 'Unconformity; An Historical Study', by S. Tomkeieff (*Proceedings of the Geologists' Association*, **73**, 383–416, 1962), which traces the concept from its use by Steno and the coining of the term by Robert Jameson in 1805 through its gradual adoption and use in the nineteenth century and the advent of distinctions; 'The Influence of Torbern Bergman (1735–1784) on Stratigraphy: A Résumé', by H. D. Hedberg (*Acta Universitatis Stockholmiensis, Stockholm Contributions in Geology*, **20**, 19–47, 1969), in which the development of stratigraphy in Europe up to Bergman's time is discussed; and 'Lavoisier's Fundamental Contribution to Stratigraphy', by A. Carozzi (*Ohio Journal of Science*, **65**, 71–85, 1965).

References to the original publications in which the fundamental principles of stratigraphy were formulated are given in 'Guiding Principles in Stratigraphy', by H. G. Schenck (*Journal of the Geological Society of India*, **2**, 1–10, 1961).

The development of stratigraphical terminology is outlined briefly in *Growth of a Prehistoric Time-scale Based on Organic Evolution*, by W. B. N. Berry (Freeman, 1968) and 'Concepts of Stratigraphical Classification and Terminology', by L. Størmer (*Earth-Science Reviews*, **1**, 5–28, 1966).

Geochronology

The controversy over the age of the earth which flared up in the early nineteenth century, partly as a result of the developments in stratigraphy, is discussed in its temporal setting in *The Age of the World: Moses to Darwin*, by F. C. Haber (Johns Hopkins Press; Oxford University Press, 1959). In spite of the book's title, Haber does not discuss Hebrew notions of time but is concerned primarily with the overthrow of Mosaic chronology in Christendom in the years between the publication of Thomas Burnet's *Sacred Theory of the Earth* (1681) and of J. B. Lamarck's *Hydrogéologie* (1802). A review of more modern aspects of the story is given in *The Exploration of Time*, by R. N. C. Bowen (Newnes, 1958). A more strictly geological review of the growth of a time-scale is given by W. B. N. Berry in his *Growth of a Prehistoric Time-scale Based on Organic Evolution* (see above), which contains a detailed history of the growth and development of the various stratigraphical units.

Mineralogy

Glimpses of mineralogy during classical and mediaeval times are given in three recent translations: (a) E. R. Caley and J. C. Richards, the former a chemist and the latter a classical scholar, have translated Theophrastus' *De Lapidus* as *History of Stones*

(Ohio State University Press, 1956) and provide an elaborate commentary of 160 pages together with a bibliography and comprehensive Greek and English indexes. (b) Theophrastus' *De Lapidus*, edited by D. E. Eichholz (Clarendon Press, 1965). The appearance of a new edition of the *De Lapidus* only nine years after the edition of Caley and Richards is a sign of the growing interest in Theophrastus' work. (c) Dorothy Wyckoff has issued a translation of Albertus Magnus' *Mineralia* or *Book of Minerals* (c.1260) (Clarendon Press, 1967) with appendices on Aristotle, lapidaries, astrology and magic, and alchemy.

The Gemological Institute of America (Los Angeles, 1950) have issued *A Roman Book on Precious Stones, Pliny's Natural History Book XXXVII* based on Philemon Holland's translation of 1601. The mediaeval lapidaries are cited and discussed in *Magical Jewels of the Middle Ages and the Renaissance*, by J. Evans (Oxford University Press, 1922); *Anglo-Norman Lapidaries*, by P. Studer and J. Evans (Paris, 1924); and *English Medieval Lapidaries*, by J. Evans and M. S. Serjeantson (Oxford University Press, 1933).

The 'classic' history of mineralogy is *Geschichte der Mineralogie von 1650–1860*, by Franz von Kobell (Gottaschen, Munich, 1864; facsimile reprint, Johnson, 1965), which contains in the first part a history of mineralogy, mineral chemistry, mineral physics and crystallography; and in the second, the origins and history of mineral names. A less ambitious but much more modern study is *Les Sciences Minéralogiques au XIXᵉ Siècle (Minéralogie, Cristallographie, Lithologie)*, by J. Orcel (Université de Paris, 1962). In addition there are the sections devoted to the subject in the standard textbooks already cited, and particularly in F. D. Adams' *The Birth and Development of the Geological Sciences* and William Whewell's *History of the Inductive Sciences*.

An extremely useful bibliography of bibliographies is provided in 'Catalogue of Topographical Mineralogies and Regional Bibliographies', by L. J. Spencer (*Mineralogical Magazine*, **28**, 303–332, 1948).

Crystallography

The two definitive histories of the science of crystallography are *La Genèse de la Science des Cristaux*, by Hélène Metzger (Alcan, Paris, 1918), which is concerned primarily with the seventeenth- and eighteenth-century speculations concerning crystalline matter which culminated in the emergence of the science of crystals in the work of René Just Hauy; and the Munich crystallographer Paul

Groth's *Entwicklungsgeschichte der Mineralogischen Wissenschaften* (Springer, 1926), which describes the positive contribution to crystallography of the host of scientists from the seventeenth century through the early years of the twentieth century who were concerned with crystals. The book, which has no illustrations of crystals, has a short but important section giving excellent, brief biographies of important mineralogists. *Origins of the Science of Crystals*, by J. G. Burke (University of California Press, 1966), is designed as a supplement to the two European works 'by bringing to light information, either not contained or little noted in them, and which bears directly on the early development of the science'. The history of crystallography is also outlined in detail in the British Museum (Natural History) handbook *An Introduction to the Study of Minerals*, by L. Fletcher (British Museum, 1924). The British contribution is emphasised in Herbert Deas' article 'Crystallography and Crystallographers in England in the Nineteenth Century' (*Centaurus*, **6**, 129–148, 1959), and a discussion of certain moments in the development of crystallography in Russia (up to 1917) in I. I. Shafranobskii's substantial *Istoriya Kristallografii v Rossii* (Izdatel'stvo Akademii Nauk SSR, 1962).

One important facet of the growth of crystallography since World War I is described in *Fifty Years of X-ray Diffraction*, by P. P. Ewald (Oosthoek for the International Union of Crystallography, 1962), and another in 'Zur Geschichte des Kristallographischen Grundgesetzes; M. v. Laue 65, Geburtstag' (*Zeitschrift für Kristallographie*, **106**, 1–4, 1954).

Petrology

The two best introductions to the history of petrology are *Gesteins- und Lagerstattenbildung im Wandal der Wissenschaftlichen Auschanung*, by W. Fischer (Schweizerbart'sche, 1961), an elaborate history of petrology and of mineral deposits; and *Vvedenie v Istoriyu Petrografii (An Introduction to the History of Petrography)*, by F. Y. Loewinson-Lessing, once director of the Petrographical Institute of Moscow (Moscow, 1936). The Russian work notes the main developments only, interwoven with the personal and often critical views of the author. It was translated as *A Historical Survey of Petrology*, by S. I. Tomkeieff (Oliver and Boyd, 1954) and at the same time the bibliography was brought up to date. The work is sub-divided into sections on the 'geological', 'petrographical', 'chemical', 'experimental' and 'synthetic' approaches to petrology.

An English-language history is given in *History of the Theory of*

Ore Deposits, by T. Crook (Murby, 1933), in which one chapter is devoted to the 'Rise of Petrology'.

Important papers outlining various aspects of the history of igneous petrology include: 'The Classification of Igneous Rocks: An Historical Approach', by S. I. Tomkeieff (*Geological Magazine*, **76**, 41–48, 1939), a discussion which starts with the appearance of the 'curious book' by the Scottish antiquarian and historian J. Pinkerton—*Petralogy. A Treatise on Rocks* (2 vols., London, 1811); 'Fifty Years of Progress in Petrography and Petrology 1876–1926', by F. Bascom (*Johns Hopkins University Studies in Geology*, **8**, 1927); 'The Rise of Petrology as a Science', by L. V. Pirsson (*American Journal of Science*, Ser. 4, **46**, 222–239, 1918); and 'The Evolution of Petrological Ideas', by J. J. H. Teall (*Quarterly Journal of the Geological Society of London*, **57**, lxii-lxxxvi, 1901).

The birth of microscopical petrography is discussed in the various papers on the work of H. C. Sorby—for example, in 'Henry Clifton Sorby, and the Birth of Microscopical Petrology', by J. W. Judd (*Geological Magazine*, Dec. 5, **5**, 193–204, 1908).

An index of the rapid and remarkable success of petrography during the last quarter of the nineteenth century can be obtained by comparing the contents of the two editions of Zirkel's *Lehrbuch der Petrographie*. The first edition, which appeared (as two volumes) in Bonn in 1866, was concerned primarily with the macroscopic methods of the older teaching, whereas the second edition (3 vols., Leipzig, 1893–94) is a frank and full acknowledgment of the petrographical reform necessitated by microscopic and microchemical methods.

Sedimentary petrology

Papers outlining aspects of the history of sedimentary petrology include: 'The Evolution of Petrological Ideas [Sedimentary Rocks; Crystalline Schists and Metamorphic Rocks]', by J. J. H. Teall (*Quarterly Journal of the Geological Society of London*, **58**, lxiii-lxxviii, 1902); 'The Birth and Development of the Concepts of Diagenesis (1866–1966)', by G. D. de Segonzac (*Earth-Science Reviews*, **4**, 153–201, 1968); and 'Oolith: zur Geschichte eines Gesteines', by Ludwig Rüger (*Hallesches Jahrbuch für Mitteldeutsche Erdgeschichte*, **1**, 184–188, 1951), which traces the history of inquiry concerning the nature and origin of oolites from the explanation offered by Pliny to the end of the eighteenth century.

Geochemistry

Brief outlines of the history of geochemistry are given in: 'Über den Historischen Charakter Geochemischer Bedingungen', by Heinz Schulz (*Deutsche Gesellschaft für Geologische Wissenschaften, Berichte*, Reihe B, **12**, 99–113, 1967); 'Half a Century of Geochemistry', by A. P. Vinogradov (*Geochemistry International*, **4**, 1027–1029, 1969); and 'The Geochemical Society', by E. Ingerson (*Geotimes*, **6**, 8–14, 1962).

Geomorphology

Two recent works provide a broad-based survey of the history of geomorphology up to the last quarter of the nineteenth century. They are: *The History of the Study of Landforms or the Development of Geomorphology. 1. Geomorphology before Davis*, by R. J. Chorley, A. J. Dunn and R. P. Beckinsale (Methuen/Wiley 1964); and *The Earth in Decay. A History of British Geomorphology 1578–1878*, by G. L. Davies (Macdonald, 1969). The emphasis in the former, in spite of its comprehensive title, is overwhelmingly on British and American geomorphological thought in the nineteenth century. The work contains many lengthy quotations, making up about half of the text, and the authors consistently interpret and evaluate the ideas of the past using the terms and the concepts of modern geomorphology. The latter work deals in greater detail than any previous work with the literature of geomorphology prior to Hutton's renowned *Theory of the Earth* (see above) and also draws attention to the intellectual background against which the scholars worked, and the influence of this background upon geomorphic thought is analysed.

Hydrology

A comprehensive history of the concept of the hydrological cycle has yet to be attempted. Three recent papers deal with the cycle itself: 'Development of the Hydrologic Cycle Concept and our Challenge in the 20th Century', by R. R. Parizek (*Mineral Industries*, **22**, 1–8, 1963); 'The Hydrologic Cycle', by A. K. Biswas (*Civil Engineering*, **35**, (4), 70–74, 1965); and 'Leonardo da Vinci and the Hydrological Cycle', by A. K. Biswas (*Civil Engineering*, **35**, (9), 73, 1965). Aspects of early hydrology are outlined in 'Hydrology during the Hellenic Civilisation', by A. K. Biswas (*Bulletin of the International Association of Scientific Hydrology*, **12**, (1), 5–14, 1967); and the over-all development in *A History of Hydrology*, by A. K. Biswas (North-Holland, 1970). The birth of modern hydrology is often taken as the appearance of *De l'Origine des Fontaines*, by P. Perrault (1611–80) (Pierre le Petit, Paris, 1674). The work is now available in a complete English translation—*On the Origin of Springs*, by Aurele La Rocque (Hafner, 1967). Although the essence of the work is given in the statement 'the waters of rains and snows are sufficient to cause the flow of all rivers in the World', almost half of the treatise is taken up by preliminary research—a lengthy discussion of the work of 21 writers from Plato to Palissy, giving the opinions of these writers along with Perrault's own comments. The origin of springs and rivers is outlined in Chapter 12 of *The Birth and Development of the Geological Sciences*, by F. D. Adams (see p. 401). The development of hydrogeology is given in Chapter 1 of *Hydrology*, by O. E. Meinzer (McGraw-Hill/Constable, 1942; Dover reprint, 1955) and in the paper 'Historical Development of Ideas Regarding the Origin of Springs and Ground-Water', by M. N. Baker and R. E. Horton (*Transactions of the American Geophysical Union*, (2), 395–400, 1936).

Marine geology and sedimentation

Apart from the brief and often historically orientated introductions to the many modern textbooks and monographs on marine geology and sedimentation, references are comparatively few. P. D. Krynine draws attention to some of the very early ideas in 'On the Antiquity of "Sedimentation" and "Hydrology" (with some moral conclusions)' (*Bulletin of the Geological Society of America*, **71**, 1721–1726, 1960) and urges the restudy of the Greek philosophers. A. A. Manten reviews the history of theories and explanations up through the Challenger Expedition (1872–76) in *Marine Geology* (**2**, 1–28, 1964); J. R. L. Allen draws attention to Sorby's work in 'Henry Clifton Sorby and the Sedimentary Structures of Sands and Sandstones in Relation to Flow Conditions' (*Geologie en Mijnbouw*, **42**, 223–228, 1963); and A. Lamont notes the 'First Use of Current-bedding to Determine Orientation of Strata' (*Nature*, **145**, 1016–1017, 1940). The development of marine studies in general is given in: 'The Major Deep-sea Expeditions and Research Vessels, 1873–1960', by G. Wust, in *Progress in Oceanography* (Pergamon, 1964); 'Founders of Marine Science in Britain: The Work of the Early Fellows of the Royal Society', by M. Deacon (*Notes and Records* *of the Royal Society*, **20**, 28–50, 1965); *Scientists and the Sea 1650–1900. A Study of Marine Science*, by M. Deacon (Academic Press, 1971); and in the more popular *Frontiers of the Sea: The Story of Oceanographic Exploration*, by R. C. Cowen (Gollancz, 1960). General problems of oceanographic history were discussed at the First International Congress on the History of Oceanography held at Monaco in 1966 (see *Geographical Journal*, **133**, 212–213, 1967).

Tectonics, geotectonics and geophysics

The origins of continents and oceans is as much a tectonic as a geomorphological consideration. Papers on the early development of such ideas include: 'The English Worldmakers of the Seventeenth Century and their Influence on the Earth Sciences' (*Geographical Review*, **38**, 104–112, 1948); and 'The Origin of Continents and Oceans: A Seventeenth Century Controversy' (*Geographical Journal*, **116**, 193–198, 1950), by E. G. R. Taylor. The latter is a study of the controversy that arose among English divines and virtuosi during the late seventeenth century concerning the origin of the major surface features of the earth, a controversy centering on Dr. Thomas Burnet's book *Telluria Sacra Theoria*, first published in Latin in 1681 (London), expanded in English in 1684, and often printed in England and abroad down to the early nineteenth century.

Aspects of the comparatively recent history of the theory of continental drift are given in W. A. J. M. van Waterschoot van der Gracht's introduction to *Theory of Continental Drift. A Symposium on The Origin and Movement of Land Masses both Inter-continental and Intra-continental, as Proposed by Alfred Wegener* (American Association of Petroleum Geologists, 1928) and P. M. S. Blackett's introduction to *A Symposium on Continental Drift. Organized for the Royal Society* . . . (The Royal Society, 1965).

A broad outline of the development of geotectonics is given in *Basic Problems in Geotectonics*, by V. V. Beloussov (McGraw-Hill, 1962), which is a translation, with many additions, of the Russian edition of 1954. A review of the birth and development of the geosynclinal concept is given in *Geosynclines*, by J. Aubouin (Elsevier, 1965), and in 'Geosynclines: A Fundamental Concept in Geology', by M. F. Glaessner and C. Teichert (*American Journal of Science*, **245**, 465–482, 571–591, 1947).

A useful chronology of tectonic discovery (1775–1893), with particular reference to the Alps, is given in *Tectonic Essays: Mainly Alpine*, by Sir Edward Bailey (Clarendon Press, 1935);

and a presentation of the three turning points in the development of the literature on Swiss Alpine geology is given in the second William Smith Lecture—'Three Stages in the Evolution of Alpine-Geology, De Saussure-Studer-Heim'—by E. de Margerie (*Quarterly Journal of the Geological Society of London*, **102**, xcvii-cxiv, 1946). A similar study on the Appalachians is given in 'Evolution of Thought on Structure of Middle and Southern Appalachians', by J. Rodgers (*Bulletin of the American Association of Petroleum Geologists*, **33**, 1643–1654, 1949).

Suess' definitive synthesis *Das Antlitz der Erde* has already been mentioned. A little-known sequel is *Terrestrial Theories. A Digest of Various Views as to the Origin and Development of the Earth and their Bearing on the Geology of Egypt*, by W. F. Hume (Geological Survey of Egypt, 1948), which contains not merely a 'Digest' but a connected and reasoned survey of earth theories culled from over 300 major and minor works on the subject in most of the languages of science.

The drawing of early geological sections is discussed in 'White Watson (1760–1835) and his Geological Sections', by T. D. Ford (*Proceedings of the Geologists' Association*, **71**, 349–363, 1960).

Charles Davison in his *The Founders of Seismology* (Cambridge University Press, 1927) provides a history of seismology as well as of seismologists, and there is a historical review in *Seismology. A Brief Historical Survey and a Catalogue of Exhibits in the Seismology Section of the Science Museum* (HMSO, 1958).

G. E. Sweet has compiled *The History of Geophysical Prospecting* (Los Angeles, 1966), and another useful review is 'Half a Century in Geophysics', by Sir Harold Jeffreys (*Advancement of Science*, **10**, 113–119, 1953).

Applied (economic) geology

Books outlining various aspects of the development of applied geology include: the series *A History of Technology*, by C. Singer (Clarendon Press, 1954–58), each volume of which contains both separate chapters and other references to topics of applied geology, including mining, quarrying, water-supply, building materials, etc.; Vol. 7 of the series *Studies in Ancient Technology*, by R. J. Forbes (Brill, 1963), which surveys the geological concepts held by ancient philosophers, geographers and historians, and the development of mining and metallurgy before the Middle Ages, with a chronological bibliography of ancient authors on geology and mining; and *Seventy-five Years Progress in the Mineral Industry*,

1871–1946, edited by A. B. Parsons (American Institute of Mining and Metallurgical Engineers, 1948).

Two relevant bibliographies are: *A Century of Oil and Gas in Books; A Descriptive Bibliography*, by E. B. Swanson (Appleton, 1960); and *The Beginnings of the Petroleum Industry: Sources and Bibliography*, by P. H. Giddens (Pennsylvania Historical Commission, 1941).

The 1970 Symposium of the International Committee on the History of Geological Sciences at Freiberg in Saxony was devoted to the 'History of Concepts on Mineral Deposits' and many of the papers were concerned with applied geology. Fifty-one of the papers presented at the symposium have been published in two special numbers of *Geologie* (*Zeitschrift für das Gesamtgebiet der Geologischen Wissenschaften*), **20**, (4/5), (6/7), 1971.

REFERENCES ARRANGED BY COUNTRY

A comprehensive list of the historically orientated books and papers on all the countries in the world is well beyond the scope of this chapter. The emphasis given in the preceding sections to work carried out in Europe (particularly Western Europe) and North America is therefore continued in this section.

Britain

There is probably more written about the development of geology in the British Isles than on any other area of comparable size. Undoubtedly the best bibliographic guide is the series of four papers in the *Annals of Science* by John Challinor: 'The Early Progress of British Geology. I. From Leland to Woodward, 1538–1728' (**9**, 124–153, 1953); 'II. From Strachey to Michell, 1719–1788' (**10**, 1–9, 1954); 'III. From Hutton to Playfair, 1788–1802' (**10**, 107–148, 1954); 'The Progress of British Geology during the Early Part of the Nineteenth Century' (**26**, 177–234, 1970). The first three papers contain extensive quotations from the primary sources and cite most of the secondary sources. Challinor has also recently issued a book entitled: *The History of British Geology: A Bibliographical Study* (David and Charles, 1971).

Continental Europe

In addition to the many papers cited under the subject classification, the following brief selection of references covers many aspects of the history of geology in Europe.

The main features of the early development and position of historical geology, palaeontology, stratigraphy and regional geology in Sweden and Finland before 1800 are given in 'On the Position of Palaeontology and Historical Geology in Sweden before 1800', by G. Regnéll (*Arkiv för Mineralogie och Geologi (Kungliga Svenska Vetenskapsakademiens*), **1**, 1–64, 1949). Pertinent quotations and translations are given from, among others, Swedenborg, Linnaeus, Wallerius and Hisinger. A biographical index and an extensive bibliography are appended. An outline of the development of the science in Finland during the nineteenth and early twentieth centuries is given in *The History of Geology and Mineralogy in Finland 1828–1918*, by H. Hansen (Societas Scientiarum Fennica, 1968).

Geologiens Historie i Danmark, by A. Garboe (2 vols., Reitzels Forlag, 1959, 1961) provides a basis for understanding why a science in a small country can have a tradition of its own despite the close connections with foreign scientists. Volume 1 of this history of the geological sciences in Denmark, subtitled 'From Myth to Science', covers the period from the earliest times to 1835 (including Norway up to 1814). Volume 2, subtitled 'Researches and Results', covers the period from 1835 to the present.

The first part of a projected history of geological research in Luxemburg appears as 'Aperçu Historique sur les Recherches Géologiques Concernant le Pays de Luxembourg', by Michel Lucius (*Bulletin du Société des Naturalistes Luxembougeois*, **51**, 130–141, 1959). It deals with the period between 1808 and the date of publication of the geological map of Luxemburg by N. Weis and P. M. Siegen.

The extensive literature on the history of geology in various parts of Germany is summarised in the comprehensive bibliography in *Geologie und Paläontology in Texten und ihrer Geschichte*, by H. Holder (Alber, 1960).

The Development of Geodesy and Geophysics in Switzerland, edited by J. C. Thams (Berichthaus, Zurich, 1967), is dedicated to the participants in the 14th General Assembly of the International Union of Geodesy and Geophysics, held in Switzerland. It contains papers on volcanology, hydrology, ice and glaciers, tectonophysics (including geology), etc. Two commemorative publications are: *Hundert Jahre Schweizerische Geologische Kommission, Organ der Schweizerischen Naturforschenden Gessellschaft 1860–1960*, by A. Buxtorf and O. P. Schwarz (Kümmerly and Frey, Bern, 1960); and 'Die Entwicklung der Geologie in den Letzten 50 Jahren', by H. Suter (*Naturforschende Gesellschaft in Zurich, Vierteljahrs-schrift*, **91**, 207–218, 1946), which outlines the progress in Swiss geology since the meetings of the International Geological Congress in Zurich in 1894.

Until 1963 there was no scholarly presentation under one cover of the influence of the geology of Italy on the development of geological ideas. The book *L'Esplorazione Naturalistica dell'Appennino*, by F. Rodelico (le Monnier, Florence, 1963) successfully fills a large part of this gap. One unusual aspect and asset of the book is the full-page photographs of landscapes and outcrops at the end of the volume accompanied by the comments they inspired from famous travellers.

Two papers on the history of geology in the Iberian peninsula are: 'Do Conhecimento Géologico de Portugal Continental', by J. Carrington da Costa (*Anais da Faculdade de Ciências do Porto*, **26**, 206–229, 1941), which outlines the history of geological study in Portugal from the appearance of the earliest scientific writings; and 'Ahora Hace Cien Años . . . ; Ojeada Retrospectiva', by P. H. Sampelayo and J. M. Ríos (*Boletin del Instituto Geológico y Minero de España*, **60**, i–xliii, 1948). The latter was issued in connection with the centenary of the Spanish edition of Charles Lyell's *Elements of Geology* and outlines the history of progress in geology, with particular reference to Spain. Brief biographical sketches of several Spanish geologists are included.

'The Development of the Geological Sciences in the USSR from Ancient Times to the Middle of the Nineteenth Century' is the title of V. V. Tikhomirov's contribution in *Toward a History of Geology*. Developments in general geology, including stratigraphy, tectonics and palaeogeography during the first half of the nineteenth century, are given in the same author's *Geologiya v Rossii Pervoi Poloviny 19 Veta [Geology in Russia in the First Half of the 19th Century]* (2 vols., Akademiya Nauk SSSR, Geologicheskii Institut, 1960, 1963).

North America

The main volumes on the history of geology and allied sciences in the USA and Canada are: *The First One Hundred Years of American Geology*, by G. P. Merrill (Yale University Press, 1924; facsimile reprint, Hafner, 1964): *A Century of Science in America with Special Reference to the American Journal of Science, 1818–1918*, edited by E. S. Dana (Yale University Press, 1918), and published to commemorate the centenary of the founding of the *Journal* by Benjamin Silliman in July 1818; and *Etudes Américaines: Géologie et Géographie*, by E. de Margerie (2 vols., Colin, Paris, 1952, 1954)—a continuation of the series *Critique et Géologie* by the same author (1943–48) and of a similar nature, being critical and historical reviews of geology.

Relevant bibliographies include: (a) *A Bibliography of American Natural History, the Pioneer Century, 1769–1865*, by M. Meisel (3 vols., Premier Publishing Co., Brooklyn, 1924–29). An annotated bibliography of the material published up to 1924 relating to the history, biography, and bibliography of American natural history and its institutions, during colonial times and the pioneer century; with a classified subject and geographical index, and a bibliography of biographies. It includes early state and regional geological surveys. (b) 'Catalogue and Index of Contributions to North American Geology, 1732–1891', by N. H. Darton (*Bulletin of the US Geological Survey*, **127**, 1896). (c) *Catalogue and Index of the Publications of the Hayden, King, Powell, and Wheeler Surveys*, by L. M. Schmeckebier (USGPO, 1904). A catalogue and index of the following surveys: Geological and Geographical Survey of the Territories; Geological Exploration of the 40th Parallel; Geographical and Geological Surveys of the Rocky Mountain Region; and Geographical Surveys West of the 100th Meridian.

A number of the papers presented at the 'Symposium on the History of American Geology' at the American Association for the Advancement of Science Meeting in 1958 are published in the *Journal of the Washington Academy of Sciences*, **49** (1959).

References to relevant manuscript material are given in 'Some Manuscript Resources in the History of Nineteenth Century American Natural Sciences', by E. Lurie (*Isis*, **44**, 363–370, 1953).

BIOGRAPHIES

The standard, and almost indispensable, work of reference for biographical information on scientists in general is *Biographisch-literarisches Handwörterbuch zur Geschichte der Exacten Wissenschaften enthaltend Nachweisungen über Lebensverhältnisse und Leistungen von Mathematikern, Astronomen, Physikern, Chemikern Mineralogen, Geologen . . .* , by J. C. Poggendorff (Barth, 1863–1904; Chemier, 1925–40; facsimile reprint, Edwards, 1945— 10 vols., vols. 1–2 to 1857; vol. 3, 1858–83; vol. 4, 1883–1904; vol. 5, 1904–22; vol. 6, 1923–31).

An alphabetical list of palaeontologists, giving brief details and references to biographical sources is given in 'Palaeontologi Catalogus bio-bibliographicus', by Kalman Lambrecht and W. Quenstedt (*Fossilium Catalogus I. Animalia*, **72**, 1935).

One of the most frequently quoted sets of biographical sketches of geologists is that in the famous *éloges* in which Baron Georges Cuvier immortalised deceased geologists (*Eloges Historiques, Précédés de l'Eloge de l'Auteur par M. Flourens*, Paris, 1835). Pen portraits of 31 French mineralogists and geologists, from Desmarest to Hauy, are given in *Figures de Savants*, by A. Lacroix (4 vols., Gauthier-Villars, 1932–38). Italian geologists are described in 'Italian Pioneers in Geology and Mineralogy', by Michele Gortani (*Cahiers d'Histoire Mondiale*, **7**, 503–519, 1963); and the work of the pioneer Russian geologists A. Karpinsky, I. V. Mushketov, S. Nikitin, A. P. Pavlov, F. N. Chernyshev and V. A. Obruchev is discussed in 'Pioniere Geologischer Forschung in Russland (Beiträge zur Geschichte der Russischen Geologie)', by N. Polutoff (*Geologische Rundschau. Zeitschrift für Allgemeine Geologie Herausgegeben von der Geologischen Vereinigung*, **31**, 457–487, 1940).

Useful lists are given in 'Biographies of Geologists', by J. W. Wells and G. W. White (*Ohio Journal of Science*, **58**, 285–298, 1958), which supplements the list given in the chapter on the history and biography of biology in *The Use of Biological Literature*, and the catalogue in the bibliographic essay in C. C. Gillispie's *Genesis and Geology* (Harvard University Press, 1951). Other useful lists are given in: *Earth for the Layman*, by M. W. Pangborn (American Geological Institute, 1957); 'Biographical Material in the *Philosophical Magazine* to 1900', by E. Scott Barr (*Isis*, **55**, 88–90, 1964); 'Nature's Scientific Worthies', by the same author (*Isis*, **56**, 354–356, 1965).

BIBLIOGRAPHIES

The importance of bibliographical studies for the historian of geology is stressed in the article 'Bibliography and the History of Science', by V. A. Eyles (*Journal of the Society for the Bibliography of Natural Science*, **3**, 63–71, 1955) and epitomised in the work of C. D. Sherborn.

Two essential bibliographical tools for the historian of science in general are *Catalogue of Scientific Papers, 1800–1900* (see p. 74) and *Catalogue of the Books, Manuscripts, Maps and Drawings in the British Museum (Natural History)*. Vol. 1 (A–D), by B. B. Woodward (British Museum (Natural History), 1903).

Two equally essential volumes for the historian of geology are *Bibliographia Zoologiae et Geologiae. A General Catalogue of all Books, Tracts, and Memoirs on Zoology and Geology*, by Louis Agassiz, corrected, enlarged and edited by H. E. Strickland and W. Jardine (see p. 74) and *Catalogue des Bibliographies Géologiques Rédigé avec le Concours des Membres de la Commission Bibliographique du Congrès*, compiled by E. de Margerie (Gauthier-Villars for Congrès Géologique International, 1896; reprint 1967), which lists over 4000 general, regional and local bibliographies of geological interest, with author and subject indexes. Although intended for geologists and not historians and designed, as the title states, as a catalogue of published geological sources rather than as a bibliography of geological literature, the work is so thorough that it is probably one of the best starting points for the student of the sources of the history of British geology.

A number of eighteenth-century bibliographies are cited by V. A. Eyles in his contribution to *Toward a History of Geology* and a very useful list of recent bibliographical publications is given in 'A Check-list of Natural History Bibliographies and Bibliographical Scholarship, 1966–1970', by G. Bridson and A. P. Harvey (*Journal of the Society for the Bibliography of Natural History*, **5**, 428–467, 1971).

Practical exercises

D. N. Wood and A. G. Myatt

The following exercises have been prepared around individual chapters of the book and are intended to be used in connection with courses of instruction on the literature of the earth sciences. Practical work is an essential part of such courses, since it is only by handling the literature that students can become familiar with its structure and skilled in its use.

The exercises are followed by lists of relevant source materials but it is recognised that these sources provide only some of the ways in which answers to the questions may be located. Bibliographic details of the source materials are to be found in the appropriate chapters.

EXERCISE 1   LIBRARIES AND THEIR USE

1. What are the UDC, DDC and L of C classification numbers for *submarine topography*? **2.** Does your library hold (a) *Tuff Lavas and Ignimbrites—A Survey of Soviet Studies*, edited by E. F. Cook (Elsevier, 1966); (b) any books on the geology of India; (c) the following periodicals: *Buletín del Instituto Antartico Argentino*; *Geological Journal*; and *Water Supply Papers, US Geological Survey*? If so, where are they located and in the case of the periodicals what is the length of the run? **3.** Does your library hold a general geological map of the UK (scale around 1:1M) and the USA (scale around 1:5M)? **4.** What is the address of the Bibliothèque de l'Ecole Nationale Supérieure des Mines, Paris?

EXERCISE 2   PRIMARY LITERATURE

1. An article by R. H. Rastall on continental drift appeared in either the *Quarterly Journal of the Geological Society* or *Geological Magazine* in the 1920s or 1930s. What is the full reference? **2.** Give details of a subject classified guide to Australian periodicals. **3.** What major British and American libraries hold extensive sets of the following periodicals: *Geologische Rundschau, Journal of the Geological Society of Tokyo, Geophysical Prospecting*? Give not more than four libraries for each title and for each quote holdings. **4.** Give details of price, publisher and frequency of the following periodicals: *Journal of Petrology, Geoderma, Bulletin of the Seismological Society of America*. **5.** The British Natural Environment Research Council published a monograph on wollastonite in 1969 or 1970; who was the author and how much does it cost? **6.** In the early part of this century the US Geological Survey issued a bulletin on the Miocene and Pleistocene foraminifera from South Carolina. Give bibliographic details. **7.** What is the latest edition of the British Regional Geology Handbook which covers the Tertiary Volcanic District? Quote its ISBN. **8.** Sometime during the 1960s the US Government issued a report on the Carboniferous cephalopods of Arkansas. Give full bibliographic details. **9.** Is *Journal of Geology* available in microform? If so, which volumes and how much does it cost? **10.** The Cold Regions Research and Engineering Laboratory in the USA has been carrying out work on deep ice cores from Greenland. Have any reports been issued on the subject? **11.** In the late 1950s and early 1960s members of the Research Institute of African Geology at the University of Leeds were studying late-Karroo igneous rocks in Southern Rhodesia. What publications and theses have been produced as a result of this work?

EXERCISE 3   SECONDARY LITERATURE—REFERENCE AND REVIEW PUBLICATIONS

1. What is the address of Yasuichi Mima, the Japanese mining executive? **2.** Is there a museum housing a collection relating to Tasmanian geology and palaeontology? Who is its director? **3.** Which organisation is concerned with hydrology and water geology in Saudi Arabia? **4.** What are the publications of the UK Water Resources Board? **5.** Who is the head of the geology department at the University of Peshawar? **6.** Locate the address and details of holdings of an American library devoted to lignite. **7.** How has Turkey's production of crude petroleum varied between 1950 and 1970? **8.** Who published *Geology of North Santo* by G. P. Robinson? **9.** What are (a) hydrargillite; (b) ferberite; (c) wohlerite?

EXERCISE 4 SECONDARY LITERATURE— BIBLIOGRAPHIES, ABSTRACTS AND INDEXES

1. James Mitchell was a nineteenth-century English geologist interested in Cretaceous stratigraphy. Locate a list of his papers. **2.** Locate some references to introductory material on selenium. **3.** Locate bibliographies on (a) bacteria, (b) anorthosite. **4.** Provide full bibliographic details of a paper by Caleb Alexander on eruptions in West River Mountain, New Hampshire. The paper was written late in the eighteenth century. **5.** Dr. V. Krokos is reported to have written an article on geochronology in a Russian journal in the mid-1930s. Can you locate the full reference? **6.** Discover as many abstracts as you can of the paper by R. T. Prider on noonkanbahite, a potassic batisite from lamproites of Western Australia in *Mineralogical Magazine*, **34**, 403–405 (1965). Note the amount of information in each abstract, the way in which it was subject-indexed and the speed with which the abstracts appeared. **7.** Using the cycling procedure, use the latest volume of *Science Citation Index* to compile a short list of recent references on continental drift. **8.** How soon after its appearance in *Journal of Geophysical Research* (1971) did P. D. Rabinowitz's article on gravity anomalies across the East African continental margin first get listed in the secondary literature.

EXERCISE 5 TRANSLATIONS AND FOREIGN LITERATURE

1. Locate a translation into English of (a) the paper 'Tarasovite— A New Dioctahedral Ordered Mixed-Layer Mineral', by E. K. Lazarenko and Yu. M. Korolev in *Zap. Vses. Mineral. Obshch. SSSR*, (2) (1970); (b) the paper, published in late 1962, by S. S. Ellern and R. N. Valeyev on 'The Principal Devonian Trench of the East European Platform'. **2.** Locate the full reference and if possible a translation into English of the paper published in 1967 by A. S. Povarennykh on crystallochemical classification of phosphates. From where can the translation be obtained? **3.** Locate

English abstracts of the following publications (a) a thesis by G. Ranchin (1971) entitled *Uranium Geochemistry and Granite Differentiation in the Uranium Bearing Province of Nord-Limousin*; (b) a paper published in 1965 by L. H. Bactsle, W. F. Maes and J. Souffriau on the transportation of radioisotopes in the soil; (c) I. A. Menyailov, 'Fumarole Gases of Pyroclastic Flows of Bezymyannyi and Katmai Volcanoes', *Vulkany Izverzheniya* (Izdatel'stvo Nauka, 1969).

EXERCISE 6   MAPS

1. Information is required on the availability of geological maps of the following areas. Where possible give bibliographic details, cost and address of source. (a) 1 : 200 000 of south-west Germany; (b) 1 : 100 000 series of New England, Australia; (c) 1 : 100 000 of Munhino, Angola; (d) 1 : 50 000 of the Avignon area of France; (e) 1 : 24 000 of the Panther quadrangle, Daviess County, Kentucky; (f) any reasonably detailed map of the Barra Isles, Outer Hebrides; (g) 1 : 500 000 of Rabat, Morocco; (h) 1 : 1 000 000 of the Mare Vaporum Quadrangle, the Moon; (i) approximately 1 : 500 000 of the area between the Mawson and Priestly Glaciers, Antarctica; (j) approximately 1 : 250 000 of the south-western part of the Nuanetsi district of Rhodesia. **2.** Locate an illustration showing the geological map by William Smith dated 1801 and thought to be the earliest geological map of England and Wales.

EXERCISE 7   STRATIGRAPHY INCLUDING REGIONAL GEOLOGY

1. What is the age of the Goodwood Limestone found near Otago in New Zealand? What fossils does it contain and where can an account of the geology of this area be found? **2.** Locate a geological excursion guide to the Assynt area of Sutherland. **3.** Locate a general account of the geology of Sao Tomé. **4.** Locate a general account of the Precambrian rocks of the Canadian Shield. **5.** Has any recent (1968 onwards) work been reported on the stratigraphy of the Mafingi Hills in Northern Malawi and Zambia?

EXERCISE 8   PALAEONTOLOGY

1. One of two known fossil sea spiders was discovered in the Lower

Devonian of Western Germany. What is its size? **2.** Locate a diagram of the skull of *Didelphis marsupialis*. **3.** A description of a new fossil genus *Pleuromeiopsis* was published in 1962. Where? **4.** A. E. Kasper Jr., working at the University of Connecticut, has recently (last five years) reported on a new genus of Devonian fossil plants from northern Maine. Locate a reference to this work which names the genus. **5.** Of what significance was the discovery, first reported in 1965, of saturated hydrocarbons (alkanes) in the Precambrian Soudan Formation of Minnesota? **6.** O. H. Schindewolf (1942) and T. G. Iljina (1965) have suggested that the coelenterate order Scleractinia evolved from the sub-family of Rugosa known as the Plerophyillinae. (a) Are translations into English of these German and Russian papers available? (b) Has any work been reported since 1965 to substantiate the relationship and if so which particular genera were studied? (c) Find out who first described these genera and where the descriptions were published.

EXERCISE 9 MINERALOGY, PETROLOGY, ETC.

1. What is the name of the mineral with the following formula: $BaBe_2Si_2O_7$? Who first described it and where is the type locality? Is any information available on its crystal structure? **2.** A meteorite landed in Duncanville, Texas in 1961. What were the co-ordinates of the fall and how much did it weigh? **3.** What are the optical properties of lawsonite? **4.** Locate data including full chemical analyses, on the composition of typical lamprophyres. **5.** A new mineral, wermlandite, was recently reported from Längban in Sweden. What is the full reference to the paper? **6.** The Royal Society of London organised a meeting in 1971 to discuss matters relating to the petrology of the ocean floor. Has any report of this meeting been published? **7.** Has any work been reported since 1970 on the structural differences between talc and steatite? **8.** According to a directory of research projects, work has been done at the University of Illinois on the engineering properties of montmorillonite. Has this work been described anywhere? **9.** Has any information been published in the last 5 years on the chlorine content of volcanic rocks and the migration of chlorine within the earth?

EXERCISE 10 STRUCTURAL GEOLOGY

1. What is 'crenulation cleavage' and when was the term first introduced? **2.** Locate an introductory account and bibliography of (a) Appalachian tectonics; (b) folding. **3.** Is there any recently published information on the Curie–Sander Symmetry Concept as applied to petrofabric analysis. **4.** Has any work recently been reported on the use of twinned calcite as an indicator of strain? **5.** Locate a review and bibliography of experimental structural geology.

EXERCISE 11   APPLIED GEOLOGY

1. What are (a) 'dilly holes'; (b) 'burgee'; (c) 'zwitter'; (d) 'phyteral'? **2.** Locate a photograph showing zoning in cassiterite in thin section. **3.** Locate a general account of the coalfields of India. **4.** What work, if any, has been published since 1969 on the water vapour permeability of building stones? **5.** Mundford in Norfolk (UK) was proposed as the site for a 300 GeV proton accelerator. This machine has stringent requirements for ground stability and deformation under load. Has any work been reported indicating how the load bearing properties of the Middle Chalk of this area were measured? **6.** Locate a general account of the recent gas discoveries off the north-west coast of Australia. **7.** A paper on tracer techniques in soil erosion studies was written by an American called McHenry (initials unknown) in 1969 or 1970. Locate the full reference.

EXERCISE 12   GEOPHYSICS

1. Locate a bibliography of publications pertaining to the IQSY programme. **2.** Locate a recent (post-1970) review on the geophysical applications of high-resolution magnetometers. **3.** Give the names and addresses of any companies in the UK manufacturing gravimetric survey equipment. **4.** Has any work been published on the basement rocks of the Northern Pennines (UK) as revealed by geophysical observations. Is there a Bouguer anomaly map of this area? **5.** The University of California (San Diego) is reported to have developed a laser earth strain meter. Has any account of this instrument been published?

EXERCISE 13   GEOMORPHOLOGY, ETC.

1. A paper by F. C. Wezel was published in *Riv. Miner. Siciliana*, **21**, entitled 'Intepretazione Dinamica della Eugeosinclinale Mesomediterranea'. Locate a lengthy English-language abstract. **2.** There is a river flow station on the River Wharfe at Ilkley. What was the highest gauged flow in m³/s in 1965–66? When was this station established and when was the highest ever flow recorded? **3.** Is there a publication (of some substance and written in the last 10 years) covering the various aspects of meteorology related to agriculture? **4.** Find some information on the various theoretical equations concerning the growth rate and thickness of sea ice. **5.** What are the average daily maximum and minimum temperatures in Tristan da Cunha in May? **6.** A portable flow-cell membrane salinometer has recently (1971) been described. Where was the account published? **7.** Locate a list of symbols used on geomorphological maps. **8.** Locate a map showing the karst regions in Romania. **9.** A meeting on the evolution of shorelines and continental shelves during the Quaternary was held during the 8th INQUA Congress in Paris in 1969. Have the proceedings been published yet? **10.** Locate a post-1967 bibliography on ocean currents.

EXERCISE 14   SOIL SCIENCE

1. Locate a fairly recent general account of the soils of Lancashire. **2.** Where has any work been published since 1966 which describes the absorption of radioactive strontium-90 from river water by natural soil? **3.** H. J. Van Rensburg wrote about some work on run-off and soil erosion tests in the Mpwapwa area of Central Tanganyika. It was published in about 1955. Locate the full reference. **4.** Where can information be found concerning the assessment of sub-soil and soil conditions associated with the pre-planning of golf courses? **5.** Locate a recent paper on the physical and chemical properties of desert sands in Antarctica.

SOURCES FOR PRACTICAL EXERCISES

EXERCISE 1
1. Published guides to the various classification systems (see Chapter 2). **2.** Author, subject and periodical catalogues maintained by the library. **3.** Map catalogue or other locally available guide. **4.** *International Library Directory*.

EXERCISE 2
1. Cumulated indexes of both journals. **2.** *Guides to Scientific Periodicals*. **3.** *BUCOP, Union List of Serials in the United States and Canada, New*

Serial Titles. **4.** *Ulrich's International Periodicals Directory, Geoserials.* **5.** *Government Publications, British National Bibliography.* **6.** *Publications of the Geological Survey 1879–1961.* **7.** *Government Publications, Sectional List 45.* **8.** *USGPO Price List 15—Geology.* **9.** *Guide to Microforms in Print.* **10.** *US Government Research and Development Reports, Government Reports Announcements.* **11.** *Index to Theses Accepted for Higher Degrees . . . , Annual Report of the Research Institute of African Geology.*

EXERCISE 3

1. *International Who's Who.* **2.** *World of Learning.* **3.** *Ocean Research Index.* **4.** *Councils, Committees and Boards, A Handbook of Advisory, Consultative, Executive and Similiar Bodies in British Public Life.* **5.** *Commonwealth Universities Yearbook.* **6.** *Directory of Special Libraries and Information Centers.* **7.** *Statesman's Yearbook.* **8.** *British Books in Print.* **9.** *Chemical Index of Minerals, Dana's System of Mineralogy.* **10.** *Dana's System of Mineralogy.*

EXERCISE 4

1. Royal Society's *Catalogue of Scientific Papers.* **2.** *Subject Guide to Books in Print, McGraw-Hill Basic Bibliography of Science and Technology,* Encyclopaedias. **3.** (a) *Bibliographic Index, Handbook of Palaeontological Techniques*; (b) *Geoscience Documentation.* **4.** *Geologic Literature on North America 1785–1918.* **5.** *Bibliographie des Sciences Géologiques.* **6.** *Bulletin Signalétique, Referativnyi Zhurnal, Bibliography and Index of Geology, Chemical Abstracts, Mineralogical Abstracts.* **7.** Locate a key early paper or book, e.g. Wegener, A., *The Origin of Continents.* **8.** *Current Contents, Physical and Chemical Sciences.*

EXERCISE 5

1. (a) *Bibliographie des Sciences de la Terre*; (b) *Technical Translations.* **2.** *Chemical Abstracts, Translations Register-Index.* **3.** (a) *Nuclear Science Abstracts*; (b) *Soils and Fertilizers*; (c) *Chemical Abstracts.*

EXERCISE 6

1. (a) *Bibliographie Cartographique Internationale*; (b) *Bibliographie Cartographique Internationale*; (c) *Bibliography and Index of Geology*; (d) *Catalogue des Publications, Bureau de Recherches Géologiques et Minières*; (e) *New Publications of the Geological Survey*; (f) *Transactions of the Royal Society of Edinburgh*; (g) *Geological Newsletter*; (h) *Geotimes*; (i) *New Zealand Journal of Geology and Geophysics*; (j) *Bibliography and Index of Geology Exclusive of North America.* **2.** *A Bibliographical Dictionary of Scientists, Proceedings of the Yorkshire Geological Society.*

EXERCISE 7

1. *Lexique Stratigraphique Internationale.* **2.** *Welsh Geological Quarterly.* **3.** *Sie Beiträge zur Regionalen Geologie der Erde.* 10. *Geology of the South Atlantic Islands.* **4.** *The Geologic Systems: the Precambrian,* Vol. 2. **5.** *Annotated Bibliography and Index of the Geology of Zambia.*

EXERCISE 8

1. *Invertebrate Fossils.* **2.** *Traité de Paléontologie.* **3.** *Traité de Paléobotanique,* Vol. 2. **4.** *Dissertation Abstracts International.* **5.** *Biological Abstracts.* **6.** (a) *Translations Register-Index*; (b) start by obtaining details of coelenterate classification from any general palaeontological textbook, then use *Bibliographie des Sciences de la Terre* and *Bibliography and Index of Geology*; (c) *Nomenclator Zoologicus.*

EXERCISE 9
1. Locate name from *Chemical Index of Minerals* or *Chemical Abstracts* and details of first description from *Dana's System of Mineralogy.* The reference to a paper describing the crystal structure can be obtained from *Chemical Abstracts* or *Structure Reports.* **2.** *Meteorites, Catalogue of Meteorites.* **3.** *Microscopic Identification of Minerals.* **4.** *Igneous and Metamorphic Petrology.* **5.** *Mineralogical Abstracts.* **6.** *Index of Conference Proceedings Received by the NLL.* **7.** *British Ceramic Abstracts.* **8.** *Dissertation Abstracts, British Ceramic Abstracts.* **9.** *Bibliography and Index of Geology.*

EXERCISE 10
1. *Memoirs, American Association of Petroleum Geologists,* 7. **2.** *Subject Guide to Books in Print.* **3.** *Mineralogical Abstracts.* **4.** *Bibliographie des Sciences de la Terre, F, Bibliography and Index of Geology.* **5.** *Earth Science Reviews.*

EXERCISE 11
1. *A Dictionary of Mining, Mineral and Related Terms.* **2.** *Atlas of the Most Important Ore Mineral Paragenesis under the Microscope.* **3.** *Coal Mining Geology.* **4.** *Building Science Abstracts.* **5.** *Engineering Index, British Technology Index.* **6.** *Gas Abstracts.* **7.** *Highway Research Abstracts, Nuclear Science Abstracts, Science Citation Index.*

EXERCISE 12
1. *Subject Guide to Books in Print.* **2.** *Handbuch der Physik.* **3.** *Kompass UK.* **4.** *Geophysical Abstracts.* **5.** *Physics Abstracts.*

EXERCISE 13
1. *Deep Sea Research, Oceanographic Abstracts and Oceanographic Bibliography Section.* **2.** *Surface Water Yearbook of Great Britain.* **3.** *Meteorological and Geoastrophysical Abstracts.* **4.** *Antarctic Bibliography.* **5.** *Tables of Temperature, Relative Humidity and Precipitation for the World.* **6.** *Oceanic Citation Journal.* **7.** *Encyclopedia of Geomorphology.* **8.** *Romanian Scientific Abstracts.* **9.** *Index of Conference Proceedings Received by the NLL.* **10.** *Ocean Currents.*

EXERCISE 14
1. *Report—Rothamsted Experimental Station.* **2.** *Water Pollution Abstracts.* **3.** *Bibliography of Soil Science.* **4.** *Engineering Index.* **5.** *Soils and Fertilizers.*

Index